Computational Medical Image Analysis

Computational Medical Image Analysis

Editor

Anando Sen

Basel • Beijing • Wuhan • Barcelona • Belgrade • Novi Sad • Cluj • Manchester

Editor
Anando Sen
John Walton Muscular
Dystrophy Research Centre
Newcastle University
Newcastle upon Tyne
United Kingdom

Editorial Office
MDPI
St. Alban-Anlage 66
4052 Basel, Switzerland

This is a reprint of articles from the Special Issue published online in the open access journal *Computation* (ISSN 2079-3197) (available at: www.mdpi.com/journal/computation/special_issues/Computational_Medical_Anatomical_Image).

For citation purposes, cite each article independently as indicated on the article page online and as indicated below:

Lastname, A.A.; Lastname, B.B. Article Title. *Journal Name* **Year**, *Volume Number*, Page Range.

ISBN 978-3-7258-1394-0 (Hbk)
ISBN 978-3-7258-1393-3 (PDF)
doi.org/10.3390/books978-3-7258-1393-3

Cover image courtesy of cottonbro studio

© 2024 by the authors. Articles in this book are Open Access and distributed under the Creative Commons Attribution (CC BY) license. The book as a whole is distributed by MDPI under the terms and conditions of the Creative Commons Attribution-NonCommercial-NoDerivs (CC BY-NC-ND) license.

Contents

About the Editor vii

Anando Sen
Computational Medical Image Analysis: A Preface
Reprinted from: *Computation* **2024**, *12*, 109, doi:10.3390/computation12060109 1

Daniel Fernando Santos and Helbert Eduardo Espitia
Secure Medical Image Transmission Scheme Using Lorenz's Attractor Applied in Computer Aided Diagnosis for the Detection of Eye Melanoma
Reprinted from: *Computation* **2022**, *10*, 158, doi:10.3390/computation10090158 4

Hassaan Haider, Jawad Ali Shah, Kushsairy Kadir and Najeeb Khan
Sparse Reconstruction Using Hyperbolic Tangent as Smooth l_1-Norm Approximation
Reprinted from: *Computation* **2023**, *11*, 7, doi:10.3390/computation11010007 20

Shubhangi A. Joshi, Anupkumar M. Bongale, P. Olof Olsson, Siddhaling Urolagin, Deepak Dharrao and Arunkumar Bongale
Enhanced Pre-Trained Xception Model Transfer Learned for Breast Cancer Detection
Reprinted from: *Computation* **2023**, *11*, 59, doi:10.3390/computation11030059 41

Baidaa Mutasher Rashed and Nirvana Popescu
Performance Investigation for Medical Image Evaluation and Diagnosis Using Machine-Learning and Deep-Learning Techniques
Reprinted from: *Computation* **2023**, *11*, 63, doi:10.3390/computation11030063 58

Juan Carlos Aguirre-Arango, Andres Alvarez-Meza and German Castellanos-Dominguez
Feet Segmentation for Regional Analgesia Monitoring Using Convolutional RFF and Layer-Wise Weighted CAM Interpretability
Reprinted from: *Computation* **2023**, *11*, 113, doi:10.3390/computation11060113 101

Nishant Chauhan and Byung-Jae Choi
Regional Contribution in Electrophysiological-Based Classifications of Attention Deficit Hyperactive Disorder (ADHD) Using Machine Learning
Reprinted from: *Computation* **2023**, *11*, 180, doi:10.3390/computation11090180 124

Catur Supriyanto, Abu Salam, Junta Zeniarja and Adi Wijaya
Two-Stage Input-Space Image Augmentation and Interpretable Technique for Accurate and Explainable Skin Cancer Diagnosis
Reprinted from: *Computation* **2023**, *11*, 246, doi:10.3390/computation11120246 138

Aravind Kolli, Qi Wei and Stephen A. Ramsey
Predicting Time-to-Healing from a Digital Wound Image: A Hybrid NeuralNetwork and Decision Tree Approach Improves Performance
Reprinted from: *Computation* **2024**, *12*, 42, doi:10.3390/computation12030042 152

Seweryn Lipiński
Creation of a Simulated Sequence of Dynamic Susceptibility Contrast—Magnetic Resonance Imaging Brain Scans as a Tool to Verify the Quality of Methods for Diagnosing Diseases Affecting Brain Tissue Perfusion
Reprinted from: *Computation* **2024**, *12*, 54, doi:10.3390/computation12030054 169

Sadiq Alinsaif
COVID-19 Image Classification: A Comparative Performance Analysis of Hand-Crafted vs. Deep Features
Reprinted from: *Computation* **2024**, *12*, 66, doi:10.3390/computation12040066 . . . . **183**

Dominika Petríková and Ivan Cimrák
Survey of Recent Deep Neural Networks with Strong Annotated Supervision in Histopathology
Reprinted from: *Computation* **2023**, *11*, 81, doi:10.3390/computation11040081 . . . . **202**

Mónica Vieira Martins, Luís Baptista, Henrique Luís, Victor Assunção, Mário-Rui Araújo and Valentim Realinho
Machine Learning in X-ray Diagnosis for Oral Health: A Review of Recent Progress
Reprinted from: *Computation* **2023**, *11*, 115, doi:10.3390/computation11060115 . . . . **219**

About the Editor

Anando Sen

Dr. Anando Sen is a Data Scientist at the John Walton Muscular Dystrophy Research Center, Newcastle University. He completed his PhD in Mathematics in 2012 at the University of Houston. He held postdoctoral positions at the University of Houston (Biomedical Engineering) and Columbia University (Biomedical Informatics). Most recently, he was a Research Investigator (Imaging Physics) at MD Anderson Cancer Center. His research interests include quantitative applications in healthcare, particularly medical imaging, clinical trials and data analysis.

Editorial

Computational Medical Image Analysis: A Preface

Anando Sen

John Walton Muscular Dystrophy Research Centre, Translational and Clinical Research Institute, Newcastle University and Newcastle Hospitals NHS Foundation Trust, Newcastle upon Tyne NE1 3BZ, UK; anando.sen@newcastle.ac.uk

There has been immense progress in medical image analysis over the past decade. Methodologies have transitioned from analytical to implementation-based to machine-learning-based techniques. This Special Issue was introduced to showcase the latest innovations in medical image analysis. The focus was on computational applications of medical images. The scope was not limited by the imaging modalities or applications presented. I envisaged a combination of modalities such as planar imaging (e.g., X-ray, planar nuclear imaging), anatomical imaging (e.g., computed tomography [CT], magnetic resonance [MR] imaging), nuclear medicine (e.g., positron emission tomography [PET]), bimodal or multi-modal imaging (e.g., PET-CT), as well as pathology. While some applications were listed as examples in the invitation, there were no restrictions on the presented applications as long as they were clinically relevant.

This Special Issue received enthusiastic responses from the academic community. With 12 successful submissions and the interest these generated, we were motivated to work on a second edition, which is already open for submissions. Among the articles, 10 are research articles while two are review articles. Further, of the 12 articles, 10 (including both the review articles) dealt with machine-learning- and deep-learning-based methods, highlighting the transition towards these methods in the past decade. I was also glad to see the geographic diversity of the published articles. The first authors of the 12 papers were based in 11 different countries and four continents.

This Special Issue contains some very important and innovative applications. A summary of the papers along with the tackled applications is provided in Table 1.

Table 1. Summary of the 12 papers published in the Special Issue 'Computational Medical Image Analysis.

First Author/Reference	Type	Description
Santos et al [1]	Research	A computer-aided diagnosis method for eye melanoma detection using Lorenz's attractor
Haider et al. [2]	Research	A compressed sensing-based based method for sparse reconstruction of MR images
Joshi et al. [3]	Research	A method for detecting breast cancer in histopathology images using convolutional neural networks (CNNs)
Rashed et al. [4]	Research	Evaluate various machine-learning and deep-learning techniques for classifying chest X-ray and dermoscopy images into normal and abnormal
Aguirre-Arango et al. [5]	Research	Technique for feet segmentation in thermal images that can be used for pain relief during pregnancy
Chauhan et al. [6]	Research	Evaluate several classifiers on electroencephalography (EEG) data to classify Attention Deficit Hyperactive Disorder (ADHD) patients and healthy controls

Citation: Sen, A. Computational Medical Image Analysis: A Preface. *Computation* **2024**, *12*, 109. https://doi.org/10.3390/computation12060109

Received: 17 May 2024
Accepted: 21 May 2024
Published: 24 May 2024

Copyright: © 2024 by the author. Licensee MDPI, Basel, Switzerland. This article is an open access article distributed under the terms and conditions of the Creative Commons Attribution (CC BY) license (https://creativecommons.org/licenses/by/4.0/).

Table 1. *Cont.*

First Author/Reference	Type	Description
Supriyanto et al. [7]	Research	Use a geometric augmentation and generative adversarial networks (GANs) to classify clinical (white-light) images of skin lesions into cancerous and non-cancerous
Kolli et al. [8]	Research	Predict the time to wound healing based on digital images of the wound
Lipinski [9]	Research	Development of simulated dynamic susceptibility contrast MR images of the brain that can be used to evaluate quality assessment methods for disease diagnosis
Alnsaif [10]	Research	Development of a classifier to differentiate between COVID-19 and non-COVID-19 chest CT scans using features extracted from pre-trained deep-learning models
Petrikova et al. [11]	Review	Provide an overview of deep-learning-based histopathology classification tasks covering a variety of applications and organs
Martins et al. [12]	Review	Investigate recent progress in machine-learning methods for the diagnosis of oral diseases using oral X-ray images

I would like to thank the MDPI team for their smooth processing of the submitted articles. The editorial team and Academic Editors ensured each article received adequate consideration. The Editorial Board was called upon a few times to provide the final decision when reviewer opinions diverged. Finally, a big thank you to all reviewers who provided scientific expertise for this Special Issue despite their busy schedules. I look forward to working with all of them for the second edition and invite all authors and readers to consider this Special Issue for the submission of their research outputs.

Conflicts of Interest: The author declares no conflict of interest.

References

1. Santos, D.F.; Espitia, H.E. Secure Medical Image Transmission Scheme Using Lorenz's Attractor Applied in Computer Aided Diagnosis for the Detection of Eye Melanoma. *Computation* **2022**, *10*, 158. [CrossRef]
2. Haider, H.; Shah, J.A.; Kadir, K.; Khan, N. Sparse Reconstruction Using Hyperbolic Tangent as Smooth l1-Norm Approximation. *Computation* **2023**, *11*, 7. [CrossRef]
3. Joshi, S.A.; Bongale, A.M.; Olsson, P.O.; Urolagin, S.; Dharrao, D.; Bongale, A. Enhanced Pre-Trained Xception Model Transfer Learned for Breast Cancer Detection. *Computation* **2023**, *11*, 59. [CrossRef]
4. Rashed, B.M.; Popescu, N. Performance Investigation for Medical Image Evaluation and Diagnosis Using Machine-Learning and Deep-Learning Techniques. *Computation* **2023**, *11*, 63. [CrossRef]
5. Aguirre-Arango, J.C.; Álvarez-Meza, A.M.; Castellanos-Dominguez, G. Feet Segmentation for Regional Analgesia Monitoring Using Convolutional RFF and Layer-Wise Weighted CAM Interpretability. *Computation* **2023**, *11*, 113. [CrossRef]
6. Chauhan, N.; Choi, B.-J. Regional Contribution in Electrophysiological-Based Classifications of Attention Deficit Hyperactive Disorder (ADHD) Using Machine Learning. *Computation* **2023**, *11*, 180. [CrossRef]
7. Supriyanto, C.; Salam, A.; Zeniarja, J.; Wijaya, A. Two-Stage Input-Space Image Augmentation and Interpretable Technique for Accurate and Explainable Skin Cancer Diagnosis. *Computation* **2023**, *11*, 246. [CrossRef]
8. Kolli, A.; Wei, Q.; Ramsey, S.A. Predicting Time-to-Healing from a Digital Wound Image: A Hybrid Neural Network and Decision Tree Approach Improves Performance. *Computation* **2024**, *12*, 42. [CrossRef]
9. Lipiński, S. Creation of a Simulated Sequence of Dynamic Susceptibility Contrast—Magnetic Resonance Imaging Brain Scans as a Tool to Verify the Quality of Methods for Diagnosing Diseases Affecting Brain Tissue Perfusion. *Computation* **2024**, *12*, 54. [CrossRef]
10. Alnsaif, S. COVID-19 Image Classification: A Comparative Performance Analysis of Hand-Crafted vs. Deep Features. *Computation* **2024**, *12*, 66. [CrossRef]

11. Petríková, D.; Cimrák, I. Survey of Recent Deep Neural Networks with Strong Annotated Supervision in Histopathology. *Computation* **2023**, *11*, 81. [CrossRef]
12. Martins, M.V.; Baptista, L.; Luís, H.; Assunção, V.; Araújo, M.-R.; Realinho, V. Machine Learning in X-ray Diagnosis for Oral Health: A Review of Recent Progress. *Computation* **2023**, *11*, 115. [CrossRef]

Disclaimer/Publisher's Note: The statements, opinions and data contained in all publications are solely those of the individual author(s) and contributor(s) and not of MDPI and/or the editor(s). MDPI and/or the editor(s) disclaim responsibility for any injury to people or property resulting from any ideas, methods, instructions or products referred to in the content.

Article

Secure Medical Image Transmission Scheme Using Lorenz's Attractor Applied in Computer Aided Diagnosis for the Detection of Eye Melanoma

Daniel Fernando Santos and Helbert Eduardo Espitia *

Facultad de Ingeniería, Universidad Distrital Francisco José de Caldas, Bogotá 110231, Colombia
* Correspondence: heespitiac@udistrital.edu.co

Abstract: Early detection of diseases is vital for patient recovery. This article explains the design and technical matters of a computer-supported diagnostic system for eye melanoma detection implementing a security approach using chaotic-based encryption to guarantee communication security. The system is intended to provide a diagnosis; it can be applied in a cooperative environment for hospitals or telemedicine and can be extended to detect other types of eye diseases. The introduced method has been tested to assess the secret key, sensitivity, histogram, correlation, Number of Pixel Change Rate (NPCR), Unified Averaged Changed Intensity (UACI), and information entropy analysis. The main contribution is to offer a proposal for a diagnostic aid system for uveal melanoma. Considering the average values for 145 processed images, the results show that near-maximum NPCR values of 0.996 are obtained along with near-safe UACI values of 0.296 and high entropy of 7.954 for the ciphered images. The presented design demonstrates an encryption technique based on chaotic attractors for image transfer through the network. In this article, important theoretical considerations for implementing this system are provided, the requirements and architecture of the system are explained, and the stages in which the diagnosis is carries out are described. Finally, the encryption process is explained and the results and conclusions are presented.

Keywords: chaotic attractors; computer vision; disease diagnosis; encryption; computer-assisted diagnosis; convolutional neural networks

Citation: Santos, D.F.; Espitia, H.E. Secure Medical Image Transmission Scheme Using Lorenz's Attractor Applied in Computer Aided Diagnosis for the Detection of Eye Melanoma. *Computation* **2022**, *10*, 158. https://doi.org/10.3390/computation10090158

Academic Editor: Demos T. Tsahalis

Received: 26 July 2022
Accepted: 11 September 2022
Published: 14 September 2022

Publisher's Note: MDPI stays neutral with regard to jurisdictional claims in published maps and institutional affiliations.

Copyright: © 2022 by the authors. Licensee MDPI, Basel, Switzerland. This article is an open access article distributed under the terms and conditions of the Creative Commons Attribution (CC BY) license (https://creativecommons.org/licenses/by/4.0/).

1. Introduction

Computer assistance in providing disease diagnoses has a broad range of applications, and the development of tools that help to reach this end is of paramount significance. Computer-Assisted Diagnosis (CAD) has been applied in many different contexts, including Digital Imaging and Communications in Medicine (DICOM); in this regard, a web application for disease diagnosis through a browser is shown in [1]. Other examples of medical images transmission can be found in [2,3]. In [4], the authors evaluated a fuzzy clustering algorithm for breast cancer detection, while [5] illustrates developments in the detection of diabetic retinopathy involving computer-aided diagnostic systems.

In Colombia, as well as in other parts of the world, access to an ophthalmologist entails several appointments and procedures, which usually lead to long waits. For this reason, it is essential to create tools to aid in timely diagnosis in order to provide adequate treatment. In [6], the authors proposed a telemedicine system to diagnose stomach diseases. Issues with developing computer-aided diagnosis systems (CADS) have been studied in [7], where the authors explained a new model along with several fundamental CADS techniques. Cloud Computing (CC), TensorFlow (TF), and Django are used as support for the construction of such systems. CC supports the storage and access of information of interest for different parties, while TF (created by Google under an open-source Apache 2.0 license [8]) provides an interface for building and running machine learning algorithms to use and run eye disease prediction models.

Regarding computer-assisted diagnosis, reference [9] presents a diagnostic system for schizophrenia using effective connectivity of resting-state electroencephalogram (EEG) data, while [10] studies the practicality of deep learning algorithms applied to chest X-ray images for COVID-19 detection. In [11], the authors presented a COVID-19 prediction applying supervised machine learning algorithms using the Waikato Environment for Knowledge Analysis (WEKA), which is an open-source software developed at the University of Waikato in New Zealand. Lastly, reference [12] proposed a system for detection of cancer cells using commercially automated microscope-based screeners. Employing supervised machine learning, the authors developed software capable of classifying Feulgen-stained nuclei within eight diagnostically important types.

Regarding eye image processing (classification), in [13], the authors described the implementation of a framework for healthy and diabetic retinopathy retinal image recognition. In [14], the authors presented a framework for eye tracking calibration where features extracted from the synthetic eyes dataset are used in a fully connected network to isolate the effect of a specific user's features. Their work was oriented towards the design of low-cost eye-tracking systems. In [15], the authors performed an Image Quality Assessment (IQA) of eye fundus images in the context of digital fundoscopy with Topological Data Analysis (TDA) and machine learning methods. IQA is a fundamental step in digital fundoscopy for clinical applications, and is considered one of the first steps in the preprocessing stages of Computer-Aided Diagnosis (CAD) systems using eye fundus images. Their research employed cubical complexes to represent the images; the grayscale version was then used to calculate a homology illustrated with persistence diagrams and thirty vectorized topological descriptors were calculated from each image for use as input to a classification algorithm. Finally, Diabetic Retinopathy (DR) is a disease that is one of the main causes of blindness around the world. Therefore, reference [16] employed retinal fundus images as diagnostic tools to screen abnormalities associated with eye diseases. In this regard, article [16] proposed an algorithm to segment and detect hemorrhages in retinal fundus images. The method they described performs preprocessing on retinal fundus images by utilizing a windowing-based adaptive threshold to segment hemorrhages. In this way, conventional features are extracted for each candidate and classified using a support vector machine.

In regards to research related to image encryption based on chaos, developed approaches include specially fractional-order chaotic systems, which exhibit more complex dynamics than integer-order chaotic systems. In [17], a fractional-order memristor was developed, analyzed, and electronically implemented. In this order, a three-dimensional (3D) fractional-order memristive chaotic system with a single unstable equilibrium point was proposed for use in an encryption system applied to grayscale images. Other related research was presented in [18], where the authors proposed using a chaotic oscillator without linear terms as a random number generator for application in biomedical image encryption. They demonstrated the physical realization of the oscillator and carried out a security and performance analysis. In [19], an oscillator with chaotic dynamics was presented and various properties of the oscillator, such as bifurcations, equilibria, and Lyapunov exponents, were studied in order to show the existence of chaotic dynamics (as the oscillator has a chaotic attractor). Using the features of the chaotic oscillator, a method for generating pseudo-random numbers was presented in the context of designing secure substitution boxes applied to an image cryptosystem. In this same orientation, in [20] the authors developed, analyzed, tested, and electronically implemented a 4D fractional-order memcapacitor that observed the nonlinear dynamic properties of a hyperchaotic system. On this basis, they proposed an encryption algorithm for color encryption based on the system's chaotic behavior in which every pixel value of the original image is incorporated into the secret key to strengthen the encryption algorithm. A related work was presented in [21] involving a hyperchaotic 4D fractional discrete Hopfield neural network system. The chaotic dynamics features were analyzed and the chaotic system was used as a pseudo-random number generator for an image encryption scheme based on a fractal-like model scrambling method. This approach was able to enhance the complexity and security of the encryption algorithm. Finally, in [22], a chaotic oscillator was presented in which the chaotic dynamics were pre-located around manifolds. After analyzing the complex dynamics of the oscillator, this approach was employed in the design of an image cryptosystem, and the results of the cryptosystem were tested while considering different metrics.

The present study proposes a system for computer-aided diagnosis, detection, and classification of eye diseases using chaotic-based encryption for image transmission. The proposal is based on the previous works in [23,24] for image encryption and [25,26] for image diagnostics.

Regarding differences with other related works, references [9–12] display various applications in computer-assisted diagnosis that are not applicable to eye image diagnosis; while [13–16] are focused specifically on eye image processing and classification. Regarding image encryption methods based on chaotic attractors, references [17–22] each describe several different developments, while [18] considers the encryption of medical images and [22] is concerned with Internet of Things (IoT) applications for remote diagnosis. The novelty of the present work is in its integration of an identification system for Uveal melanoma with an encryption mechanism in order to obtain a computer-assisted diagnosis system that can help professionals in improving the diagnostic process. As such, a computer-aided diagnosis system is presented here that integrates an image processing (classification) system based on a convolutional neural network and a mechanism for image transmission using an encryption method based on a chaotic attractor.

The rest of this paper is constructed as follows. Section 2 displays the design of the proposed system. Section 3 specifies the encryption framework and reviews the chaotic Lorenz attractor and its relevant properties for encryption. Section 4 presents the implementation results for the encryption system. Finally, Sections 5 and 6 respectively present the discussion and conclusions.

2. Proposed Diagnosis and Encryption System

This section presents the system architecture and the process used to diagnose the graphic user interface, then explains the system operation employing Convolutional Neural Network (CNN).

The structure of the application is similar to the one proposed in [27] divided into three layers: presentation, domain logic, and data access. The presentation layer comprises the patient and doctor user interfaces and all actions that a user can carry out. The domain logic contains the business module and the process of transferring and ciphering images over the network; this frame uses the data access layer, which stores the data the systems need to operate. Figure 1 shows a graphic of the doctor user interface (presentation layer), providing an example of a result after diagnosis. The specialist interface provides a diagnostic from all the eye images received. Contiguous to each eye, there is a textbox where the specialist can formulate the diagnostic; additionally, it offers options to run the automatic CNN diagnostic model. A button to send the diagnostic is provided. When clicking on the image of the eye, the specialist performs the operations described in Table 1. This interface has a responsive design, allowing it to be used from a smartphone.

The system allows different operations to be executed on the eye; these are shown in Table 1. These operations are aimed at modifying eye the image according to user needs, for instance, zooming in on a particular region or removing noise.

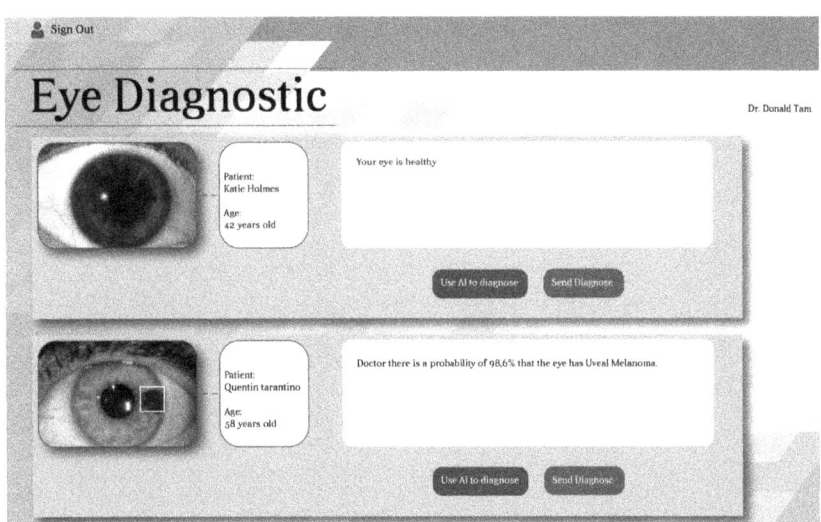

Figure 1. Specialist's interface with different diagnostic options.

Table 1. Operations that the specialist carries out on the images. These operations allow the doctor to perform a more reliable diagnostic.

Operation	Description
Grayscale transformation	Conversion of the $I[x, y, z]$ color image to grayscale $O_1[x, y]$ to reduce noise and improve the performance of the following stages. The conversion of a color image to a grayscale image consists of converting RGB values (24 bits) to grayscale values (8 bits).
Apply median filter	This process is performed to apply smoothing, which is achieved by sliding a window over the image, thus suppressing the higher frequencies. It can be seen as a change of the brightness of the input image.
Apply thresholding	For the present project, this utilizes mean adaptive thresholding and Gaussian adaptive thresholding to clearly define the borders. The main objective of this step is to provide better definition of the edges.
Dilate the image	By applying a morphological operation to reduce noise, dilation allows objects to be expanded, thus potentially filling small holes, in this case reducing pepper noise.
Rotation	The image is rotated at a predefined or random angle. In the case of the iris, 360 different rotations can be performed.
Zooming	This technique creates new versions of an image with different zoom views, in many cases focusing on the region of interest. The resulting images are enlarged or reduced according to a predefined range.

The user takes a picture of the eye and sends it using the application; this image is saved in the server through a request, then the specialist receives this photo and writes a diagnosis. This process can be seen in Figure 2. Before transmitting over the network, images are ciphered using chaotic encryption to maintain privacy. The Diffie–Hellman algorithm shares the initialization conditions for ciphering and deciphering the server and the client.

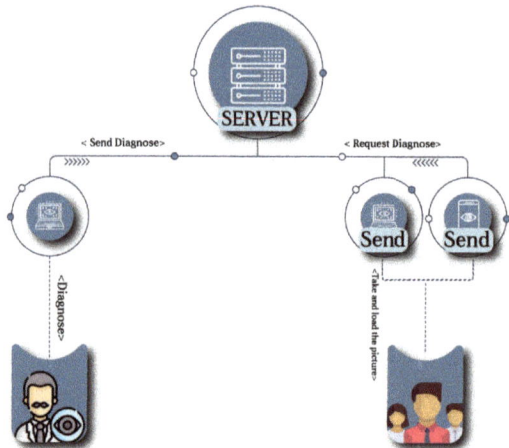

Figure 2. Model explaining communication between the patient and specialist. The patient can send images through their cellphone or computer.

Diagnostic Using Deep Convolutional Neural Network

For implementing the layer corresponding to the diagnostic, preliminary tests were conducted recognizing fuzzy systems neural networks and neuro-fuzzy systems as shown in [25,26]. However, better results are attained utilizing convolutional neural networks. Thus, the diagnostic system employs a famous pre-trained model, Resnet18, which recognizes abnormalities with an accuracy of 99%. Figure 3 presents the Resnet18 architecture.

In order to train the CNN, a data augmentation process was employed obtaining a dataset of 2048 figures, consisting of 1024 healthy and 1024 unhealthy. The original database is taken from [28], consisting of images of 150 healthy and 33 unhealthy patients. Additional details related to the design and training of the CNN can be found in [29].

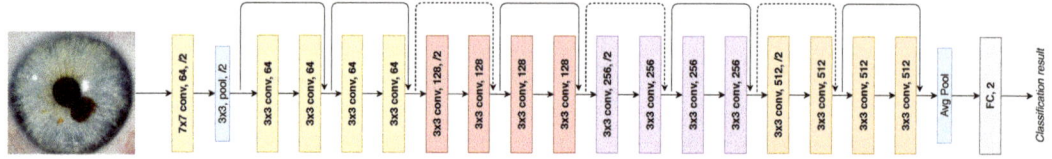

Figure 3. Resnet18 architecture.

The model trained with the Resnet18 architecture receives an image (iris image) and produces a value corresponding to the classification; in addition, there is a second model that produces a bounding box with the location of the detected abnormality. The model is used when the specialist clicks the "Use AI to Diagnose" button. In [29], different convolutional neural networks (CNNs) were used to detect ocular abnormalities with an illustrative case of uveal melanoma (UM), a type of ocular cancer. Thus, this work is a complement to that research, seeking to implement a CAD.

Resnet is a well-known convolutional neural network architecture that allows the training of hundreds or thousands of layers and achieves excellent performance. The biggest advantage of Resnet is its ability to reduce the vanishing gradient problem [30]. Before Resnet, a deep network was hard to train, as the gradients need to back-propagate through an enormous number of layers in a deep network, which makes the gradient infinitely small. Resnet has solved this problem, as it can skip the backwards connections between layers and create identity shortcuts in the gradients' path, that allows the gradients to flow faster to the initial layer [31].

The model was trained with a cross-entropy loss function to minimize the distance between predicted and ground-truth probabilities. This is defined in Equation (1), where p_i and q_i are the ground-truth and predicted probability, respectively. The loss function was minimized utilizing the Adam optimization algorithm, as it is computationally efficient and works well with noisy or sparse gradient problems.

$$L = -\sum_{i=1}^{N} p_i \log(q_i) \tag{1}$$

3. Security Techniques

This section reviews the chaotic Lorenz attractor and its relevant properties for encryption, and describes notable security techniques.

Privacy is essential for systems that hold patient information, and is indispensable in speeding up the diagnostic process. Various techniques have been introduced, including data encryption standard (DES), Rivest–Shamir–Adleman (RSA), and chaos, among others. Chaos provides high sensitivity to initial conditions and unpredictability. For instance, reference [32] used chaotic Arnold Maps (AM) to randomize the original position of the pixels, causing the image to become noisy. In [33,34], the authors proposed a system for encrypting color using the advantage of chaotic maps. The idea behind these maps is to distribute the pixels with a transformation such that the correlation of adjacent pixels can be reduced. Using compression and security features, this scheme can be applied in public networks. In [35], a feasible system for image encryption was presented using techniques applicable for real-time image transmission and encryption. However, applications in medical image transfer are relatively scarce; one of the few that has been found is the use of Arnold maps for the diffusion stage in a system that allows the encoding of pixels of biometric data [36]. This system uses a chaotic Chen system to change the statistical properties and resist attacks of the same type, achieving a robust system capable of resisting brute force attacks and thus demonstrating that this system is applicable for the transmission of biometric data over open and shared networks. However, AM is not sufficient to protect against statistical attacks. This is why a second phase of encryption is needed using Lorenz' system, as used in different works such as [37,38]. The Lorenz system is a model of thermally induced fluid convection in the atmosphere, which has properties that make it ideal for ciphering images. It is defined by the following set of equations:

$$x_1 = a(x_2 - x_1) \tag{2}$$
$$x_2 = cx_1 + x_2 - x_1x_3 - x_4 \tag{3}$$
$$x_3 = x_1x_2 - bx_3 \tag{4}$$
$$x_4 = kx_2x_3 \tag{5}$$

Figure 4 shows the Lorenz system for x_1, x_2, and x_3, with initial conditions $x_1 = 2.7$, $x_2 = 1.3$, $x_3 = -1.7$, and $x_4 = -5$.

Equations (2)–(5) correspond to the 4D hyperchaotic Lorenz system, where $a, b, c, k > 0$ are the control parameters. Using suitable values can obtain the desired chaotic behavior. In this work, we employ the encryption scheme presented in [23] that utilizes this Lorenz model. In addition, we consider [24], where the 3D Lorenz classical model is used for iris image encryption.

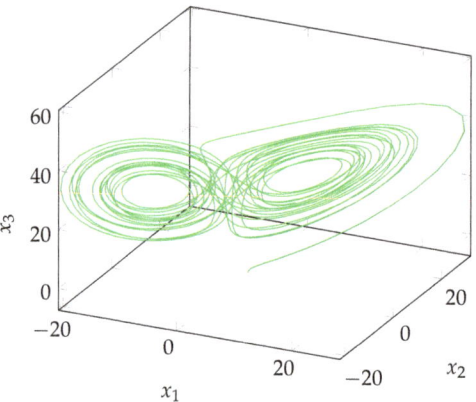

Figure 4. Lorenz system with initial conditions $x_1 = 2.7$, $x_2 = 1.3$, $x_3 = -1.7$, and $x_4 = -5$. Plot of x_1, x_2, and x_3.

Encryption Process

The introduced encryption algorithm uses RGB images of the eye and bases its operation on two means, permutation and diffusion; the first is performed through Arnold's chaotic map, while the second is accomplished through the numerical solutions 4th of Lorenz's system generated by Runge Kutta. In the permutation phase, each pixel is re-positioned with one-to-one correspondence, i.e., all pixels composing the permuted image correspond to the group of pixels of the original image, making it possible to recover the actual image without any distortion. Different techniques can be applied in this situation; Arnold's Chaotic Map provides easy and efficient implementation and shows consistent results in terms of the metric used to establish how much the pixels have moved from the original position [39].

To describe the encryption process, the width w and height of the image h are obtained, and three arrays arr_r, arr_g, and arr_b of cardinality $w \times h + \delta$ are generated with values obtained from the Lorenz map using R4. The value δ is a natural number representing the amount of iterations required for the values of x_1, x_2, x_3, and x_4 to enter the chaotic system.

Figure 5 illustrates the image ciphering process and Figure 6 is a subset of images of the CASIA dataset used to perform statistical analysis. First, the image histogram is observed without carrying out the encryption process. Later, the respective encryption allows for observation of cases in which image transmission is susceptible to a "digital attack". In this context, a "digital attack" attempts to reconstruct the transmitted image without authorization.

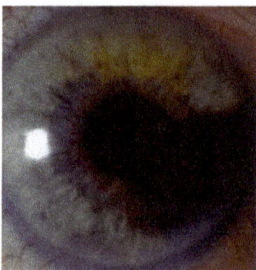

Figure 5. Image with an abnormality in the bottom right part of the iris (image used for testing).

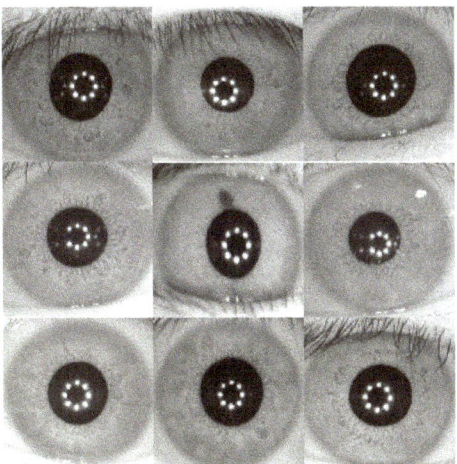

Figure 6. Subset of images from the CASIA Iris dataset used to perform statistical analysis of the proposed ciphering scheme.

The more uniform the histograms are, the more secure the ciphering is against statistical attacks; the histograms of the real image can be seen in Figure 7a–c. If permutation is not carried out, it is feasible that an attacker could gain insights into the original image.

An example of a process of ciphering without using permutation is presented in Figure 8. it can be seen that the ciphering process without a permutation procedure causes the circular structure of the iris to be somewhat recognizable.

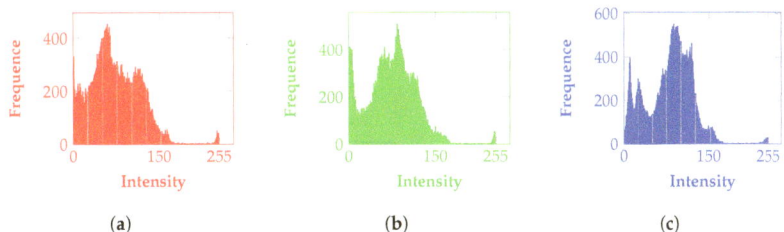

Figure 7. Histograms for RGB channels of the original image (Figure 5): (**a**) red image histogram, (**b**) green image histogram, (**c**) blue image histogram.

Figure 8. Ciphering process without a permutation process causes the circular structure of the iris to be recognizable.

The patterns displayed in the images permuted with Arnold's map are based on the number of iterations. For instance, Figure 9 shows the images in Figure 6, and with these results Figure 10 displays the diffused images with Lorenz.

Figure 9. Nine of the ciphered images after Arnold's map.

Figure 10. Nine ciphered images with Lorenz' attractor and Arnold's map.

4. Results

A simulated example of a diagnosis carried out over two eyes of the test dataset can be seen in Figure 1. The top image shows a healthy picture corresponding to the diagnosis "Your eye is healthy" and at the bottom, the image contains a potential abnormality, an unhealthy eye. In this case, the option "Use AI to diagnose" was used to produce the results, which triggers the CNN network and produces a number with a probability, which in this case is "Doctor, there is a 98.6% probability" that the eye has Uveal Melanoma. Next, considering Figure 5 to illustrate the process of image ciphering, the module receives an eye image which is permuted using Arnold's Map to produce Figure 11a. Finally, this image is diffused using Lorenz' attractor, generating Figure 11b, which can then be transmitted over the network. The original database used here is taken from [28].

(a)   (b)

Figure 11. Results of the permutation and diffusion processes. The input of the diffusion process is the result of the permutation stage. As can be seen, the shape of the iris is no longer presented. (**a**) Image permuted with seven iterations using Arnold's map. (**b**) Image permuted ciphered using Lorenz' attractor.

As explained in [40], there are different kinds of correlation attacks. Correlation analysis techniques include the Mean Difference Method (MDM) and Pearson Correlation Coefficient Method (PCCM); producing lower values of correlation makes the system more robust against these attacks. Hence, it is a crucial statistical analysis tool based on the frequency distribution of the encrypted pixels illustrated in the histograms (visual representation of such distribution, plotting the number of pixels at each level). After performing the proposed scheme, the results of the correlation can be seen in Table 2, exposing small correlation values for the transmitted images. In addition, the histograms can be seen in Figure 12a–c, showing that the ciphered images have a uniform distribution in the intensity of each color component, which in plain sight would result in an inconsistent or meaningless image. Thus, the possibility of an attacker obtaining the actual image is noticeably low, as the pixels of each color component of the encrypted image are distributed without providing any indication to use in statistical analysis to obtain a possible image.

Table 2. Correlation values for Red, Green, and Blue channels for different images in the encryption process.

Image	Red	Green	Blue
Figure 5 (Image without ciphering)	0.996	0.996	0.995
Figure 11a (Image after Arnold's map)	0.314	0.276	0.274
Figure 11b (Image after Arnold's map and Lorenz' attractor)	0.002	0.001	0.002

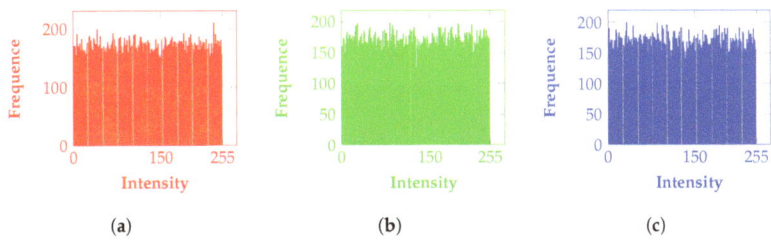

(a)   (b)   (c)

Figure 12. Histograms of RGB channels for a given ciphered image: (**a**) histogram red channel, (**b**) histogram green channel., (**c**) histogram blue channel.

4.1. Sensibility in the Key

As mentioned above, in a chaotic system the initial conditions significantly affect the performance of the Chen system. A security system must be sensitive to a wrong key to ensure that data cannot be obtained without the proper key. To provide an example, the following are the initial conditions: $[x_1 = 2.7, x_2 = 1.3, x_3 = -1.7, x_4 = -5]$, and to measure to sensitivity, these values are slightly changed to $[x'_1 = 2.71, x'_2 = 1.3, x'_3 = -1.71, x'_4 = -5.08]$. This slight alteration produces large changes, confirming the high sensitivity present in the Lorenz system. In order to observe the sensibility of the initial conditions, Figure 13 shows the values for x_1, x_2, x_3 and x_4. The values of (x_1, x_2, x_3) are used to cipher each of the RGB pixels. Therefore, a small alteration in any of the initial conditions produces remarkably different results.

Sensitivity to initial conditions is one of the requirements defined by Shannon [41] for confusion and diffusion in cryptography; the problem with these systems is that they can be broken due to their small key space [42,43], as the most important part of any encryption algorithm is the key that defines whether the system is sufficiently strong against attacks. However, as shown in [44], the Lorenz system can be used to generate keys that successfully pass the National Institute of Standards and Technology (NIST) statistical test suite.

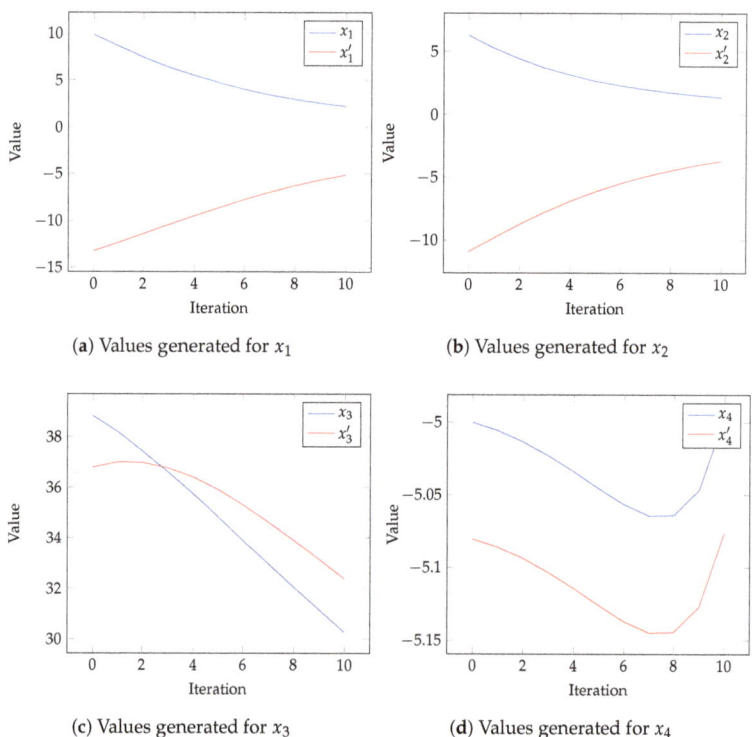

Figure 13. Values generated with initial conditions for x_1, x_2, x_3, and x_4.

4.2. Metrics

According to [45], the NPCR and UACI statistical tests are employed when dealing with base chaos encryption; for example, references [46,47] employed these metrics. The Number of Pixel Change Rate (NPCR), which computes the pixel difference ratio between ciphered and original images, is calculated with Equation (6); in this equation, $D(i, j)$ corresponds to Equation (7):

$$\text{NCPR} = \frac{1}{M \times N} \sum_{i=1}^{M} \sum_{j=1}^{N} D(i, j) \times 100\% \qquad (6)$$

$$D(i, j) = \begin{cases} 0, & \text{if } A(i, j) = B(i, j) \\ 1, & \text{if otherwise} \end{cases} \qquad (7)$$

where $A(ij)$ is the pixel value of the original image, $B(i,j)$ is the pixel value of the encrypted image, and (M, N) corresponds to image dimensions. A higher NPCR displays better algorithm performance. The range of NPCR values is $[0, 1]$.

Meanwhile, UACI computes the difference between the ciphered image and the original image, allowing the strength of the encryption algorithm to be observed. UACI measures the average change in a pixel's values between the ciphered and original images by employing Equation (8):

$$\text{UACI} = \frac{1}{M \times N} \sum_{i=1}^{M} \sum_{j=1}^{N} \frac{|A(i,j) - B(i,j)|}{255} \times 100\%. \tag{8}$$

A high UACI indicate that the systems is resistant against different attacks. The state of the art shows that a UACI of 0.33 is a secure value [24].

The entropy metric relies on the probability of pixel values and computes the degree of randomness; this metric is calculated using Equation (9), where $P(i)$ is the probability of pixel value i and is computed by Equation (10):

$$E = \sum_{i=0}^{255} \left(P(i) \log_2 \left(\frac{1}{P(i)} \right) \right), \tag{9}$$

$$P(i) = \frac{\text{Frequency of the pixel value } i}{\text{Total number of image pixels}}. \tag{10}$$

The efficiency of the encrypted image is superior if the entropy value is greater. The maximum entropy value is 8.

In this work, the results of these metrics for RGB images are shown in Table 3, in which it can be seen that the NPCR, UACI, and Entropy in the ciphered images have high values of security acceptable for use in image transmission.

Table 3. Results of NPCR, UACI, and Entropy tests for a RGB image after permutation and diffusion.

Measure	Red	Green	Blue
NPCR	0.996	0.996	0.996
UACI	0.310	0.328	0.334
Entropy	7.996	7.996	7.995

In order to ensure that the obtained result was not an outlier, we used the Chinese Academy of Sciences—Institute of Automation (CASIA) database, taking several iris images in grayscale to obtain a dataset that allows a statistical analysis to be performed. For iris recognition research, CASIA contains a free access database; the images were captured using a uniform illumination to obtain an adequate iris image. This database of free access images can be found in [48].

Example images and the results of the iterations are shown in Figures 6, 9 and 10, while the metrics obtained after this process are shown in Table 4 for 145 images. These results show that near-maximum values of NPCR are obtained, as are near-safe values of UACI and high entropy in the different channels for the ciphered images. These values are sufficiently high for encrypted images, and therefore can be considered to have strong resistance to differential attacks.

Table 4. Metrics: median, standard deviation, minimum and maximum of 145 images for NPCR, UACI, and Entropy using CASIA dataset.

Measure	Median	Standard Deviation	Min	Max
NPCR	0.996	0.0004	0.994	0.997
UACI	0.296	0.0196	0.266	0.357
Entropy	7.954	0.0022	7.949	7.961

5. Discussion

Although this article describes a system for medical diagnosis, the main results aim to showing the various aspects of the encryption process; consequently, a user test is outside the scope of this work. This is due to the difficulty of establishing a group of professionals to carry out such tests. Additionally, the details of the convolutional neural networks used for the classification system can be consulted in [29].

As mentioned, there are several different works related to the proposal made in this document, included those on computer-aided diagnosis systems, image classification, and encryption systems. Several references cited in the introduction section were considered here, as follows:

- Computer Aided Diagnosis [9–12];
- Eye Image Classification [13–16];
- Chaotic Encryption [17–22].

It should be noted that [18] considered the encryption of medical images and [22] considered internet of things applications which included the possibility of remote diagnosis.

As observed, a comparison with related works can be made considering different approaches. In this respect, a comparison consists of the process of image encryption taking similar works as reference. Then, considering the average values for the implementations made in other related works, Table 5 displays the NPCR, UACI, and entropy values.

Although all cited works present better values in the metrics considered, the results obtained are close to those reported in [17–22], taking into account that the best values of the indicators are close to 1 for NPCR, around 0.33 for UACI, and 8 for entropy. In addition, in this comparison it should be considered that different numbers of figures with several sizes and features were used to carry out the tests. As displayed in Table 5, in this work 145 figures were used to validate the encryption process, which is more than in the other works. Therefore, to make a uniform comparison, in future works a benchmark must be defined considering standard figures according to the application in consideration.

Table 5. Performance comparison with other related works.

Research	Images Used	NPCR	UACI	Entropy
This work	145	0.9960	0.2960	7.9540
Reference [17]	1	0.9987	0.4996	7.9951
Reference [18]	3	-	-	7.9957
Reference [19]	4	0.9961	0.3347	7.9027–7.9999
Reference [20]	1	0.9981	0.3362	7.9996
Reference [21]	3–10	0.9961	0.3344	7.9983
Reference [22]	4	0.9962	0.3345	7.9993

6. Conclusions

In this paper, we have presented a system for medical image diagnosis using chaotic-base encryption with a particular case for Uveal Melanoma diagnosis. This cipher scheme was assessed using several statistical tests, including entropy test analysis, key sensitivity test, correlation properties, and randomization tests using UACI and NPCR. Although the encryption confirms that the original and encrypted images have no visual correspondence, statistical analysis through histograms reveals uniform distributions. Nonetheless, the correlation coefficients of adjacent pixels are low enough to guarantee that the original image cannot be easily recovered from the image resulting from the encryption process without knowledge of the initial conditions.

It should be noted that the Arnold maps with the Lorenz system are an encryption scheme with suitable results for the transmission of images over public networks, which requires the confidentiality, integrity, and privacy of the message.

The results display adequate performance of the encryption system, with high values obtained for NPCR (0.994 to 0.997), near-safe values for UACI (0.266 to 0.357), and high entropy of 7.949 to 7.961 for the ciphered images.

Considering the results in Table 4 for the 145 images, NPCR describes the lowest variation (standard deviation) of 0.0004 for the tests performed, followed by entropy with 0.0022, and finally UACI at 0.0196. This shows that the experiments carried out do not present greater variation when encrypting the largest number of processed figures in the same way.

In subsequent work, we intend to carry out user tests in order to improve the computer-aided diagnosis system.

Author Contributions: Conceptualization, D.F.S. and H.E.E.; Methodology, D.F.S. and H.E.E.; Project administration, H.E.E.; Supervision, H.E.E.; Validation, D.F.S.; Writing—original draft, D.F.S.; Writing—review and editing, D.F.S. and H.E.E. All authors have read and agreed to the published version of the manuscript.

Funding: This research received no external funding. The study was self-financed.

Institutional Review Board Statement: In this work, direct tests were not carried out on individuals (humans). The historical data used for this study can be found in [28].

Informed Consent Statement: The data used were requested from [28].

Data Availability Statement: The original database can be found at [28].

Acknowledgments: We would like to thank Paul T. Finger, the New York Eye Cancer Center, and the Institute of Automation of the Chinese Academy of Sciences for providing the data for the present study. We wish to thank the Universidad Distrital Francisco José de Caldas, the École Nationale Supérieure Mines-Télécom Atlantique Bretagne-Pays and the University of Nantes for encouraging this research.

Conflicts of Interest: The authors declare no conflict of interest.

References

1. Wang, L.; Wang, X.L.; Yuan, K.H. Design and implementation of remote medical image reading and diagnosis system based on cloud services. In Proceedings of the 2013 IEEE International Conference on Medical Imaging Physics and Engineering, Shenyang, China, 19–20 October 2013; pp. 341–347. [CrossRef]
2. Coatrieux, G.; Puentes, J.; Roux, C.; Lamard, M.; Daccache, W. A Low Distorsion and Reversible Watermark: Application to Angiographic Images of the Retina. In Proceedings of the 2005 IEEE Engineering in Medicine and Biology 27th Annual Conference, Shanghai, China, 17–18 January 2005; pp. 2224–2227. [CrossRef]
3. Coatrieux, G.; Le Guillou, C.; Cauvin, J.M.; Roux, C. Reversible Watermarking for Knowledge Digest Embedding and Reliability Control in Medical Images. *IEEE Trans. Inf. Technol. Biomed.* **2009**, *13*, 158–165. [CrossRef] [PubMed]
4. Ahmad, A. Evaluation of Modified Categorical Data Fuzzy Clustering Algorithm on the Wisconsin Breast Cancer Dataset. *Scientifica* **2016**, *2016*, 4273813. [CrossRef] [PubMed]
5. Amin, J.; Sharif, M.; Yasmin, M. A Review on Recent Developments for Detection of Diabetic Retinopathy. *Scientifica* **2016**, *2016*, 6838976. [CrossRef]

6. Lu, X. A Cooperative Telemedicine Environment for Stomatological Medical Diagnosis. In Proceedings of the 2006 IEEE International Conference on Mechatronics and Automation, Luoyang, China, 25–28 June 2006.
7. Ouyang, H.B.; Liu, S.; You, L.; Huang, W.H.; Zhong, S.Z. Study on the new design of computer-aided diagnosis system. In Proceedings of the 2009 IEEE International Symposium on IT in Medicine Education, Jinan, China, 14–16 August 2009; Volume 1, pp. 50–55. [CrossRef]
8. Abadi, M.; Agarwal, A.; Barham, P.; Brevdo, E.; Chen, Z.; Citro, C.; Corrado, G.; Davis, A.; Dean, J.; Devin, M.; et al. TensorFlow: Large-Scale Machine Learning on Heterogeneous Distributed Systems. *arXiv* **2015**, arXiv:1603.04467.
9. Ciprian, C.; Masychev, K.; Ravan, M.; Manimaran, A.; Deshmukh, A. Diagnosing Schizophrenia Using Effective Connectivity of Resting-State EEG Data. *Algorithms* **2021**, *14*, 139. [CrossRef]
10. Alorf, A. The Practicality of Deep Learning Algorithms in COVID-19 Detection: Application to Chest X-ray Images. *Algorithms* **2021**, *14*, 183. [CrossRef]
11. Villavicencio, C.N.; Macrohon, J.J.E.; Inbaraj, X.A.; Jeng, J.H.; Hsieh, J.G. COVID-19 Prediction Applying Supervised Machine Learning Algorithms with Comparative Analysis Using WEKA. *Algorithms* **2021**, *14*, 201. [CrossRef]
12. Böcking, A.; Friedrich, D.; Schramm, M.; Palcic, B.; Erbeznik, G. DNA Karyometry for Automated Detection of Cancer Cells. *Cancers* **2022**, *14*, 4210. [CrossRef]
13. Akande, O.N.; Abikoye, O.C.; Kayode, A.A.; Lamari, Y. Implementation of a Framework for Healthy and Diabetic Retinopathy Retinal Image Recognition. *Scientifica* **2020**, *2020*, 4972527. [CrossRef]
14. Garde, G.; Larumbe-Bergera, A.; Bossavit, B.; Porta, S.; Cabeza, R.; Villanueva, A. Low-Cost Eye Tracking Calibration: A Knowledge-Based Study. *Sensors* **2021**, *21*, 5109. [CrossRef]
15. Avilés-Rodríguez, G.J.; Nieto-Hipólito, J.I.; Cosío-León, M.d.l.A.; Romo-Cárdenas, G.S.; Sánchez-López, J.d.D.; Radilla-Chávez, P.; Vázquez-Briseño, M. Topological Data Analysis for Eye Fundus Image Quality Assessment. *Diagnostics* **2021**, *11*, 1322. [CrossRef] [PubMed]
16. Aziz, T.; Ilesanmi, A.E.; Charoenlarpnopparut, C. Efficient and Accurate Hemorrhages Detection in Retinal Fundus Images Using Smart Window Features. *Appl. Sci.* **2021**, *11*, 6391. [CrossRef]
17. Rahman, Z.A.S.A.; Jasim, B.H.; Al-Yasir, Y.I.A.; Abd-Alhameed, R.A. High-Security Image Encryption Based on a Novel Simple Fractional-Order Memristive Chaotic System with a Single Unstable Equilibrium Point. *Electronics* **2021**, *10*, 3130. [CrossRef]
18. Almatroud, O.A.; Tamba, V.K.; Grassi, G.; Pham, V.T. An Oscillator without Linear Terms: Infinite Equilibria, Chaos, Realization, and Application. *Mathematics* **2021**, *9*, 3315. [CrossRef]
19. El-Latif, A.A.A.; Ramadoss, J.; Abd-El-Atty, B.; Khalifa, H.S.; Nazarimehr, F. A Novel Chaos-Based Cryptography Algorithm and Its Performance Analysis. *Mathematics* **2022**, *10*, 2434. [CrossRef]
20. Rahman, Z.A.S.A.; Jasim, B.H.; Al-Yasir, Y.I.A.; Abd-Alhameed, R.A. Efficient Colour Image Encryption Algorithm Using a New Fractional-Order Memcapacitive Hyperchaotic System. *Electronics* **2022**, *11*, 1505. [CrossRef]
21. Liu, Z.; Li, J.; Di, X. A New Hyperchaotic 4D-FDHNN System with Four Positive Lyapunov Exponents and Its Application in Image Encryption. *Entropy* **2022**, *24*, 900. [CrossRef]
22. Li, L.; Abd El-Latif, A.A.; Jafari, S.; Rajagopal, K.; Nazarimehr, F.; Abd-El-Atty, B. Multimedia Cryptosystem for IoT Applications Based on a Novel Chaotic System around a Predefined Manifold. *Sensors* **2022**, *22*, 334. [CrossRef]
23. Santos, D.F.; Barrera Amaya, I.; Suárez Parra, C.A. Encryption algorithm for color Images based on chaotic systems. *Ingeniería* **2020**, *25*, 144–161. [CrossRef]
24. Santos, D.F. Chaos-based Digital Image Encryption Using Unique Iris Features. *Int. J. Appl. Eng. Res.* **2020**, *15*, 358–363. [CrossRef]
25. Santos, D.F.; Espitia, H.E. Detection of Uveal Melanoma using fuzzy and neural networks classifiers. *Telkomnika* **2020**, *18*, 2213–2223. [CrossRef]
26. Santos, D.F.; Espitia, H.E. Proposal for a Neuro-Fuzzy System for Uveal Melanoma Detection. *J. Eng. Appl. Sci.* **2021**, *16*, 523–531.
27. Sodhi, B.; Agrawal, A.; Prabhakar, T.V. Appification of web applications: Architectural aspects. In Proceedings of the 2012 1st IEEE International Conference on Communications in China Workshops (ICCC), Beijing, China, 15–17 August 2012; pp. 1–7. [CrossRef]
28. New York Eye Cancer Center. Iris Tumors. Available online: https://eyecancer.com/eye-cancer/image-galleries/iris-tumors/ (accessed on 1 February 2021).
29. Daniel-Fernando, S.B.; Binh-Minh, N.; Helbert-Eduardo, E. Towards automated eye cancer classification via VGG and ResNet networks using transfer learning. *Eng. Sci. Technol. Int. J.* **2022**, in press. [CrossRef]
30. Goceri, E. Analysis of Deep Networks with Residual Blocks and Different Activation Functions: Classification of Skin Diseases. In Proceedings of the 2019 Ninth International Conference on Image Processing Theory, Tools and Applications (IPTA), Istanbul, Turkey, 6–9 November 2019. [CrossRef]
31. He, K.; Zhang, X.; Ren, S.; Sun, J. Deep Residual Learning for Image Recognition. In Proceedings of the 2016 IEEE Conference on Computer Vision and Pattern Recognition (CVPR), Las Vegas, NV, USA, 27–30 June 2016; pp. 770–778. [CrossRef]
32. Umamageswari, A.; Suresh, G. Security in medical image communication with arnold's cat map method and reversible watermarking. In Proceedings of the 2013 International Conference on Circuits, Power and Computing Technologies (ICCPCT), Nagercoil, India, 20–21 March 2013; pp. 1116–1121. [CrossRef]

33. Fu, C.; Tang, J.; Zhou, W.; Liu, W.; Wang, D. A symmetric color image encryption scheme based on chaotic maps. In Proceedings of the 2013 15th IEEE International Conference on Communication Technology, Guilin, China, 17–19 November 2013; pp. 712–716. [CrossRef]
34. Peng, J.; Jin, S.; Liu, Y. Design and Analysis of an Image Encryption Scheme Based on Chaotic Maps. In Proceedings of the 2010 International Conference on Intelligent Computation Technology and Automation, Changsha, China, 11–12 May 2010. [CrossRef]
35. Chen, D. A Feasible Chaotic Encryption Scheme for Image. In Proceedings of the 2009 International Workshop on Chaos-Fractals Theories and Applications, Shenyang, China, 6–8 November 2009. [CrossRef]
36. Mehta, G.; Dutta, M.K.; SooKim, P. Biometric data encryption using 3-D chaotic system. In Proceedings of the 2016 2nd International Conference on Communication Control and Intelligent Systems (CCIS), Mathura, India, 18–20 November 2016. [CrossRef]
37. Zou, C.; Zhang, Q.; Wei, X.; Liu, C. Image Encryption Based on Improved Lorenz System. *IEEE Access* **2020**, *8*, 75728–75740. [CrossRef]
38. Cellk, K.; Kurt, E. A new image encryption algorithm based on lorenz system. In Proceedings of the 2016 8th International Conference on Electronics, Computers and Artificial Intelligence (ECAI), Ploiesti, Romania, 30 June–2 July 2016; pp. 1–6. [CrossRef]
39. Abd-El-Hafiz, S.K.; AbdElHaleem, S.H.; Radwan, A.G. Permutation techniques based on discrete chaos and their utilization in image encryption. In Proceedings of the 2016 13th International Conference on Electrical Engineering/Electronics, Computer, Telecommunications and Information Technology (ECTI-CON), Chiang Mai, Thailand, 28 June–1 July 2016. [CrossRef]
40. Fei, H.; Daheng, G. Two kinds of correlation analysis method attack on implementations of Advanced Encryption Standard software running inside STC89C52 microprocessor. In Proceedings of the 2016 2nd IEEE International Conference on Computer and Communications (ICCC), Chengdu, China, 14–17 October 2016; pp. 1265–1269. [CrossRef]
41. Patidar, V.; Pareek, N.; Purohit, G.; Sud, K. A robust and secure chaotic standard map based pseudorandom permutation-substitution scheme for image encryption. *Opt. Commun.* **2011**, *284*, 4331–4339. [CrossRef]
42. Ye, R.; Guo, W. An Image Encryption Scheme Based on Chaotic Systems with Changeable Parameters. *Int. J. Comput. Netw. Inf. Secur.* **2014**, *6*, 37–45. [CrossRef]
43. Guo, W.; Wang, X.; He, D.; Cao, Y. Cryptanalysis on a parallel keyed hash function based on chaotic maps. *Phys. Lett. A* **2009**, *373*, 3201–3206. [CrossRef]
44. Oğraş, H.; Türk, M. A Robust Chaos-Based Image Cryptosystem with an Improved Key Generator and Plain Image Sensitivity Mechanism. *J. Inf. Secur.* **2017**, *8*, 23–41. [CrossRef]
45. Özkaynak, F. Role of NPCR and UACI tests in security problems of chaos based image encryption algorithms and possible solution proposals. In Proceedings of the 2017 International Conference on Computer Science and Engineering (UBMK), Antalya, Turkey, 5–8 October 2017; pp. 621–624. [CrossRef]
46. Shah, D.; Haq, T.U.; Shah, T. Image Encryption Based on Action of Projective General Linear Group on a Galois Field GF(28). In Proceedings of the 2018 International Conference on Applied and Engineering Mathematics (ICAEM), London, UK, 4–6 July 2018; pp. 38–41. [CrossRef]
47. Elkamchouchi, H.M.; Shawky, M.A.; Takieldeen, A.E.; Fouda, I.; Khalil, M.; Elkomy, A.; AbdElrasol, A. A New Image Encryption Algorithm Combining the Meaning of Location with Output Feedback Mode. In Proceedings of the 2018 10th International Conference on Communication Software and Networks (ICCSN), Chengdu, China, 6–9 July 2018; pp. 521–525. [CrossRef]
48. CASIA. Iris Database. Available online: http://forensics.idealtest.org (accessed on 1 February 2021).

Article

Sparse Reconstruction Using Hyperbolic Tangent as Smooth l_1-Norm Approximation

Hassaan Haider [1], Jawad Ali Shah [1,*], Kushsairy Kadir [2] and Najeeb Khan [3]

[1] Department of Electrical and Computer Engineering, Faculty of Engineering and Technology, International Islamic University, Islamabad 44000, Pakistan
[2] Electrical Section, British Malaysian Institute, Universiti Kuala Lumpur, Gombak 53100, Malaysia
[3] Department of Electrical and Computer Engineering, University of British Columbia, Vancouver, BC V6T 1Z4, Canada
* Correspondence: jawad.shah@iiu.edu.pk; Tel.: +92-333-5274448

Abstract: In the Compressed Sensing (CS) framework, the underdetermined system of linear equation (USLE) can have infinitely many possible solutions. However, we intend to find the sparsest possible solution, which is l_0-norm minimization. However, finding an l_0 norm solution out of infinitely many possible solutions is NP-hard problem that becomes non-convex optimization problem. It has been a practically proven fact that l_0 norm penalty can be adequately estimated by l_1 norm, which recasts a non-convex minimization problem to a convex problem. However, l_1 norm non-differentiable and gradient-based minimization algorithms are not applicable, due to this very reason there is a need to approximate l_1 norm by its smooth approximation. Iterative shrinkage algorithms provide an efficient method to numerically minimize l_1-regularized least square optimization problem. These algorithms are required to induce sparsity in their solutions to meet the CS recovery requirement. In this research article, we have developed a novel recovery method that uses hyperbolic tangent function to recover undersampled signal/images in CS framework. In our work, l_1 norm and soft thresholding are both approximated with the hyperbolic tangent functions. We have also proposed the criteria to tune optimization parameters to get optimal results. The error bounds for the proposed l_1 norm approximation are evaluated. To evaluate performance of our proposed method, we have utilized a dataset comprised of 1-D sparse signal, compressively sampled MR image and cardiac cine MRI. The MRI is an important imaging modality for assessing cardiac vascular function. It provides the ejection fraction and cardiac output of the heart. However, this advantage comes at the cost of a slow acquisition process. Hence, it is essential to speed up the acquisition process to take the full benefits of cardiac cine MRI. Numerical results based on performance metrics, such as Structural Similarity (SSIM), Peak Signal to Noise Ratio (PSNR) and Root Mean Square Error (RMSE) show that the proposed tangent hyperbolic based CS recovery offers a much better performance as compared to the traditional Iterative Soft Thresholding (IST) recovery methods.

Keywords: compressed sensing; holography cardiac cine MRI; l_1-norm smooth approximations; hyperbolic tangent function; soft thresholding

Citation: Haider, H.; Shah, J.A.; Kadir, K.; Khan, N. Sparse Reconstruction Using Hyperbolic Tangent as Smooth l_1-Norm Approximation. *Computation* **2023**, *11*, 7. https://doi.org/10.3390/computation11010007

Academic Editors: Anando Sen and Demos T. Tsahalis

Received: 11 October 2022
Revised: 29 November 2022
Accepted: 16 December 2022
Published: 4 January 2023

Copyright: © 2023 by the authors. Licensee MDPI, Basel, Switzerland. This article is an open access article distributed under the terms and conditions of the Creative Commons Attribution (CC BY) license (https://creativecommons.org/licenses/by/4.0/).

1. Introduction

Compressed Sensing (CS) exploits the sparsity of signals in a certain domain to find a near-optimal solution to the underdetermined system of linear equations. In CS, the sampling of signals depends on the information rate rather than its bandwidth. CS technique facilitates simultaneous acquisition and compression of compressible or sparse signals that potentially reduce the acquisition time. The CS is a data acquisition method that allows for the reconstruction of a signal from very few measurements if the signal is transformed in a sparsifying domain, and these measurements are highly incoherent with respect to its sparsifying transform. Unfortunately, most of the reconstruction techniques of compressively sampled signals are computationally expensive and non-linear [1–3].

CS has recently been used to reconstruct under-sampled biomedical images by exploiting the sparsity of biomedical images in the sparsifying domain. The Fourier-encoded nature of the MR image scanning process and the existence of suitable sparsifying transform domains, i.e., Wavelets, Contourlets, total variation, etc., make the MRI a potentially suitable application of CS [4]. Incoherent sampling, which is another important requirement of CS, can be accomplished with the variable density k-space sampling method to introduce noise-like random aliasing artefacts during the MR image recovery. Variable density k-space under-sampling pattern samples with high density from the center of the k-space that contains maximum energy of the MR images and undersamples the outer k-space region with lower density to efficiently reduce the MR image scanning time [5].

Cardiac cine magnetic resonance imaging (MRI) is an emerging medical imaging modality to evaluate the growth of Cardiac-Vascular Disease (CVD). It is useful in evaluating the cardiac wall thickness and motion in CVD patients [6–8]. Further, cardiac cine MRI aids in performing the quantitative study of ejection fraction and cardiac output of the heart. The ejection fraction is the percentage of blood that is ejected out of the ventricles with each contraction. This amount is used to determine heart failures and other types of heart diseases [9]. Cardiac output measures the amount of blood pumped by the heart per minute. However, these advantages are limited by the lengthy acquisition process of cardiac cine MRI that requires multiple breaths-holds of the patient and extended patient engagement in MRI scanner. Therefore, it is essential to accelerate the image acquisition in cardiac cine MRI by using fast pulse sequences and/or by reducing the number of samples taken during data acquisition [10,11]. As the former approach is inherently limited by different constraints, much research interest is moved to the latter approach. CS can be applied successfully to the cardiac cine MRI, where sparsity is exploited in the temporal dimension [4,12]. However, improving speed and efficiency of CS recovery methods is an active area of interest for researchers working in medical imaging especially MRI. The key conditions for the CS framework to work are sparsity, non-linear reconstruction and incoherent undersampling.

In MR imaging, sparsity can be accomplished by transforming the image in its sparse representation. To fulfil the condition of incoherent sampling in MR imaging, various undersampling patterns can be utilized, such as radial lines sampling and variable density sampling [4]. Non-linear reconstruction numerical techniques involve l_1-norm regularization in order to find sparse solutions to the least-squares optimization problem. l_2-norm based regularization provides the linear and simplest solution to under-determined system problem. However, it minimizes the energy of the error and distributes it over all solution set that results in non-sparse solution that does not fits in CS framework. Similarly l_p-norm $(1 < p < \infty)$ based regularization, as value of p starts growing it tends to penalize only the largest parameter, such as max function, and some bad parameters may hide under the largest parameters, which results in less-sparse solution. For this very reason l_1-norm is the preferred regularization, as it promotes sparsity that perfectly fits in CS framework [1–3]. However, l_1-norm penalty is non-differentiable, so applying efficient optimization methods that involve derivative are not feasible. Therefore, various methods have been proposed to resolve the l_1-norm regularization problem. The IST based recovery methods have successfully been utilized to efficiently reconstruct images from under-sampled data in the CS framework [5,13]. An iterative hard thresholding-based recovery method is proposed for the compressed sensing problem [14]. However, this algorithm has limited performance as compared to the soft thresholding-based methods. Random filters for compressive sampling [15], Bregman iterative algorithms for compressed sensing [16], and a weighted l_1 minimization recovery algorithm [17] are proposed to solve the compressed sensing recovery problem. Lately, smooth l_1-norm penalty based sparse signal reconstruction method was evolved for approximation of l_1-norm that uses a hyperbolic tangent function [18]. The research shows that this technique can be used for the reconstruction of undersampled MR images from fewer acquired samples, which allows fast imaging without compromising spatial resolution. Jawad et al. proposed that the wavelet thresholding can be implemented using hyperbolic tangent function. It was explained that the differentiable hyperbolic function provides a much more accurate recovery than IST techniques [19–21]. It was experimentally shown that the hyperbolic tangent function performs a much improved signal recovery as compared to the hard threshold, soft threshold and Garrote threshold functions [22,23].

In this paper, CS technique is applied to reduce MRI scanning time, CS exploits the sparsity of MRI in Fourier domain that enables us to take fewer samples without compromising on the quality of image recovered from undersampled MRI. We propose a novel and more efficient CS recovery algorithm based on the hyperbolic tangent function for approximating l_1-norm and shrinkage operation for accurate recovery of compressively sampled sparse signals, MR images and the cardiac cine MRI. We introduce smooth approximation of l_1-norm, the hyperbolic tangent function, where steepest descent algorithm is applicable for minimization of objective function. The error bounds for the proposed l_1-norm penalty are presented in this paper. We have used the soft thresholding technique based on the hyperbolic tangent function that is inspired by the maximum a posteriori (MAP) noise estimator. In this work, we also recommend the efficient criteria for the tuning parameters. Performance analysis of the proposed method is shown using simulations; to recover random 1-D sparse signal, 2-D MR image and clinical cardiac cine MRI. Several quantitative performance measures are used apart from qualitative depiction, i.e., Mean Square Error (MSE), Root Mean Squared Error (RMSE), Signal to Noise Ratio (SNR), Peak Signal to Noise Ratio (PSNR), Improved Signal to Noise Ratio (ISNR), correlation, fitness, and Structural Similarity (SSIM) in order to prove the supremacy of proposed method over existing recovery techniques.

2. Materials and Methods

Reconstruction of undersampled signal through CS is an optimization problem, which promotes sparsity in our solution by minimizing the l_1-norm.

2.1. Proposed Method

Let $z \epsilon \mathbb{R}^n$ be the signal in a vector form and $y \epsilon \mathbb{C}^m$ be the undersampled measurements. Then, the CS recovery function is written as:

$$f(z) = \frac{1}{2}\|y - \Phi\Psi^H z\|_2^2 + \lambda \|z\|_1 \qquad (1)$$

where Φ is the sampling domain of the signal z, whereas Ψ represents sparsifying transform. The tuning parameter λ in Equation (1) provides an important trade-off parameter between fidelity and sparsity. The performance of our algorithm is dependent on proper threshold level selection. We have employed the fixed value expression, depending on the signal dimensions and its noise variance [13].

$$\lambda = \sigma_v \sqrt{2 \ln(n)} \qquad (2)$$

where σ_v is the noise standard variance and n is the length of the sparse signal.

Since the hyperbolic tangent function has properties, such as non-convex, odd, smooth analytical bounded function that is monotonically increasing, the slope of the function at the origin can be tuned to any desired value [21]. So, our proposed approximation for the l_1 norm in Equation (1) is defined as:

$$\|z\|_1 \cong \sum_{i=1}^{n} z_i tanh(\gamma z_i) \qquad (3)$$

Since the hyperbolic tangent function is used as a smooth and differentiable approximation to l_1-norm. Therefore, the value of γ is taken quite high to make it closer to l_1-norm, as shown in Figure 1. It is also providing the benefits of the smoothness and differentiability. Equation (1) can now be written as:

$$f(z) = \frac{1}{2}\|y - \Phi\Psi^H z\|_2^2 + \lambda \sum_{i=1}^{n} z_i tanh(\gamma z_i) \qquad (4)$$

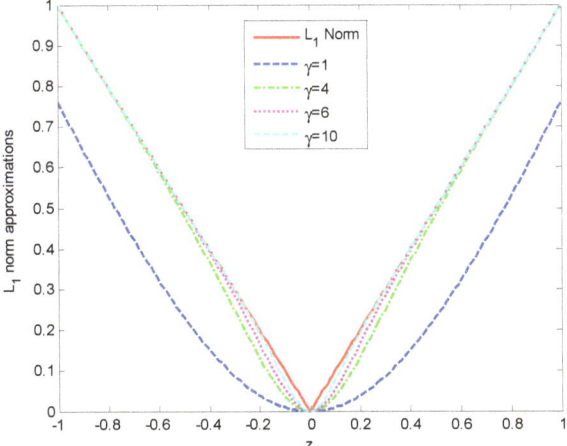

Figure 1. l_1 norm approximation using the hyperbolic tangent function for different values of $\gamma = (1, 4, 6,$ and $10)$. As the value of gamma continues to increase and the approximation is closer to the actual l_1 norm, however, it is less smooth. The proposed technique gives us the flexibility to choose between the level of smoothness and accuracy.

For steepest descent algorithm, vector differentiation can not be used. Hence, it can be rewritten as Equation (4) in element form to find the partial derivative. Let $\mathbf{A} = \Phi\Psi^H$, then the element-wise equation is defined as:

$$f(z) = \frac{1}{2}\sum_i (\mathbf{A}z - y)_i (\mathbf{A}z - y)_i + \lambda z_i tanh(\gamma z_i) \qquad (5)$$

Let $\mathbf{A} = \phi\Psi^H$, then partial derivative of Equation (4) in element form is formulated as:

$$\frac{\partial f(z)}{\partial z_l} = \sum_{ij} A_{ij} A_{il} z_j - \sum_i y_i A_{il} + \lambda \left(\tanh(\gamma z_l) + z_l \gamma \left(1 - \tanh^2(\gamma z_l)\right) \right) \qquad (6)$$

Hence, the steepest descent algorithm for lth update is:

$$(\Delta z)_l = -\eta \frac{\partial f(z)}{\partial z_l} \qquad (7)$$

Equation (7) is used to find a solution using the steepest descent algorithm.

2.2. Error Bounds for Proposed Smooth l_1-Norm

The error bounds for the proposed smooth l_1 norm approximation defined by Equation (2) are derived in this section [24]. The l_1 norm approximation is proposed based on the following two principles.

1. $|z| = (z)_+ + (-z)_+$, where $(z)_+ = \max\{z, 0\}$ is the plus function;
2. This plus function can be smoothly approximated as:

$$(z)_+ \approx p(z, \gamma) = \frac{1}{2}[z + z.\tanh(\gamma z)] \qquad (8)$$

From Equation (8), we can write a smooth approximation of l_1 norm:

$$\begin{aligned}
\|z\|_1 &= (z)_+ + (-z)_+ \approx p(z,\gamma) + p(-z,\gamma) \\
&= \tfrac{z}{2}[1+\tanh(\gamma z)] - \tfrac{z}{2}[1+\tanh(-\gamma z)] \\
&= \tfrac{z}{2}\tanh(\gamma z) + \tfrac{z}{2}\tanh(\gamma z) \\
&= z\,\tanh(\gamma z) \\
&= \|z\|_\gamma
\end{aligned} \qquad (9)$$

Equation (9) represents the γ approximation of the l_1 norm, as shown in Figure 1.

Unlike the l_1 norm, we can apply the unconstraint optimization techniques, where gradient needs to be calculated and the proposed approximation is twice differentiable and the 1st and 2nd order gradients of the proposed l_1 norm are shown in Equations (10) and (11), respectively.

$$\nabla(\|z\|) \approx \tanh(\gamma z) - \gamma z\left(tanh(\gamma z)^2 - 1\right) \qquad (10)$$

$$\nabla^2(\|z\|) \approx 2\gamma(\gamma z\,tanh(\gamma z) - 1)\left(tanh(\gamma z)^2 - 1\right) \qquad (11)$$

As the value of γ approaches infinity the error between $\|\mathbf{z}\|_1$ and $\|\mathbf{z}\|_\gamma$ approaches zero. We here propose the simple lemma to determine the error bounds for $\|\mathbf{z}\|$ and $\|z\|_\gamma$.

Lemma 1. *The proposed smooth function of l_1 norm $f(z) = z\,\tanh(\gamma z)$ fulfils the sufficient and necessary convexity condition in the interval $z \in [-1, 1]$ as its derivative $f\prime(z)$ defined by Equation (9) is monotonically non-decreasing and its second derivative $f''(z)$ defined by Equation (10) is nonnegative for $0 < \gamma \leq 1$;*

Lemma 2. *l_1 norm approximation error bounds for any $z \in \mathbb{R}$ and $\gamma > 0$.*

$$\left|\|z\|_1 - \|z\|_\gamma\right| \leq \frac{1}{2\gamma} \qquad (12)$$

Proof. Let us consider two cases, first case for $z > 0$,

$$\begin{aligned}
p(z,\gamma) - (z)_+ &= \tfrac{z}{2}(1 + \tanh(\gamma z)) - z \\
&= \tfrac{z}{2}(1 + \tanh(\gamma z)) - z \\
&= \tfrac{z}{2}(\tanh(\gamma z) - 1)
\end{aligned} \qquad (13)$$

Now, we can find the maximum value of $\tanh(\gamma z)$ to find the upper bound for Equation (13). As we know that the maximum value of $\tanh(\gamma x)$ is 1, so we can write:

$$\underset{z}{maxima}\ \tanh(\gamma z) = \frac{e^{\gamma z} - e^{-\gamma z}}{e^{\gamma z} + e^{-\gamma z}} = 1 \qquad (14)$$

Using Equation (14), the relationship between γ and z can be easily derived as:

$$z = \frac{1}{2\gamma} \qquad (15)$$

By inserting the value of z from Equation (15) in Equation (14)

$$p(z,\gamma) - (z)_+ \leq \frac{1}{4\gamma} \qquad (16)$$

□

Figure 2 shows the proposed method obeys error bounds define by Equation (16).

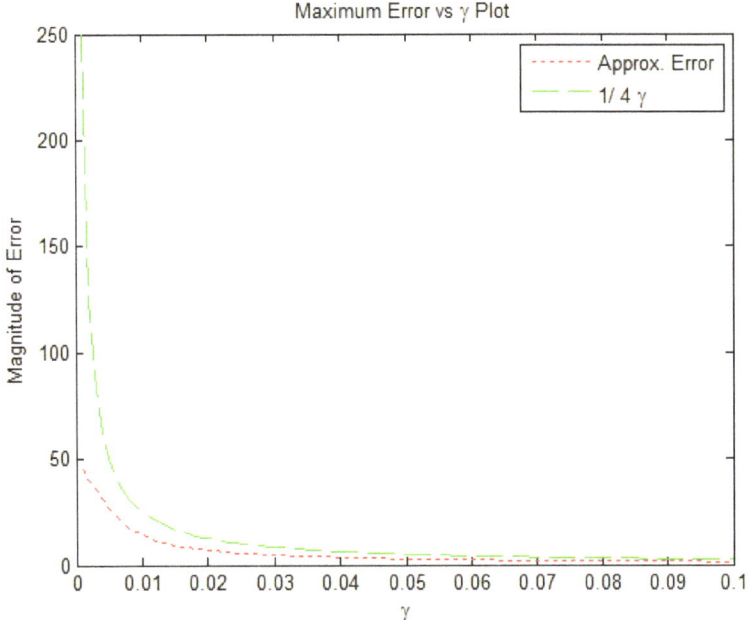

Figure 2. l_1 norm approximation error bounds for $z > 0$, the green line shows the upper bound proved mathematically in Equation (16), whereas the dotted red line shows the actual error between the proposed l_1 norm smooth approximation and actual non-differentiable l_1 norm. The error is maximum at approximately zero and approaches zero as $\gamma \to \infty$.

Figure 3 shows graphically that our proposed approximation function obeys the bounds defined by Equation (19).

For $z \leq 0$,

$$0 \leq p(z,\gamma) - (z)_+ = p(z,\gamma) \leq p(0,\gamma) \\ = \tfrac{z}{2}(\tanh(\gamma z) - 1) \leq 0 \\ = \tfrac{1}{4\gamma} \tag{17}$$

As p is the monotonically increasing function. Hence, from Equations (17) and (18), $p(z,\gamma)$ will dominate $(z)_+$, so

$$|p(z,\gamma) - (z)_+| \leq \frac{1}{4\gamma} \tag{18}$$

From Equation (8), we can insert $\|z\| = (z)_+ + (-z)_+$

$$\left|\|z\|_1 - \|z\|_\gamma\right| = |p(z,\gamma) + p(-z,\gamma) - ((z)_+ + (-z)_+)| \\ \leq |p(z,\gamma) - (z)_+| + |p(-z,\gamma) - (-z)_+| \\ \leq \tfrac{1}{4\gamma} + \tfrac{1}{4\gamma} = \tfrac{1}{2\gamma} \tag{19}$$

Figure 3 shows the error bounds versus error in the smooth approximation.

Let us define $\|z\|_{(1,\gamma)}$ as a smooth approximation to the l_1 norm function $\|z\|_1$ for a vector $z \in \mathbb{R}^n$ as:

$$\|z\|_{(1,\gamma)} = \sum_i^n \|z_i\|_\gamma \\ \left|\|z\|_{(1,\gamma)} - \|z\|_1\right| \leq 2n\tfrac{1}{4\gamma} = \tfrac{n}{2\gamma} \tag{20}$$

Hence, we can conclude that:

$$\lim_{\gamma \to \infty} \|z\|_{(1,\gamma)} = \|z\|_1 \ \forall z \in \mathbb{R}^n \tag{21}$$

Let $L : \mathbb{R}^n \to \mathbb{R}$ by any continuous cost function and defined by $f(z) = L(z) + \|z\|_1$ and $f_\gamma(z) = L(z) + \|z\|_{(1,\gamma)}$. If we define $\bar{z} = \underset{z}{argmin}\, f(z)$ and $\bar{z}_\gamma = \underset{z}{argmin}\, f_\gamma(z)$. By definition of f and f_γ and from Equation (20), it can be concluded that

$$\lim_{\gamma \to \infty} f_\gamma(z) = f(z) \ \forall z \in \mathbb{R}^n \tag{22}$$

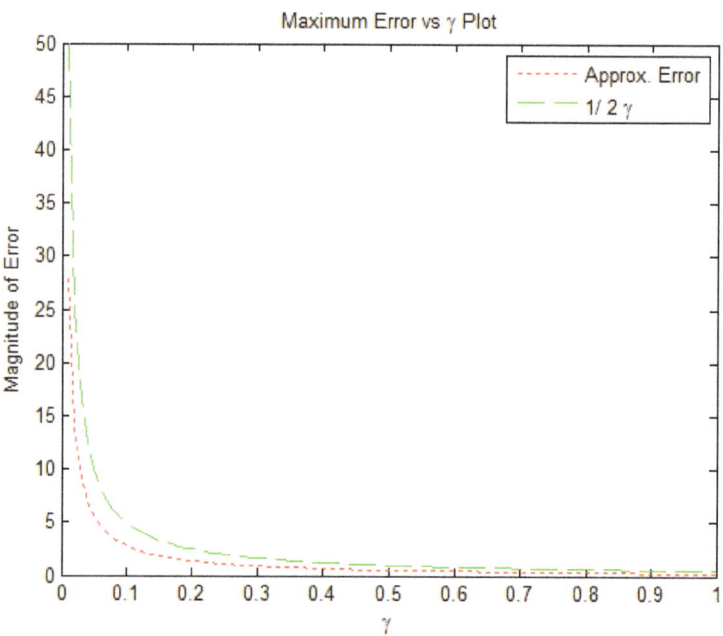

Figure 3. l_1 norm approximation error bounds for $z < 0$, the green line shows the upper bound proved mathematically in Equation (19), whereas the dotted red line shows the actual error between proposed l_1 norm smooth approximation and actual non-differentiable l_1 norm. The error is maximum at approximately zero and approaches zero as $\gamma \to \infty$.

In addition, it is a known fact that $f(\bar{z}) \leq f(z) \forall z$. In particular $f(\bar{z}) \leq f(\bar{z}_\gamma)$, then:

$$\begin{aligned} f(\bar{z}) \leq f(\bar{z}_\gamma) &= L(\bar{z}_\gamma) + \|\bar{z}_\gamma\|_1 \\ &= L(\bar{z}_\gamma) + \|\bar{z}_\gamma\|_1 + \|\bar{z}_\gamma\|_{(1,\gamma)} - \|\bar{z}_\gamma\|_{(1,\gamma)} \\ &= \left(L(\bar{z}_\gamma) + \|\bar{z}_\gamma\|_{(1,\gamma)}\right) + \left(\|\bar{z}_\gamma\|_1 - \|\bar{z}_\gamma\|_{(1,\gamma)}\right) \\ &= f_\gamma(\bar{z}_\gamma) + \left(\|\bar{z}_\gamma\|_1 - \|\bar{z}_\gamma\|_{(1,\gamma)}\right) \end{aligned} \tag{23}$$

This implies that $f(\bar{z}) - f_\gamma(\bar{z}_\gamma) \geq -\frac{n}{2\gamma}$ from Equation (21), similarly $(\bar{z}) - f_\gamma(\bar{z}_\gamma) \leq \frac{n}{2\gamma}$, hence proved that $\lim_{\gamma \to \infty} f_\gamma(\bar{z}) = f(\bar{z})$.

It can be further stated that:

$$\begin{aligned} |f(\bar{z}_\gamma) - f(\bar{z})| &= |f(\bar{z}_\gamma) - f(\bar{z}) - f_\gamma(\bar{z}_\gamma) + f_\gamma(\bar{z}_\gamma)| \\ &\leq |f(\bar{z}_\gamma) - f_\gamma(\bar{z}_\gamma)| + |f_\gamma(\bar{z}_\gamma) - f(\bar{z})| \end{aligned} \tag{24}$$

Hence, it proved that $\lim_{\gamma \to \infty} f(\bar{z}_\gamma) = f(\bar{z})$. Moreover, if L is strictly convex, it can be easily proven that: $\lim_{\gamma \to \infty} z_\gamma = \bar{z}$.

3. The Map Estimator and Proposed Thresholding Mechanism

Conventionally l_1-norm minimization has inherent soft thresholding [13]. However, when we approximate the l_1-norm by the hyperbolic tangent function, thresholding is not done implicitly. The hard thresholding operator proposed in [14] can be defined by the following equation.

$$S_\beta(z) = \begin{cases} z & |z| > \beta \\ 0 & otherwise \end{cases} \qquad (25)$$

We have used a new thresholding function based on the tangent hyperbolic function. Therefore, β is an important parameter in controlling the under-sampling noise, which has Gaussian distribution [21]. To obtain the optimum value of β, the thresholding depends upon the undersampling noise. Therefore, the following data-driven thresholding parameter β is used [13,25].

$$\beta = \frac{\sigma_v^2}{\sigma_z} \qquad (26)$$

With σ_z is the standard deviation of sparse signal and σ_v the standard deviation of Gaussian-like noise produced due to under-sampling.

To enhance the performance under different scenarios, different mathematical thresholding operators could be found in the literature [26–28]. The main idea in this approach is mapping the values nearer to the origin to zero and those that are further away from the origin are shrunk towards zero.

The basic denoising technique aims to find the estimate of original image or signal from its perturbed set of observations, as shown in Equation (27).

$$y = z + v \qquad (27)$$

where $y \in \mathbb{R}^n$ is the noisy image, $z \in \mathbb{R}^n$ is the original signal and v is the zero-mean Gaussian noise with probability distribution function (pdf) given by:

$$p_v(\theta) = \frac{1}{\sqrt{2\pi\sigma_v^2}} \exp\left(\frac{\|\theta\|_2^2}{2\sigma_v^2}\right) \qquad (28)$$

By taking the Wavelet transform of Equation (27), we get:

$$q = s + v \qquad (29)$$

where $q = \Psi y$ and $s = \Psi z$ represent the sparsifying domain for noisy image and the original image, respectively. As Wavelet transform is the linear operator, therefore the zero-mean Gaussian noise v after transformation will not change. The MAP estimation of random vector s is given by:

$$\hat{s} = \max_{s \in \mathbb{R}^n} p(s|q) \qquad (30)$$

By using Bayes' rule, one can ignore $p(q)$ as it is independent of s, MAP estimator can be written as:

$$\hat{s} = \max_{s \in \mathbb{R}^n} p(q|s) p_s(s) \qquad (31)$$

The problem defined in Equation (31) can be further simplified by taking $p(q|s) = p_v(q-s)$:

$$\begin{aligned} \hat{s} &= \max_s [p_v(q-s)] p_s(s) \\ &= \max_s [ln p_v(q-s) + ln p_s(s)] \\ &= \max_s \left[ln\left\{ \frac{1}{\sqrt{2\pi\sigma_v^2}} \exp\left(-\frac{\|q-s\|_2^2}{2\sigma_v^2}\right) \right\} + ln p_s(s) \right] \\ &= \max_s \left[ln\left\{ \left(\frac{1}{\sqrt{2\pi\sigma_v^2}}\right)^n \exp\left(-\frac{\|q-s\|_2^2}{2\sigma_v^2}\right) \right\} + ln p_s(s) \right] \\ &= \max_s \left[-\frac{\|q-s\|_2^2}{2\sigma_v^2} + f(s) \right] \end{aligned} \qquad (32)$$

where $f(s) = \ln p_s(s)$. By differentiating the argument of Equation (32) w.r.t. s and equating the result to zero, we can calculate the MAP estimator for Wavelet coefficients of the noise-free image as:

$$\frac{(q_i - \hat{s}_i)}{\sigma_v^2} + f'(\hat{s}_i) = 0, \quad 1 \leq i \leq n \tag{33}$$

The pdf of biomedical images are more peaked at the center than Gaussian, so Laplacian can better estimate the distribution of Wavelet domain coefficients, i.e.,

$$p_s(s_i) = \frac{1}{\sqrt{2}\sigma_v} \exp\left(\frac{\sqrt{2}}{\sigma_v}|s_i|\right) \tag{34}$$

gives $f'(\hat{s}_i) = -\frac{\sqrt{2}}{\sigma} sig(\hat{s}_i)$. Solving Equation (33) will result in

$$q_i = \hat{s}_i + \sqrt{2}\, sig(\hat{s}_i) \tag{35}$$

Let $\beta = \sqrt{2}\sigma_v^2$ and solve Equation (35) for $\hat{s}_i$ to formulate the nonlinear shrinkage:

$$\hat{s}_i = S_\beta(q) = max\{|q| - \beta, 0\}.sig(q) \tag{36}$$

Equation (36) can be further elaborated as:

$$S_\beta(q) \cong \begin{cases} sgn(q)(|q| - \beta) & |q| > \beta \\ 0 & otherwise \end{cases} \tag{37}$$

In this paper, novel thresholding approach has been proposed, which is used on the hyperbolic tangent, as the hyperbolic tangent function slope can be adjusted from the origin and it is a bounded function that makes it an suitable surrogate function for soft thresholding. Hence, hyperbolic tangent based soft thresholding can be described mathematically by following equation:

$$S_\beta(q) \cong \begin{cases} cz\{tanh(\alpha(|q| - \beta))\} & |q| > \beta \\ 0 & otherwise \end{cases} \tag{38}$$

where β is a thresholding parameter and parameter α is used to control the shape of the hyperbolic tangent function. If α is closer to zero, Equation (38) approximately changes into the soft thresholding function. When α approaches ∞, Equation (38) changes to the hard thresholding function, as shown in Figure 4. Our proposed Algorithm 1 starts as a soft thresholding function and smoothly changes to a hard threshold at higher iterations. The proposed approximation of this soft thresholding results in a better reconstruction as compared to the conventional soft thresholding method [21] as illustrated in Figure 4.

Algorithm 1. Proposed Algorithms.

Inputs:
Sensing matrix $\mathcal{F}_u$, measurement vector $y \epsilon \mathbb{C}^m$, parameters γ, λ and β,
Output:
A k-sparse vector $\hat{x} \in R^n$
Initialization: Initialize x_0, Index $i = 0$
Step-1 (**Sparse Representation**): $z_i = \Psi x_i$
Step-1 (**Gradient Computation**): Find $\nabla f(z_i)$ using Equation (5)
Step-2 (**Solution Update**): Compute the update using Equations (6) and (7).
Step-3 (**Shrinkage**): Estimate Solution using Equation (38), i.e., $\hat{z}_{i+1} = S_\beta(z_{i+1})$
Step-4 (**Repeat**): If stopping criterion is not met, $i = i + 1$ & go to step 1
Output: $\hat{x} = \Psi^H \hat{z}_i$

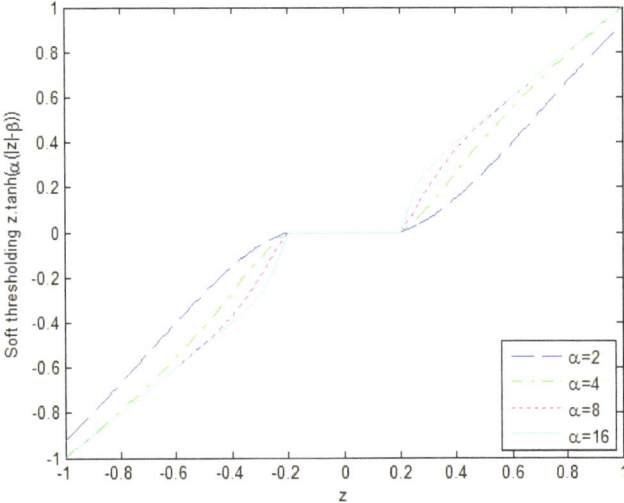

Figure 4. Hyperbolic tangent function based thresholding for alpha $\alpha = (2, 4, 8, 16)$, the value of α determines the slope of soft thresholding. The proposed method gives us the flexibility to shape the curves using α depending upon its application.

4. Results and Discussions

In order to gauge the performance of proposed algorithm, we have applied our algorithm to 1-D sparse signal, Compressively sampled MR image and Cine Cardiac MRI. MRXCAT simulator is used to evaluate the proficiency of recovery algorithms in the field of Cardiac MRI. We have evaluated the performance of our proposed technique quantitatively and qualitatively. The performance measures that are used in this research article are: pictorial depiction of under-sampling artefacts, Structural Similarity (SSIM), Peak Signal to Noise Ratio (PSNR) and Root Mean Square Error (RMSE).

4.1. 1-D Sparse Signal Recovery

The proposed algorithm is applied for the recovery of the 1-D sparse signal recovery, where the random sparse signal of length $n = 512$ is created in MATLAB and the support for the sparse signal was generated randomly with K = 85 non-zero elements. The random sparse signal is compressively sampled using a random measurement matrix $A \in R^{256 \times 512}$ with only $m = 256$ measurements.

Figure 5 shows the fitness achieved by the proposed algorithm and soft-thresholding method. The proposed method achieved faster convergence as compared to soft thresholding. Figure 6 shows sparsity effect on successful recovery achieved by the soft thresholding and proposed algorithm. The proposed algorithm performs much better even with a higher sparsity level as compared to the soft thresholding technique. Similarly, in Figure 7, the proposed method recovered the sparse signal with great accuracy, whereas the soft thresholding technique failed to accurately recover the sparse signal. The accuracy of the proposed technique was also measured against performance measures, such as SNR, MSE and correlation, as shown in Table 1. The proposed algorithm performed much better than the soft thresholding method against all these performance measures. The time comparison for proposed algorithm is 1.57 s as compared to 1.34 s by conventional soft thresholding.

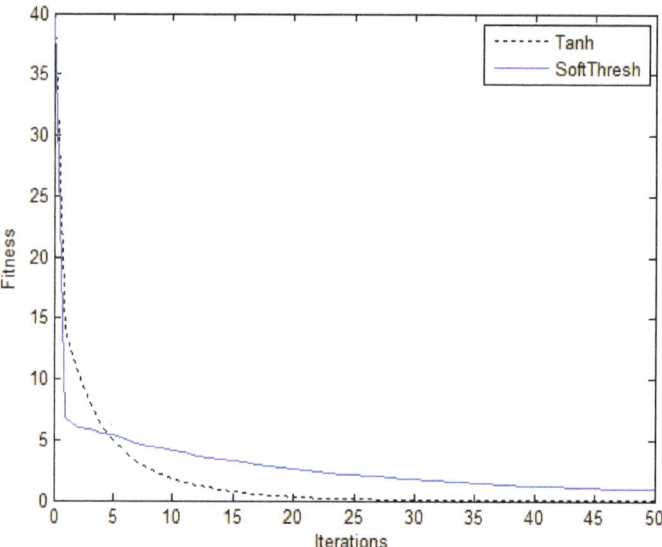

Figure 5. Fitness achieved by the soft thresholding and proposed algorithm. The proposed algorithm converges rapidly as compared to the soft thresholding technique.

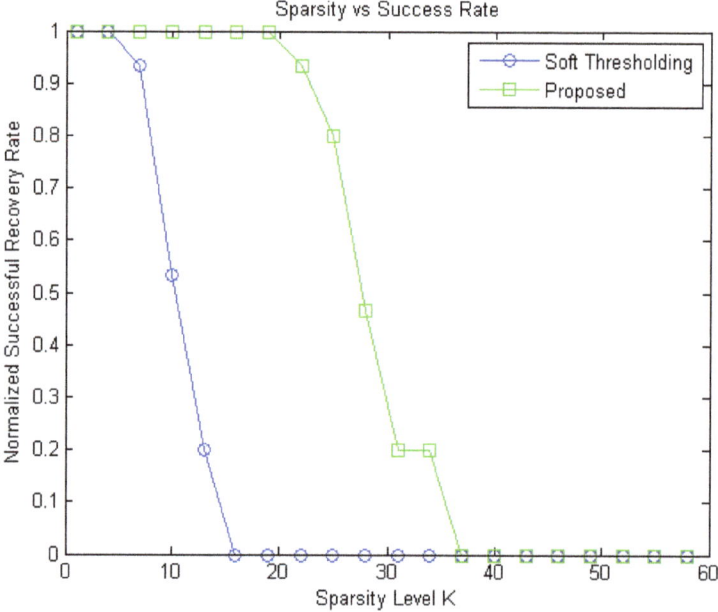

Figure 6. Sparsity effect on successful recovery achieved by the soft thresholding and proposed algorithm. The proposed algorithm performs much better even with higher sparsity level as compared to the soft thresholding technique.

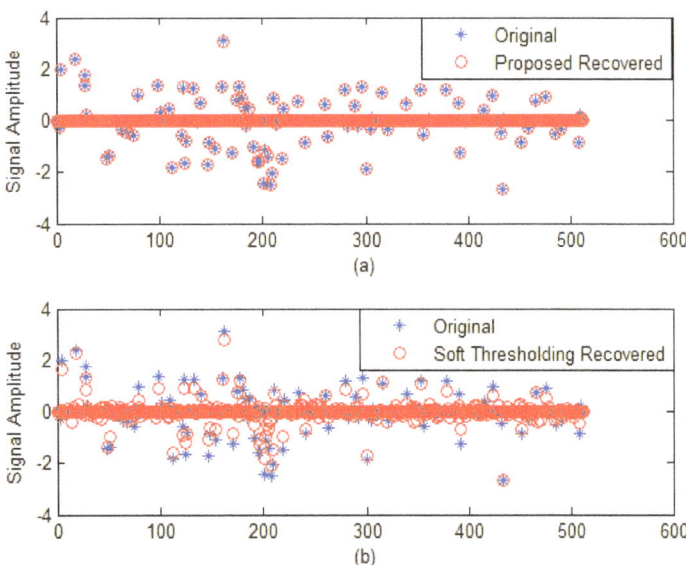

Figure 7. (a) The recovered sparse signal from the proposed algorithm; (b) The recovered sparse signal from soft thresholding.

Table 1. Performance comparison of different sparsity transforms using mean squared error in the transform domain. Temporal FFT performs better in cardiac cine MRI.

Performance Metrics	Soft Thresholding	Proposed Algorithm
MSE	1.00×10^{-2}	1.61×10^{-4}
Fitness	0.8664	0.0224
SNR	12.6712	30.6259
Correlation	0.9787	0.9995

4.2. 2-D Compressively Sampled MR Image Recovery

The random sampling at the CS image acquisition produces incoherent and noise-like artefacts in its sparsifying domain. In case of MR imaging or similar Fourier domain encoded biomedical imaging, where the MR image is in the spatial domain, the linear reconstruction (where, missing Fourier data points are replaced by zero and the resultant image inverse Fourier transform is taken) produces artifacts similar to additive Gaussian noise. The type of noise produced by subsampling is governed by undersampling patterns [29]. In order to recover an image, the compressed sensing recovery essentially becomes an image denoising problem. Using this analogy of CS encoding and noisy image, the first step in recovering the original image is to estimate a noise, this is achieved by maximum a posteriori (MAP) estimator. The proposed algorithm is also implemented to recover a 2-D Compressively sampled real human brain MR image of size 256 × 256. The human brain MR image is a fully sampled scanned image by a 1.5 Tesla GE-HDxt-MRI scanner with Gradient Echo (GE) sequence and 8 channels head coils with the specifications, i.e., TE = 10 msec, flip angle = 90°, bandwidth = 31.25 KHz, slice thickness=3 mm, TR = 55, and image dimensions = 256 × 256, at St. Mary's Hospital, London, UK. This MR image is compressively sampled by taking only 25% samples in k-space.

Figure 8 shows the performance of the proposed method with respect to Structural Similarity (SSIM). The proposed method achieved much better SSIM as compared to soft thresholding. The Peak Signal to Noise Ratio (PSNR) accomplished by proposed method is shown in Figure 9. Figure 10 shows the (a) Original 2D Brain MR Image,

(b) Conventional Soft Thresholding based recovered 2D MRI, (c) 2D Brain MR Image recovered from undersampled image, (d) Difference of original and soft thresholding image, (e) Difference of proposed recovery method image with original image. The Difference is scaled up by 1000 in order to enhance its visibility. Table 2 shows the proposed method has outperformed soft thresholding method in terms of PSNR and SSIM. Table 3 shows the performance of proposed algorithm and soft thresholding in terms of Mean Square Error (MSE), Improved Signal to Noise Ratio (ISNR), Correlation, SSIM, SNR, PSNR after the 15 iterations of soft thresholding and the proposed algorithm. The results show that proposed method achieved much better results in as compared to soft thresholding.

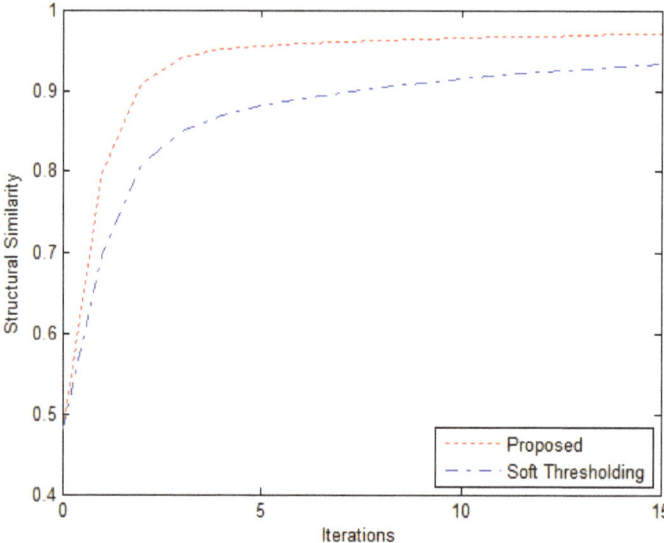

Figure 8. Structural Similarity of proposed and soft thresholding algorithm for recovery of compressively sampled MR image against each iteration.

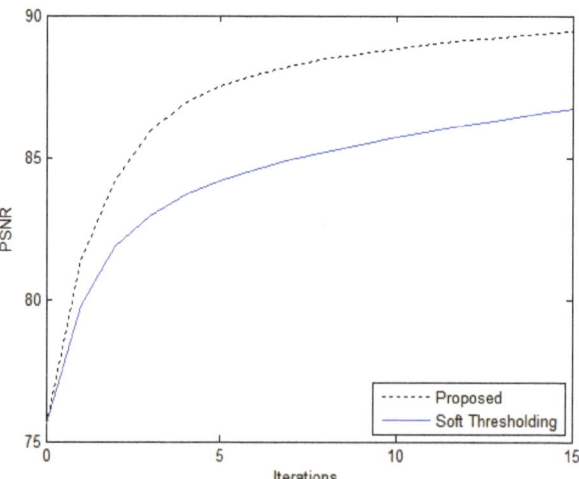

Figure 9. Correlation of proposed and soft thresholding algorithm of recovered compressively sample MR image.

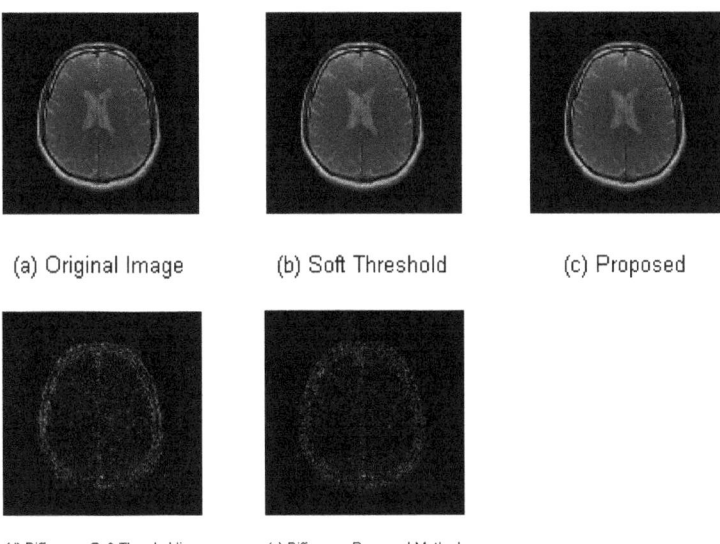

Figure 10. (**a**) Original 2D Brain MR Image, (**b**) Conventional Soft Thresholding based recovered 2D MRI, (**c**) 2D Brain MR Image recovered from undersampled image, (**d**) Difference of original and soft thresholding image, (**e**) Difference of proposed recovery method image with original image. The Difference is scaled up by the factor of 1000 in order to enhance its visibility.

Table 2. Performance comparison of conventional soft thresholding and proposed method with different compression levels, i.e., 5% to 50% of subsampling of the original 2-D MR image. These results show that the proposed method achieves better results in terms of SSIM and PSNR at varying compression ratios.

Compression Ratio	Soft Thresholding		Proposed Algorithm	
	SSIM	PSNR	SSIM	PSNR
5 %	0.6843	75.9056	0.7048	76.1609
10%	0.7786	78.9320	0.8175	79.6580
20%	0.8994	82.0316	0.8472	83.7628
30%	0.9407	87.3535	0.9790	91.1620
40%	0.9724	91.2540	0.9920	96.1281
50%	0.9884	95.4245	0.9955	99.5496

Table 3. Performance comparison of different sparsity transforms using mean squared error in the transform domain. Temporal FFT performs better in cardiac Cine MRI.

Performance Metrics	Soft Thresholding	Proposed Algorithm
MSE	1.38×10^{-4}	0.73×10^{-4}
PSNR	86.7195	89.4497
ISNR	28.3832	31.1135
SSIM	0.9346	0.9711
SNR	26.0298	28.7491
Correlation	0.9980	0.9989

4.3. Cardiac Cine Magnetic Resonance Imaging Recovery

The proposed algorithm is applied to MRXCAT, which produces breath-held undersampled cardiac cine MR image data. For MRXCAT, the following parameters were set:

recovery matrix size: 256 × 256 with 24 cardiac phases, with an image resolution set as 1 × 1 × 1 mm³, TR = 3 ms, TE = 1.5 ms. Five different acceleration rates R = (2, 4, 8, 12, 20) were used to assess the performance of the proposed method. For in vivo data, the following parameters were used: reconstruction matrix size: 256 × 256, 25 cardiac phases, with FOV of 375 mm. TE = 1 ms, TR = 3 ms and flip angle = 600. Five acceleration rates are $R = (2, 4, 8, 12, 20)$ are used to evaluate the performance of the proposed method. The reconstructed images are matched with the fully sampled original generated cardiac cine MRI as shown in Figure 11. All images are recovered in MATLAB by the proposed algorithm.

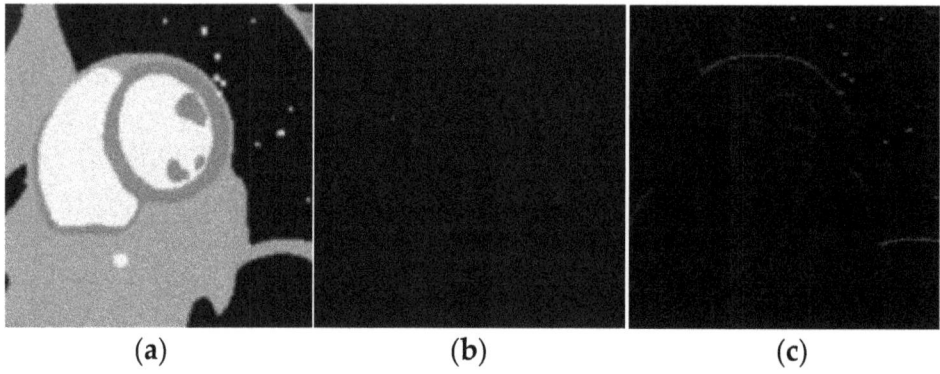

Figure 11. (a) Short axis cardiac cine MRI with completely sampled diastolic frame. (b) Sparsifying transform of cine cardiac MRI diastolic frame with temporal Fourier transform (Ψ), which results in sparse representation, (c) Another sparse representation of cardiac cine MR image (diastolic frame) using total variation transform (Ψ).

To gauge the efficiency of proposed algorithm, we use MRXCAT simulator software. It is designed for the analysis of reconstruction algorithms performance in the area of cardiac cine MRI. MRXCAT simulator is used to evaluate the proficiency of recovery algorithms in the field of Cardiac MRI. We have evaluated the performance of our proposed technique quantitatively and qualitatively. The performance measures that are used in this research article are: pictorial depiction of under-sampling artefacts, Structural Similarity (SSIM), Peak Signal to Noise Ratio (PSNR) and Root Mean Square Error (RMSE).

To evaluate the performance of proposed recovery technique qualitatively, we have experimentally depicted the recovered diastolic and systolic frames using acceleration rates of (R = 2, 4, 8, 12, 20). The quantitative assessment of the proposed algorithm is done using RMSE, PSNR and SSIM. Comparison between proposed algorithm and traditional soft thresholding technique is also performed. Figure 11 depicts the proficiency of the proposed algorithm at various acceleration rates while comparing with the soft thresholding. The first column shows the diastolic frame at different acceleration rates of cine cardiac MR image and the second column represents the systolic frame of cine MRI. The top row depicts the results of traditional IST algorithm, while the bottom row depicts proposed method results.

Table 4 shows the performance comparison of different sparsity transform using mean squared error in the transform domain. It measures the average of the error squares between the reconstructed and the acquired coefficients in the sparse domain. The proposed tangent hyperbolic based approximation performs well in temporal FFT as compared to the other sparse transforms. In particular, at higher acceleration rates, the tangent hyperbolic tangent based proposed technique shows much improved recovery of CS images.

Table 4. Performance comparison of different sparsity transforms using mean squared error in the transform domain. Temporal FFT performs better in cardiac cine MRI.

Acceleration Rates	Spatial Domain	Total Variation	Temporal FFT
2	0.1096	0.1123	0.0728
4	0.2321	0.1849	0.0848
8	0.2810	0.2438	0.0948
12	0.3533	0.2684	0.1043
20	0.4756	0.2982	0.1150

Figure 12 shows simulated data where (a) compares proposed method at bottom row with IST at top row with acceleration rate of 2. The arrow in (a) depicts the very minute presence of artefacts, while (b) depicts the performance of proposed method at acceleration rate equal to 4. The arrow in (b) depicts the presence of artefacts (c) shows the results of both algorithms with acceleration rate set at 8. The artefacts due to subsampling become gradually more visible in IST results as highlighted by arrow mark (d) depicts the results when acceleration rate is set at 12. Both techniques depicts the artefacts, however these artefacts are visible in the IST as mentioned by the white arrow in the figure (e) shows very much degraded image quality of IST, while comparing it with the proposed method. The subsampling artefacts dominate the traditional IST result when acceleration rate R is set at 20.

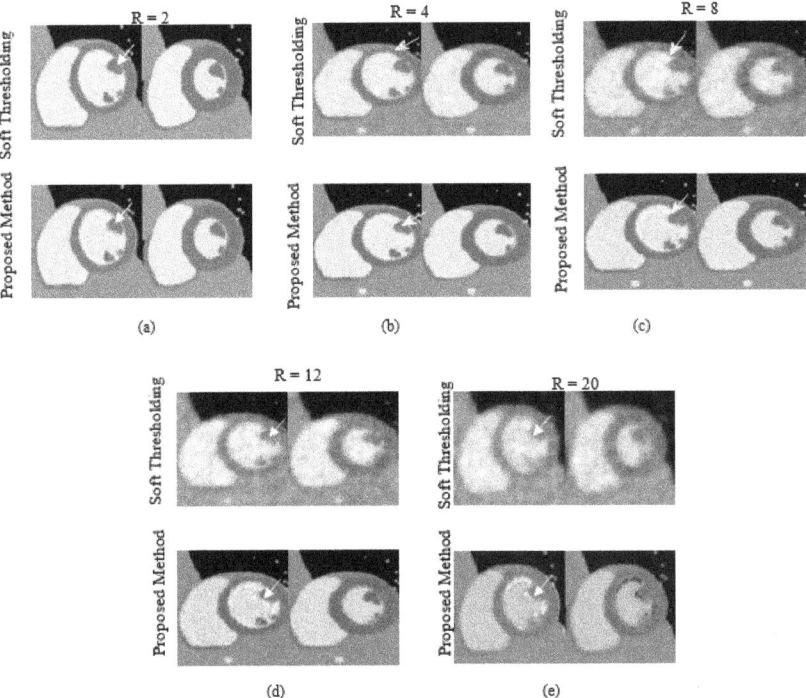

Figure 12. Simulated data (a) Compares proposed method at bottom row with IST at top row with acceleration rate of 2. The arrow in (a) depicts the very minute presence of artefacts. Here (b) depicts the performance of proposed method at acceleration rate equal to 4. The arrow in (b) depicts the presence of artefacts (c) shows the results of both algorithms with acceleration rate set at 8. (d) depicts the results when acceleration rate is set at 12. These artefacts are visible in the IST as mentioned by the white arrow in the figure (e) shows very much degraded image quality of IST, while comparing it with the proposed method.

To evaluate the recovered images quantitatively, we have used SSIM to compare the proposed technique with the IST technique. Figure 13 depicts the SSIM of our proposed algorithm, iterative soft thresholding (IST) and undersampled images. The quality of undersampled images is visibly quite poor. The efficiency of our proposed technique and IST technique is almost similar at low acceleration rates. However, the visible quality of the soft thresholding based recovered images decreases as acceleration rates are increased, as compared to the proposed method.

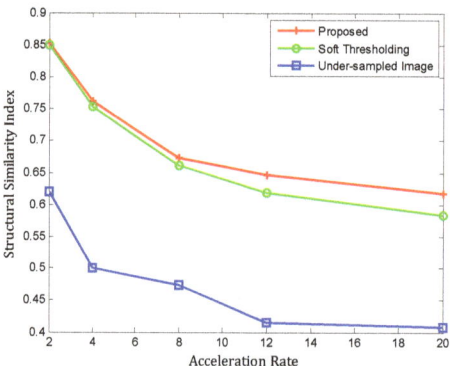

Figure 13. This figure depicts the efficiency of proposed algorithm by means of SSIM index. As acceleration rate increases, the SSIM of proposed algorithm degrades slowly while comparing it with IST algorithm.

Figure 14 shows the requisite iterations for the image reconstruction in both the methods. The proposed method solves the problem in six iterations, while the IST recovery technique takes ten iterations to reach the optimal solution. In this result, the data consistency is used to show the performance of proposed method.

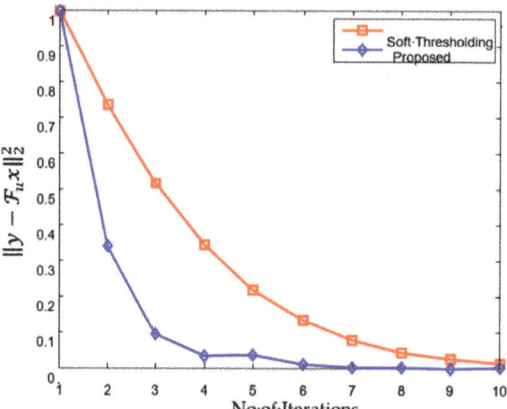

Figure 14. Comparison of cardiac cine MRI recovery at number of iterations; it can be seen from results that proposed method converges to an optimal solution in lesser iterations as compared to traditional thresholding.

To evaluate the efficiency of our recovered algorithm quantitatively, the results are shown the reconstruction results using PSNR at various acceleration rates (R = 2, 4, 8, 12, 20). We have compared our method with the traditional iterative soft thresholding technique. Figure 14 depicts the efficiency of our proposed technique at various acceleration rates as compared to IST algorithm. Red line shows the results of our method at the different acceleration rates. Figure 15 shows the PSNR of soft thresholding method and under-sampled data are shown with the green and blue lines, respectively.

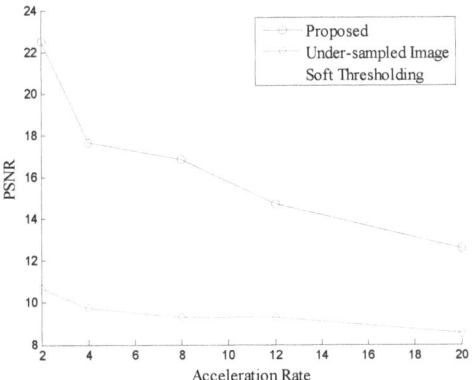

Figure 15. The performance using the peak signal-to-noise ratio (PSNR). PSNR of our method is better at all acceleration rates as compared to the soft thresholding method.

Table 5 elaborates the comparison of performance using root mean squared error (RMSE). It measures the error squares average between the recovered samples and the actual samples. The proposed method performance is much superior as compared to the IST technique. In particular, while operating at higher acceleration rates, the tangent hyperbolic method performs much better in recovering the images. However, the efficiency of traditional IST algorithm degrades at higher acceleration rates.

Table 5. Comparison of proposed method with conventional IST algorithm with RMSE. Proposed method performance is much better as acceleration rates are increased.

	Acceleration Rates	Undersampled Image	Iterative Soft Thresholding	Proposed Method
Simulated Data	2	0.081	0.0365	0.0353
	4	0.1218	0.0472	0.0372
	8	0.1498	0.0702	0.0419
	12	0.1583	0.0775	0.0485
	20	0.1782	0.0941	0.0606
In vivo Data	2	0.085	0.0099	0.0056
	4	0.106	0.0241	0.0172
	8	0.1170	0.0495	0.0206
	12	0.120	0.0567	0.0338
	20	0.1398	0.0585	0.0551

To evaluate the performance of reconstructed images qualitatively, using in vivo data, we have shown a comparison between the proposed method and the soft thresholding method in Figure 16. We have used five acceleration rates R = (2, 4, 8, 12, and 20) to show the comparison between the proposed method and the soft thresholding method. In this figure, the performance of the proposed technique and IST algorithm is similar at lower acceleration rates. However, the blurring artefacts in IST based recovered images are more prominent at higher acceleration rates as compared to the proposed algorithm as indicated by the white arrows.

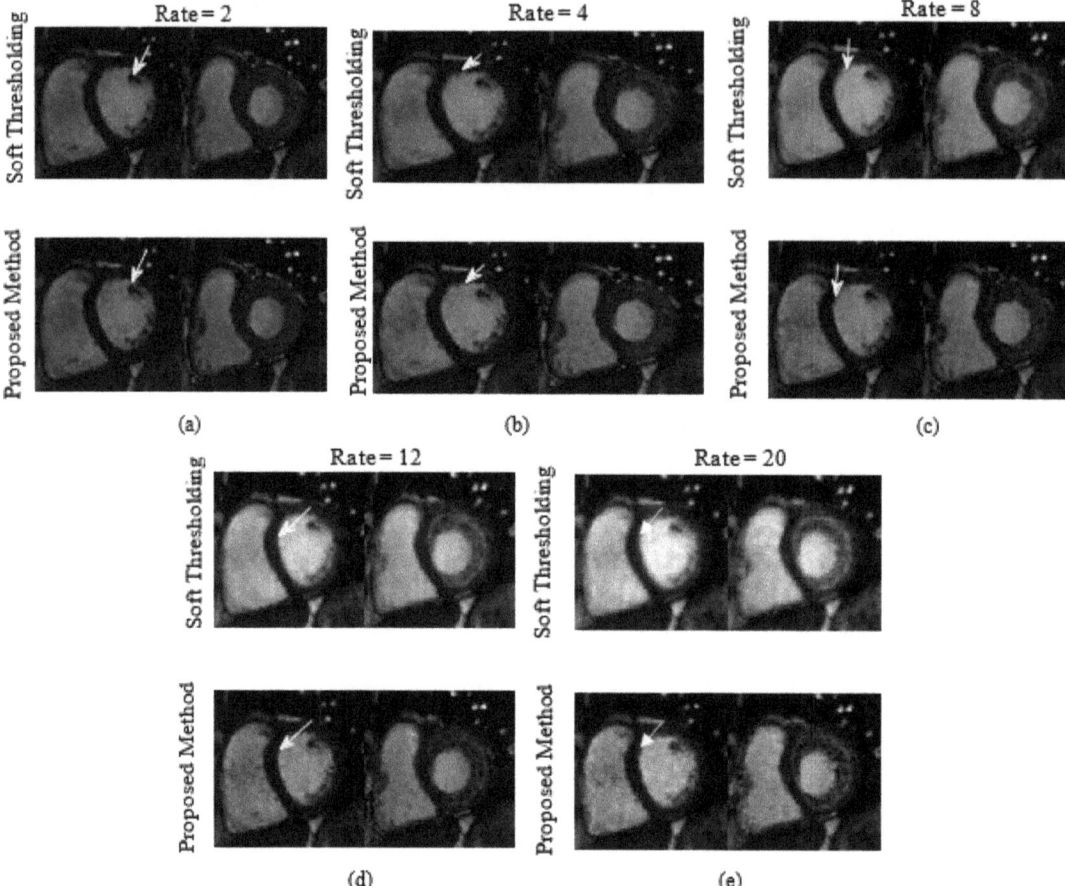

Figure 16. In real vivo data (**a**) Compares proposed method at bottom row with IST at top row with acceleration rate of 2. The arrow in (**a**) shows very minute artifacts. (**b**) depicts the performance of proposed method at acceleration rate equal to 4. The arrow in (**b**) depicts the presence of artefacts (**c**) shows the results of both algorithms with acceleration rate set at 8. The artefacts due to subsampling become gradually more visible in IST results as highlighted by arrow mark (**d**) depicts the results when acceleration rate is set at 12. Both techniques depicts the artefacts, however these artefacts are visible in the IST as mentioned by the white arrow in the figure (**e**) show very much degraded image quality of IST, while comparing it with the proposed method. The subsampling artefacts dominate the traditional IST result when acceleration rate R is set at 20.

5. Conclusions

In this paper, the novel CS recovery algorithm is proposed for compressively sampled sparse signals and biomedical images. The proposed method is applied in the 1-D sparse signal, 2-D real human brain MRI and cardiac cine MRI. In our proposed algorithm, we have introduced a hyperbolic tangent smooth approximation of non-differentiable l_1-norm and shrinkage. The experimental results quantitative analysis based on SSIM, PSNR, RMSE of recovered sparse signal and MR images have outperformed the conventional IST algorithm. The qualitative observations show significant improvement in the proposed method, especially at higher acceleration rates on Cine Cardiac MR images. In future, this research work can be further enhanced to incorporate machine learning techniques using large compressively sampled MRI datasets to restore accurate images, and the proposed method can be implemented on CS MRI scanners to reduce patient anxiety.

Author Contributions: Conceptualization, H.H. and J.A.S.; methodology; software, H.H.; validation, K.K., N.K. and J.A.S.; formal analysis, H.H.; investigation, H.H. and J.A.S.; resources, J.A.S.; writing—original draft preparation, H.H.; writing—review and editing, J.A.S. and N.K.; visualization, H.H. and J.A.S.; supervision, J.A.S.; project administration, J.A.S. and K.K.; funding acquisition, K.K. All authors have read and agreed to the published version of the manuscript.

Funding: This research was partially funded by British Malaysia Intitute, Universiti of Kaul Lumpur, including publication of this research article.

Data Availability Statement: 1-D dataset was generated using MATLAB. 2-D MR image was acquired from St. Mary's Hospital, London, England. While breath-held under-sampled cardiac cine dataset is generated using MRXCAT software.

Acknowledgments: We acknowledge the support given by University of Kaula Lumpur, British Malaysian Institute for funding the research publications. We also acknowledge Higher Education Commission Pakistan for providing support in terms of hardware for experiments.

Conflicts of Interest: The authors declare no conflict of interest.

References

1. Donoho, D.L. Compressed sensing. *IEEE Trans. Inf. Theory* **2006**, *52*, 1289–1306. [CrossRef]
2. Candes, E.J.; Wakin, M.B. An Introduction To Compressive Sampling. *IEEE Signal Process. Mag.* **2008**, *25*, 21–30. [CrossRef]
3. Candes, E.J.; Tao, T. Near-Optimal Signal Recovery From Random Projections: Universal Encoding Strategies? *IEEE Trans. Inf. Theory* **2006**, *52*, 5406–5425. [CrossRef]
4. Lustig, M.; Donoho, D.; Pauly, J.M. Sparse MRI: The application of compressed sensing for rapid MR imaging. *Magn. Reson. Med.* **2007**, *58*, 1182–1195. [CrossRef]
5. Lustig, M.; Donoho, D.L.; Santos, J.M.; Pauly, J.M. Compressed Sensing MRI. *IEEE Signal Process. Mag.* **2008**, *25*, 72–82. [CrossRef]
6. Auger, D.A.; Bilchick, K.C.; Gonzalez, J.A.; Cui, S.X.; Holmes, J.W.; Kramer, C.M.; Salerno, M.; Epstein, F.H. Imaging left-ventricular mechanical activation in heart failure patients using cine DENSE MRI: Validation and implications for cardiac resynchronization therapy. *J. Magn. Reson. Imaging* **2017**, *46*, 887–896. [CrossRef]
7. Liu, J.; Feng, L.; Shen, H.-W.; Zhu, C.; Wang, Y.; Mukai, K.; Brooks, G.C.; Ordovas, K.; Saloner, D. Highly-accelerated self-gated free-breathing 3D cardiac cine MRI: Validation in assessment of left ventricular function. *Magn. Reson. Mater. Phys. Biol. Med.* **2017**, *30*, 337–346. [CrossRef]
8. van Amerom, J.F.P.; Lloyd, D.F.A.; Price, A.N.; Kuklisova Murgasova, M.; Aljabar, P.; Malik, S.J.; Lohezic, M.; Rutherford, M.A.; Pushparajah, K.; Razavi, R.; et al. Fetal cardiac cine imaging using highly accelerated dynamic MRI with retrospective motion correction and outlier rejection. *Magn. Reson. Med.* **2018**, *79*, 327–338. [CrossRef]
9. Paulus, W.J.; Tschöpe, C.; Sanderson, J.E.; Rusconi, C.; Flachskampf, F.A.; Rademakers, F.E.; Marino, P.; Smiseth, O.A.; De Keulenaer, G.; Leite-Moreira, A.F.; et al. How to diagnose diastolic heart failure: A consensus statement on the diagnosis of heart failure with normal left ventricular ejection fraction by the Heart Failure and Echocardiography Associations of the European Society of Cardiology. *Eur. Heart J.* **2007**, *28*, 2539–2550. [CrossRef]
10. Yerly, J.; Gubian, D.; Knebel, J.-F.; Schenk, A.; Chaptinel, J.; Ginami, G.; Stuber, M. A phantom study to determine the theoretical accuracy and precision of radial MRI to measure cross-sectional area differences for the application of coronary endothelial function assessment. *Magn. Reson. Med.* **2018**, *79*, 108–120. [CrossRef]
11. Ahmed, A.H.; Qureshi, I.M.; Shah, J.A.; Zaheer, M. Motion correction based reconstruction method for compressively sampled cardiac MR imaging. *Magn. Reson. Imaging* **2017**, *36*, 159–166. [CrossRef]
12. Gamper, U.; Boesiger, P.; Kozerke, S. Compressed sensing in dynamic MRI. *Magn. Reson. Med.* **2008**, *59*, 365–373. [CrossRef] [PubMed]
13. Donoho, D.L. De-noising by soft-thresholding. *IEEE Trans. Inf. Theory* **1995**, *41*, 613–627. [CrossRef]
14. Blumensath, T.; Davies, M.E. Iterative hard thresholding for compressed sensing. *Appl. Comput. Harmon. Anal.* **2009**, *27*, 265–274. [CrossRef]
15. Tropp, J.A.; Wakin, M.B.; Duarte, M.F.; Baron, D.; Baraniuk, R.G. Random Filters for Compressive Sampling and Reconstruction. In Proceedings of the 2006 IEEE International Conference on Acoustics Speech and Signal Processing Proceedings, Toulouse, France, 14–19 May 2006.
16. Yin, W.; Osher, S.; Goldfarb, D.; Darbon, J. Bregman Iterative Algorithms for ℓ_1-Minimization with Applications to Compressed Sensing. *SIAM J. Imaging Sci.* **2008**, *1*, 143–168. [CrossRef]
17. Khajehnejad, M.A.; Xu, W.; Avestimehr, A.S.; Hassibi, B. Weighted ℓ_1 minimization for sparse recovery with prior information. In Proceedings of the 2009 IEEE International Symposium on Information Theory, Seoul, Korea, 28 June–3 July 2009; pp. 483–487.

18. Shah, J.; Qureshi, I.; Omer, H.; Khaliq, A. A modified POCS-based reconstruction method for compressively sampled MR imaging. *Int. J. Imaging Syst. Technol.* **2014**, *24*, 203–207. [CrossRef]
19. Shah, J.A.; Qureshi, I.M.; Omer, H.; Khaliq, A.A.; Deng, Y. Compressively sampled magnetic resonance image reconstruction using separable surrogate functional method. *Concepts Magn. Reson. Part A* **2014**, *43*, 157–165. [CrossRef]
20. Bilal, M.; Shah, J.A.; Qureshi, I.M.; Kadir, K. Respiratory Motion Correction for Compressively Sampled Free Breathing Cardiac MRI Using Smooth l(1)-Norm Approximation. *Int. J. Biomed. Imaging* **2018**, *2018*, 7803067. [CrossRef]
21. Shah, J.; Qureshi, I.M.; Deng, Y.; Kadir, K. Reconstruction of Sparse Signals and Compressively Sampled Images Based on Smooth l1-Norm Approximation. *J. Signal Process. Syst.* **2017**, *88*, 333–344. [CrossRef]
22. He, C.; Xing, J.; Li, J.; Yang, Q.; Wang, R. A New Wavelet Thresholding Function Based on Hyperbolic Tangent Function. *Math. Probl. Eng.* **2015**, *2015*, 528656. [CrossRef]
23. Lu, J.-y.; Lin, H.; Ye, D.; Zhang, Y.-s. A New Wavelet Threshold Function and Denoising Application. *Math. Probl. Eng.* **2016**, *2016*, 3195492. [CrossRef]
24. Schmidt, M.; Fung, G.; Rosales, R. Fast Optimization Methods for L1 Regularization: A Comparative Study and Two New Approaches. In Proceedings of the Machine Learning: ECML 2007, Berlin/Heidelberg, Germany, 17–21 September 2007; pp. 286–297.
25. Chang, S.G.; Bin, Y.; Vetterli, M. Spatially adaptive wavelet thresholding with context modeling for image denoising. *IEEE Trans. Image Process.* **2000**, *9*, 1522–1531. [CrossRef]
26. Sendur, L.; Selesnick, I.W. Bivariate shrinkage functions for wavelet-based denoising exploiting interscale dependency. *IEEE Trans. Signal Process.* **2002**, *50*, 2744–2756. [CrossRef]
27. Nasri, M.; Nezamabadi-Pour, H. Image denoising in the wavelet domain using a new adaptive thresholding function. *Neurocomputing* **2009**, *72*, 1012–1025. [CrossRef]
28. Prinosil, J.; Smekal, Z.; Bartusek, K. Wavelet Thresholding Techniques in MRI Domain. In Proceedings of the 2010 International Conference on Biosciences, Online, 7–13 March 2010; pp. 58–63.
29. Jin, J.; Yang, B.; Liang, K.; Wang, X. General image denoising framework based on compressive sensing theory. *Comput. Graph.* **2014**, *38*, 382–391. [CrossRef]

Disclaimer/Publisher's Note: The statements, opinions and data contained in all publications are solely those of the individual author(s) and contributor(s) and not of MDPI and/or the editor(s). MDPI and/or the editor(s) disclaim responsibility for any injury to people or property resulting from any ideas, methods, instructions or products referred to in the content.

Article

Enhanced Pre-Trained Xception Model Transfer Learned for Breast Cancer Detection

Shubhangi A. Joshi [1,†], Anupkumar M. Bongale [2,*,†], P. Olof Olsson [3,*,†], Siddhaling Urolagin [4,*,†], Deepak Dharrao [5,†] and Arunkumar Bongale [1,†]

1. Symbiosis Institute of Technology (SIT), Symbiosis International (Deemed University), Lavale, Pune 412 115, Maharashtra, India
2. Department of Artificial Intelligence and Machine Learning, Symbiosis Institute of Technology (SIT), Symbiosis International (Deemed University), Lavale, Pune 412 115, Maharashtra, India
3. Fujairah Genetics Center, Fujairah, United Arab Emirates
4. Department of Computer Science, Birla Institute of Technology & Science, Pilani, Dubai International Academic City, Dubai P.O. Box 345055, United Arab Emirates
5. Department of Computer Science and Engineering, Symbiosis Institute of Technology (SIT), Symbiosis International (Deemed University), Lavale, Pune 412 115, Maharashtra, India
* Correspondence: anupkumar.bongale@sitpune.edu.in or ambongale@gmail.com (A.M.B.); olof@fujairahgenetics.ae (P.O.O.); siddhaling@dubai.bits-pilani.ac.in (S.U.)
† These authors contributed equally to this work.

Abstract: Early detection and timely breast cancer treatment improve survival rates and patients' quality of life. Hence, many computer-assisted techniques based on artificial intelligence are being introduced into the traditional diagnostic workflow. This inclusion of automatic diagnostic systems speeds up diagnosis and helps medical professionals by relieving their work pressure. This study proposes a breast cancer detection framework based on a deep convolutional neural network. To mine useful information about breast cancer through breast histopathology images of the 40× magnification factor that are publicly available, the BreakHis dataset and IDC(Invasive ductal carcinoma) dataset are used. Pre-trained convolutional neural network (CNN) models EfficientNetB0, ResNet50, and Xception are tested for this study. The top layers of these architectures are replaced by custom layers to make the whole architecture specific to the breast cancer detection task. It is seen that the customized Xception model outperformed other frameworks. It gave an accuracy of 93.33% for the 40× zoom images of the BreakHis dataset. The networks are trained using 70% data consisting of BreakHis 40× histopathological images as training data and validated on 30% of the total 40× images as unseen testing and validation data. The histopathology image set is augmented by performing various image transforms. Dropout and batch normalization are used as regularization techniques. Further, the proposed model with enhanced pre-trained Xception CNN is fine-tuned and tested on a part of the IDC dataset. For the IDC dataset training, validation, and testing percentages are kept as 60%, 20%, and 20%, respectively. It obtained an accuracy of 88.08% for the IDC dataset for recognizing invasive ductal carcinoma from H&E-stained histopathological tissue samples of breast tissues. Weights learned during training on the BreakHis dataset are kept the same while training the model on IDC dataset. Thus, this study enhances and customizes functionality of pre-trained model as per the task of classification on the BreakHis and IDC datasets. This study also tries to apply the transfer learning approach for the designed model to another similar classification task.

Keywords: breast cancer detection; magnification dependent; histopathology; BreakHis; IDC; Xception model; ResNet50 model; EfficientNetB0; 40×

1. Introduction

Breast cancer is one of the most predominant types of malignancy seen in the worldwide population of woman. Early diagnosis, prognosis, and correct treatment of breast cancer can improve patients' life expectancy. According to a fact sheet by World Health Organization (WHO), in 2020 around 2.3 million women were diagnosed with breast cancer, and there were approximately 685,000 deaths globally due to this disease. Breast cancer is one of the most prevalent kinds of cancer, causing around 10.7% of deaths among all major cancer types in 2020 [1].

Due to digitalization, there has been a tremendous advancement in the types of modalities used for screening of breast cancer. The most common modalities used for breast cancer screening are mammography, magnetic resonance imaging, breast ultrasound, positron emission tomography, and histopathological analysis of breast tissue [2]. Mammography is X-ray imaging of human breasts and is usually prescribed when a suspicious mass in the breast is suspected as a result of a physical examination of breasts. Breast ultrasound is a non-invasive method for breast screening. It is usually helpful in the case of dense breast tissues. Breast magnetic resonance imaging also takes several images and usually is prescribed along with other diagnostic tests such as mammography or ultrasound. Breast cancer is also detected by positron emission tomography (PET) examination. PET is an imaging test that uses radio-active substance for the detection and localization of cancerous growth of cells. It is seen that the initial screening method, as mammography helps to find breast cancer at an early stage [3,4]. Magnetic resonance imaging is also one of the important screening modalities for breast cancer early detection. Magnetic resonance imaging (MRI) screening shows higher sensitivity for breast cancer with genetics [5].

Even though mammography is the most common technique for the initial screening of breast cancer, it has certain drawbacks such as low sensitivity in the case of dense breast analysis and low specificity [6]. Hence, morphological analysis of hematoxylin and eosin-stained (H&E-stained) breast tissue is the gold standard for breast malignancy detection with a very high confidence level [2]. Breast cancer arises due to the uncontrollable growth of breast cells. It starts when ductal–lobular cells inside the breast glands grow abnormally. There are two basic types of breast cancers: ductal carcinoma in situ (DCIS) and invasive carcinoma. When the cancerous growth is confined within the ductal–lobular structure of the breast, it is called ductal carcinoma in situ (DCIS). When these cells of DCIS break through the ductal–lobular system and invade the rest of the breast parenchyma, then it is called invasive breast cancer.

Machine learning-based computer-aided diagnosis is becoming popular in breast cancer screening as these techniques promise new knowledge insights with near-human performance. Pure machine learning-based systems involving traditional classifiers used for diagnostic applications comprise various stages such as data preprocessing, image enhancements, segmentation, feature engineering, and then classification. With the tremendous research and advancements in deep learning and computer vision, neural network-based systems are predominantly designed for diagnostic workflows. The tedious task of feature engineering is automatically performed by neural network architecture in a deep learning model. In deep learning, a more advanced trend called transfer learning dominates the field of artificial intelligence (AI) research. The pre-trained architectures of deep learning models are used in the transfer learning method to save the training time and cost of design. Breast cancer biopsy images consist of many essential features which are used as a basis for disease diagnosis by expert histopathologists. These features include nuclear pleomorphism, i.e., how distinctively different the shapes and sizes of nuclei are in tubule formation and metastasis information. In supervised machine learning, a labeled dataset is shown to the machine learning model during training. The machine learning model's learning algorithm then adjusts the model's parameters such that it learns from the dataset. Once training is completed, completely unknown testing data is shown to the model to classify the unseen information correctly. Artificial intelligence-based studies aimed to help in breast cancer diagnostic workflows, and not only comprise malignancy detection but also of the detection of breast cancer grade, intrinsic molecular subtype, lymph node status, and metastasis occurrence. Breast cancer consists of various subtypes according to the origin of occurrence. The grade of breast cancer depicts aggressiveness of the disease. The intrinsic molecular subtype depends on the presence of hormones or proteins on the cancer cell surface. Intrinsic molecular subtype identification is a crucial prognostic evaluator in cancer treatment [7].

This work is focused on comparing the performances of deep convolutional neural network models which use the backbone of pre-trained architectures. Binary classification is conducted on histopathological images having a magnification factor of 40×. These biopsy images are taken from the publicly available BreakHis breast cancer dataset and IDC (Invasive ductal carcinoma) dataset. The best model with good evaluation parameters is chosen as the final classification model. Comparative analysis between the three classification models is also presented. Pre-trained models trained on the ImageNet challenge dataset are used for this study as the backbone of a deep convolutional neural network. Custom layers are added to pre-trained models. The backbone network finds important lower-order features in the histopathology images, while the custom layers are responsible for finding higher-order abstract features leading to the task of binary classification. As these histopathology images are rotation and scale invariant; data augmentation is used to increase the size of the dataset. Two different datasets, namely BreakHis and IDC are considered for this work. The flowchart of the proposed work is given in Figure 1. As evident from the flowchart, the BreakHis 40× image dataset is first augmented using several image transforms such as rotation, flipping, and scaling. The enhanced models are trained, validated, and tested on the BreakHis dataset. Depending on the best accuracy, the enhanced customized pre-trained Xception model is finalized for the classification task. The weights learned during BreakHis learning are stored for the Xception model. The same enhanced Xception model with BreakHis weights is used for tuning on the IDC dataset.

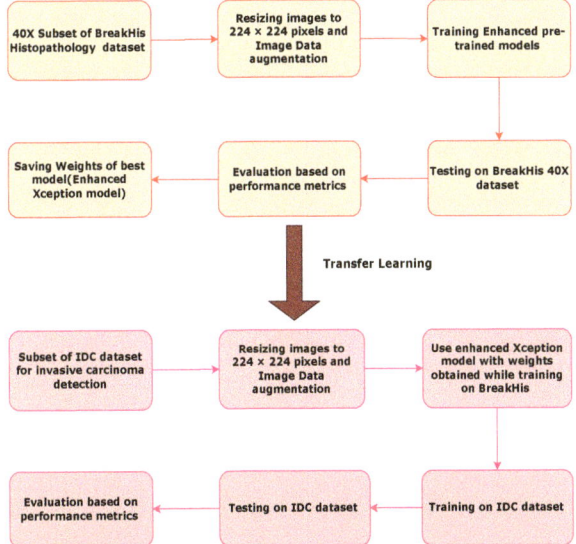

Figure 1. Flowchart of the proposed work.

The contributions of this study are summarized below.

- Rescaling and augmentation of the BreakHis 40× dataset using Keras functionality.
- Customization of pre-trained models, namely EfficientNetB0, ResNet50, and Xception, by removing the last fully connected layers and appending series of convolution as well as max pooling layers along with flattening and dense connections.
- Experimenting on the BreakHis 40× dataset by using enhanced pre-trained models for binary classification task of cancer detection.

- Using previously trained weights during tuning on the BreakHis dataset of the best model (customized Xception) to train, test, and validate a subset of the IDC dataset. Thus, this validates the potential of the Xception model trained on the BreakHis dataset to perform another similar task of invasive carcinoma detection.
- Importing of weights also exhibits ability of enhanced Xception model to transfer learn from the BreakHis dataset to IDC dataset

The rest of the article is arranged as follows. Section 2 describes similar work performed in breast cancer diagnosis using publicly available datasets. The methodology of the convolutional deep learning technique used for this binary classification problem is discussed in Section 3. In Section 4, experiments, results, and analyses are presented. Finally, the paper is concluded in Section 5.

2. Related Work

Deep learning systems ease the diagnostic classification task as deep neural networks automatically find features and patterns in the given dataset. Along with the availability of good hardware processing architectures such as graphics processing units (GPU) and tensor processing units (TPU), different open-source programming frameworks such as Tensorflow-enabled Keras and PyTorch are also responsible for the trend in using deep learning for classification problems. In addition to these programming frameworks, pre-trained architectures of deep neural nets such as visual geometry group (VGG), AlexNet, inception, ResNet, Xception, and several other models are also available for use in a deep computer vision model.

A recurrent, residual neural network was used for semantic segmentation of medical images [8]. In one of the studies, an improved version of U-Net-based architecture called IRU-Net was used to segment images of patients' tissue slides. This IRU-Net method, which was designed to detect the presence of bacteria and immune cells in tissue images, used several scaled layers of residual blocks, inception blocks, and skipped connections [9]. A multilevel semantic adaptation method was used for diverse modalities for a few-shot segmentation on cardiac image sequences. The method proved effective even under limited labels for the dataset [10].

MRI and near infrared spectral tomography-related wearable system was designed and developed in one of the research studies [11]. In dynamic contrast-enhanced magnetic resonance imaging (DCE-MRI) imaging, various features were extracted using different machine learning methods in some of the research studies [12].

In 2016, Spanhol constructed a benchmark dataset of hematoxylin and eosin (H&E)-stained histopathological images for breast cancer which were extracted using fine needle aspiration cytology (FNAC) procedure. Various feature extractors such as local phase quantization (LPQ), local binary pattern (LBP), and completed local binary pattern (CLBP) were used to obtain textural and morphometric features from H&E-stained images. Performance comparison of various classifiers was made for this task [13]. Classification of H&E-stained histopathological images was performed with features extracted from the graph run length matrix (GRLM) and gray level co-occurrence matrix (GLCM). A predictive breast cancer diagnostic model based on a modified weight assignment technique was developed [14]. Contrast limited adaptive histogram equalization method was used for contrast improvement along with several classifiers on the BreakHis dataset [15]. DNA repair deficiency (DRD) status was found from histopathology images by one of the researchers by making use of deep learning [16]. One of the researchers used several pre-trained architectures of convolutional neural networks in parallel and aggregated the results for detecting breast cancer using histopathology images with Vahadane transform for stain normalization [17]. In one magnification-dependent breast cancer binary classification study, the researcher used the transfer learning model using AlexNet [18]. An anomaly detection mechanism using a generative adversarial network is performed to find mislabeled patches of the BreakHis dataset. DenseNet121 is used for binary classification after that [19]. A deep convolutional neural network(CNN) model is used in [20–22] to classify histopathology images. Color normalization and data augmentation techniques are applied to H&E-stained histopathology slides in order to perform breast cancer classification. Fully trained and fine-tuned versions of VGG architectures were used for the task [23]. Tumor and healthy regions are segmented in one of the research studies, which used CNN for area-based annotation procedure on histopathology images [24]. The deep convolutional neural network was also designed for finding the status of cancer metastasis in lymph nodes [25,26]. Breast cancer is detected using dual-modality images such as ultrasound and histopathology in one of the studies [27]. Using transfer learned multi-head self-attention, one of the studies performed multi-class classification on the BreakHis dataset [28]. Breast cancer histopathology images are classified into four classes as normal, begin, in situ carcinoma, and invasive carcinoma in one research study using parallel combination of convolutional neural network(CNN) and recurrent neural network(RNN) [29]. GoogleNet-based hybrid CNN module was used with bagging strategy to classify breast cancer [30]. DenseNet-161 and ResNet-50 models were used for invasive carcinoma detection in IDC (invasive ductal carcinoma) dataset along with validation on the BreakHis dataset [31]. Deep learning architecture named DKS-DoubleU-Net was used to segment tubules in the breast tissue in H&E-stained images on BRACS dataset [32]. In one more research work, binary and multiclass classification on BreakHis dataset is performed using customized pre-trained models of DenseNet and ResNet networks. In this study, the researchers are able to attend maximum accuracy of 100 percent for binary classification task on 40X magnification factor [33]. In the proposed research work accuracy obtained is 93.33% on BreakHis 40× dataset. However, the proposed model is again trained on a new dataset called IDC. The IDC dataset consists of altogether different information from BreakHis dataset as IDC focuses on invasive ductal carcinoma cases. The model is again fine tuned on IDC dataset by just keeping the weights obtained during BreakHis training constant. Instead of using random weights or ImageNet weights, we have used previously trained weights (BreakHis weights). Thus, the proposed model is also able to learn new patterns in different dataset. Thus, this study tries to apply transfer learning concept. This transfer learning approach improves training and testing timing requirements and ensure generalization and robustness of the CNN model.

Deep CNN using the ResNet model was used for the binary classification of breast histopathology images. This method also used Wavelets of packet decomposition and histogram of oriented gradients [34]. Several multiple instance learning algorithms with the deep convolutional neural network experimented on the BreakHis dataset [35]. Fusion of different deep learning models has been tested on the BreakHis and ICIAR 2018 datasets in one of the research studies [36]. A fully automated pipeline using deep learning architecture for breast cancer analysis is presented in one of the research studies [37]. Transfer learning-based deep neural network was used for the breast cancer classification task in one more research study [38]. Spatial features are extracted using CapsuleNet for the breast cancer classification task in some research studies [39,40]. Color deconvolution and transformer architecture are used for histopathological image analysis in one of the research studies [41]. A pure transformer is used as a backbone to extract global features for histopathological image classification [42].

Apart from the binary or multi-class classification of breast cancer according to its subtypes, deep learning techniques are also used in grade detection and intrinsic subtype classification of breast cancer. In one of the recent studies, a deep learning model named DeepGrade was used to find grades of breast cancer from whole slide images. The prognostic evaluation of patients depending on their histology grades was also carried out [43]. In one more study, faster region convolutional neural network and deep convolutional neural network were used for mitotic activity detection [44]. Intrinsic molecular subtypes of breast cancer are leading diagnostic factors for precision therapy. Depending on whether breast cancer cells show the presence of estrogen hormone, progesterone hormone, or HER2(Human Epidermal Growth factor Receptor2) protein, there are various subtypes of breast cancer. These are called intrinsic or molecular subtypes of breast cancer. The hormonal status of histopathology images is found by one study which utilizes deep learning techniques on different datasets [45]. A generative adversarial network for stain normalization and deep learning framework for intrinsic molecular subtyping of breast tissue was used in one more research study [46]. One group of researchers used deep CNN and region of interest-based annotations on whole slide images to predict HER2 positivity in breast cancer tissue [47]. Deep learning was used in one of the works to find the status of various genomic bookmarks [48]. A deep learning image-based classifier was developed to predict intrinsic subtypes of breast cancer, and survival analysis was also conducted [49]. Even though there is much research in breast cancer detection, there is a need to conduct more studies on multi-centric data. This multi-centric data from different institutions or using different datasets can guarantee the generalization of underlying architectures and the possibility of real-life use for improving clinical outcomes. Hence, this research study tries to use network models on two different datasets for a similar type of breast cancer classification task.

3. Materials and Methods: For Breast Cancer Detection

The proposed classification algorithm uses the Keras library on top of the TensorFlow environment. The Tensor processing unit with extended RAM facilities is used from the Google Colab PRO environment. In the proposed work, the Xception model, which is already pre-trained on the ImageNet dataset, is used for low-level, deep feature extraction on H&E-stained breast histopathology images with a magnification factor of $40\times$. The other two pre-trained models were used for performance comparison. The output of the backbone Xception network is connected to custom CNN layers. These custom layers are responsible for higher-level feature abstraction and classification. The motivation behind using pre-trained models as a backbone is that they had shown successful outputs for the ImageNet challenge. Pre-trained models of the ImageNet challenge are already trained on millions of images. Furthermore, hence, they have weights already trained for one of the image-related computer vision application problems. The bottom layers of the pre-trained network are used as they are. The top layers are changed. The last fully connected layers of pre-trained CNN, which are specific to the ImageNet classification task, are removed, and custom CNN layers are appended to the pre-trained model.

3.1. Datasets

Two different datasets, namely the BreakHis and IDC (invasive ductal carcinoma) were used for experimental analysis in this study. The final network trained on the BreakHis dataset with $40\times$ zoom images also exhibited transfer learning capabilities, wherein the same model was used for fine-tuning and testing on a different IDC dataset. In a real-life scenario, for histopathologists analyzing H&E-stained slides on the computer screen, the magnification option proves beneficial as they can zoom in or out of the digitized slides to make inferences about the disease. However, for neural networks, we can provide slides of the same magnification factor as micro-level features can easily be extracted by controlling kernel sizes in the convolution process.

3.1.1. BreakHis Histopathological Image Dataset

BreakHis dataset consists of a total of 7909 breast histopathology images. These are images of patches extracted from whole slide tissue images. The dataset consists of tissue images of four different magnification factors, which are 40X, 100X, 200X, and 400X. In this study, 40X zoom images are considered for evaluating CNN architecture. Besides labeling images as benign or malignant, the dataset also gives information about other histological subtypes of breast cancer. The class imbalance problem is seen in the BreakHis dataset, where one class sample outnumbers others. For the 40X magnification factor, there are 652 benign and 1370 malignant samples in the BreakHis dataset [13].

3.1.2. IDC Breast Histopathological Image Dataset

The invasive ductal carcinoma (IDC) dataset has breast cancer histopathology images in the form of patches of dimension 50 × 50 pixels. These patches contain two categories, namely IDC positive images and IDC negative images. There are a total of 277,524 images, out of which 78,786 images are IDC +ve, and 198,738 images are IDC −ve. All these patches are of 40X magnification factor. A subset of the IDC dataset is taken in this study. As this a huge dataset exhibiting class imbalance, 10,000 samples of each category are considered in the proposed work [50].

3.2. Image Rescaling and Augmentation

The original image size of histopathology images in the BreakHis dataset is 700 × 460 pixels. All images with the 40X magnification are resized to 224 × 224 pixels. Most of the medical datasets consisting of histopathology images suffer due to limited number of training samples. To strengthen the dataset, more images are added to the 40X BreakHis dataset using various image transforms. Histopathology images being tissue images, are rotation invariant and shift invariant. The image transform operations include horizontal and vertical flips of the image, rotation of the image, zooming operation, etc. Augmented images are generated on the fly during training time. Hence, no separate memory storage is required for augmented images, resulting in better memory management. Different image transforms for the data augmentation method are shown in Figure 2.

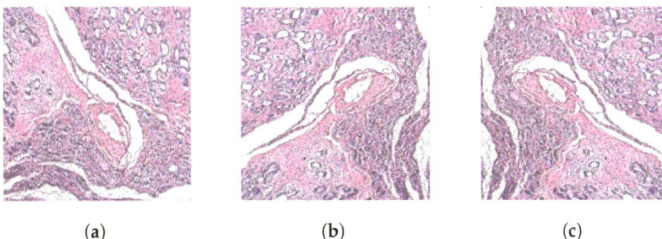

(a) (b) (c)

Figure 2. Rotation and flipping of H&E-stained tissue images. (a) Original H&E-stained tissue image. (b) Rotated tissue image. (c) Flipped tissue image.

3.3. Architecture of Proposed CNN Model

In this work, experimentation is carried out with three different pre-trained deep models as the backbone. Out of EfficientNetB0, ResNet50, and Xception, the Xception model gave the best accuracy. To increase the model's accuracy and extract higher-level features in histopathological images, custom layers in the form of three convolutional and three max-pooling layers are added to the pre-trained framework. The Xception model's weights are initialized to the ImageNet weights. Custom convolutional layers have a uniform kernel size of 3 × 3 with ReLu activation function blocks. ReLu activation function is chosen as it provides constant gradient which minimizes the vanishing gradient problem. The flattening operation converts the obtained feature map into a single-dimensional vector. Dropout is used to help reduction in overfitting, and batch normalization is used as a regularization technique. The final sigmoid activation function gives output in terms of class probability in the range from 0 to 1. Then, the whole CNN architecture consisting of a pre-trained model(except the top fully connected layers) and custom layers is trained on an augmented BreakHis dataset with the 40X magnification factor. Thus, custom CNN model layers are added on the top of the backbone Xception model. The backbone Xception model works as a potential feature extractor here. The proposed architecture is given in Figure 3.

Loss objective function used in this case is binary cross-entropy loss. During the backward pass or backpropagation pass of the neural network derivative of loss with respect to different weights is found using the chain rule. Trainable parameters of the network, meaning the weights and the biases, are updated during backpropagation.

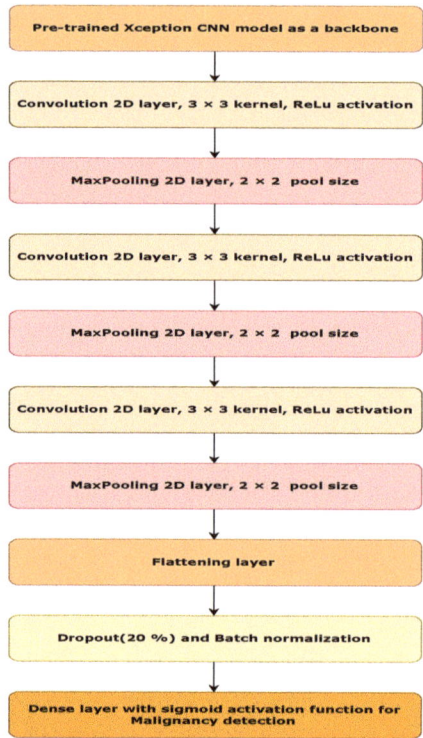

Figure 3. Architecture of proposed CNN network.

In CNNs an optimizer is also used for the optimization of trainable parameters of the network to minimize the cost objective function. In this study, the optimizer used is Adam optimizer as it gave better accuracy than other counterparts such as Stochastic gradient descent, RMSProp, AdaDelta, etc. The learning rate decides the convergence time for the model. Two learning rates were tried for this algorithm as 0.001 and 0.0001. The best suited learning rate is kept at 0.0001. The loss function used is binary cross_entropy. The cost function of binary cross_entropy has two main terms, which consist of actual labels and network predictions. This loss calculation is performed over a minibatch of a dataset.

3.4. Enhanced Pre-Trained Models

EfficientNet models are depth, width, and resolution scaled networks. This scaling is performed to obtain more detailed feature extraction. All these dimensions are scaled up with a uniform ratio. The baseline EfficientNet model is EfficientNet B0. EfficientNet is a family of models ranging from baseline models to several scaled-up versions. Thus, EfficientNet provides multiple scaling for improving accuracy [51]. ResNet50 or Residual Network 50 is the model which is also validated on the ImageNet large-scale visual recognition challenge dataset. For ease of optimization and better performance, the residual building blocks are included in ResNet models [52]. ResNet50 architecture consists of 50 such residual blocks. EfficientNet, ResNet, and Xception models are convolutional neural networks with many properties such as parameter or weight sharing, location invariance, etc., making them efficient for computer vision applications. Depth-wise separable convolution is the prominent feature used in Xception architectures. It is an updated version of inception architectures. In inception architecture, the convolutions are performed spatially and over the input depth of the image. To reduce dimensions, in Xception architecture, 1×1 convolution blocks are run across the depth. In the standard practice of convolution operation, the application of filters across all input channels corresponding to colors and a combination of these values is performed in a single execution step. Depth-wise separable convolution divides this complete process into two parts depth-wise or channel-wise convolution and point-wise convolution. In-depth wise process, a convolution filter is applied to a single input channel at a time. All output values of these separate kernels, which work on a single channel, are stacked together, and then point-wise convolution is performed on channel-wise outputs of filters [53].

3.5. Classifier Details

The proposed classifier used in this work for the breast cancer classification task is enhanced pre-trained Xception CNN. The Xception CNN model trained on the ImageNet dataset is used by excluding its final fully connected layers. Weight initialization is performed based on ImageNet weights only. After removing the last layers, additional convolutional and max pooling layers are added to the network, along with flattening and dense layers. Since pre-trained models are generally trained on larger and generalized datasets, adding custom layers help the modified models to adapt to specific tasks. Adding custom layers also helps to train the models faster. Image sizes are rescaled to 224×224 pixels. For an enhanced pre-trained Xception model, drop-out and batch normalization are used as regularization techniques. Batch normalization scales output feature values to the next layer into a standard uniform scale. Batch normalization reduces internal covariance shift to a significant extent and also helps weights to converge faster.

3.6. Training and Testing of Model

The time required for training the enhanced Xception model on the IDC dataset is 1 h 50 min, while the time taken for testing on the IDC dataset is 67 s.

The CNN architecture consisting of a pre-trained model stacked with custom CNN layers is trained on the 40X zoom images of the BreakHis dataset. A training, validation, and testing split ratio of 70%, 15%, and 15% is maintained. The model is trained for 50 epochs. Augmented images obtained from image transforms are used for training purpose. Iterations are performed for a batch size of 16 images. Overfitting or high variance problems can easily creep in for such deep architectures. Generally, neural nets trained on small datasets tend to overfit. Techniques to address the problem of overfitting or underfitting neural networks are called regularization techniques. To mitigate this overfitting issue, dropout can be inserted after any hidden layer with a different dropout factor. A dropout ratio of 0.2 is used in this work. This dropout of 20% is used after the flatten layer. Adding a dropout increases the performance of the neural network. Dropout makes some of the activation outputs of hidden neurons in hidden layers as zero. The random deactivation of neuron connections happens for every training cycle. However, the percentage of dropping remains the same. Dropout ensures minimized biasing and help to prohibit neurons from learning minute redundant details in training samples and thus eventually enhance generalization capability.

To speed up the training process and to increase its performance and stability, regularization in terms of batch normalization is used. The batch normalization process also stabilizes the weight parameters of the network. In this process, the normalized output of the previous layer is fed to the next layer. It is called batch normalization because, during the training process, the layer's inputs are normalized by using the current batch's variance and mean values. A batch normalization layer is added before the final dense layer.

Once the network is trained on the BreakHis dataset with an enhanced pre-trained Xception model, the same model is used to for fine-tuning on another dataset called IDC. A subset of the IDC dataset is used for this comparative analysis. The network is tuned and tested for IDC with weights borrowed while training on the BreakHis dataset. This new dataset considers ten thousand samples of IDC positive class and 10,000 samples of IDC negative class.

4. Experiments and Results

Among all the architectures, the Xception pre-trained model stacked with custom CNN layers exhibited the highest accuracy of 93.33% on the augmented BreakHis dataset for the 40X magnification factor. The same Xception pre-trained model stacked with custom CNN layers gave an accuracy of 88.08% on the IDC dataset.

The deep CNN models are frequently evaluated based on the parameters such as accuracy, recall, precision, and f1-score. The confusion matrix gives us parameters such as true positive values (TP), true negative values (TN), false positive values (FP), and false negative values (FN). True positives (TP) are the output values that are predicted by the classifier as positive outputs and are originally positive. In breast cancer classification, this corresponds to correctly diagnosed patients with breast cancer. These patients have breast cancer and are also correctly diagnosed as cancer-positive patients. False positives (FP) are the output values predicted by the classifier as positive outputs but are originally negative. In the context of breast cancer diagnosis, these are the patients wrongly classified as cancer-positive, whereas, in reality, these are cancer-free patients. True negatives (TN) are output values classified by the deep learning framework as negative and originally negative. These are correctly diagnosed cancer-free patients. False negatives (FN) are output values classified by the network as negatives but are positives in reality. For breast cancer diagnosis, this parameter is very sensitive because the patient having breast cancer but diagnosed as breast cancer negative is represented by a false negative. The choice of deep neural network architecture largely depends on a network providing a minimum value of false negatives.

Accuracy is a critical performance evaluator. It is the ratio of correctly predicted outputs of a deep neural network by a total number of samples. For symmetric datasets, accuracy can provide a good measure for analyzing the performance of deep CNN. The accuracy of deep CNN is given in Equation (1).

$$Accuracy = \frac{TP + TN}{TP + TN + FP + FN} \quad (1)$$

Equation (2) represents precision which is a ratio of the correct positive predictions of the network to the total positive predictions of the network.

$$Precision = \frac{TP}{TP + FP} \quad (2)$$

The recall parameter for the evaluation is given in Equation (3). In deep learning-assisted diagnostic tools, recall is a very sensitive evaluation measure as recall increases with a decrease in the number of false negative cases.

$$Recall = \frac{TP}{TP + FN} \quad (3)$$

The f1-score is given in Equation (4). F1-score, the harmonic mean of recall and precision, gives a measure of the goodness of deep CNN model for the given dataset.

$$F1 - Score = \frac{2 * Precision * Recall}{Precision + Recall} = \frac{2 * TP}{2 * TP + FP + FN} \quad (4)$$

The comparative results of performance parameters for different models are presented in Table 1.

The classification output is in terms of 0 or 1 for benign and malignant patches of BreakHis 40× dataset. For IDC dataset, the classification result is in terms of IDC or non-IDC patch. The result of classification on BreakHis dataset is shown in Figure 4.

Table 1. Custom CNN model evaluation parameters.

Backbone Pre-Trained Model with Custom Layers	Accuracy on Testing Data	Precision	Recall	f1-Score
EfficientNetB0 with custom CNN layers	89.67%	88.61%	89.32%	88.94%
ResNet50 with custom CNN layers	90.66%	89.33%	88.87%	89.09%
Xception with custom CNN layers	93.33%	92.20%	91.63%	91.91%

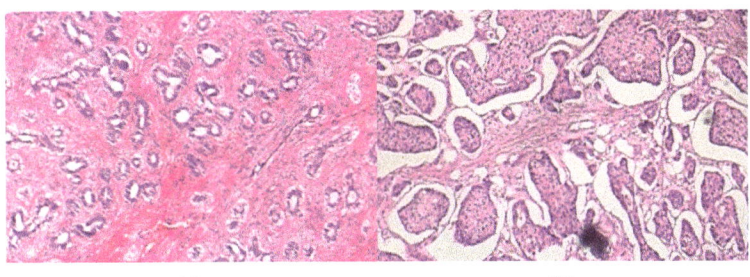

(a)   (b)

Figure 4. Classification result on patches of BreakHis40× test dataset. (**a**) Classified as benign patch. (**b**) Classified as malignant patch.

4.1. Confusion Matrices

The confusion matrix presentation shows an evaluation of the performance of the machine learning classification model. It represents a number of actual predicted output values against corresponding class labels. The heat-map representations of the confusion matrices are shown in Figure 5. As it is evident from confusion matrices, the maximum true positive value is given by CNN with Xception pre-trained model. The model with the minimum number of false negative cases in cancer diagnostics is considered more satisfactory. The false negative numbers are the number of patients who have breast cancer but are diagnosed as cancer-free. This false negative value is minimum for the Xception-based CNN model. Thus, CNN model with Xception as a backbone network performs better in the context of cancer diagnostics.

Confusion Matrix for Evaluation on IDC Dataset

The enhanced pre-trained Xception model was also trained and tested on a sub-part of the IDC (invasive ductal carcinoma) dataset. It consisted of 10,000 samples of each category, namely benign and malignant. The confusion matrix of CNN with Xception as a backbone for the IDC dataset is given in Figure 6.

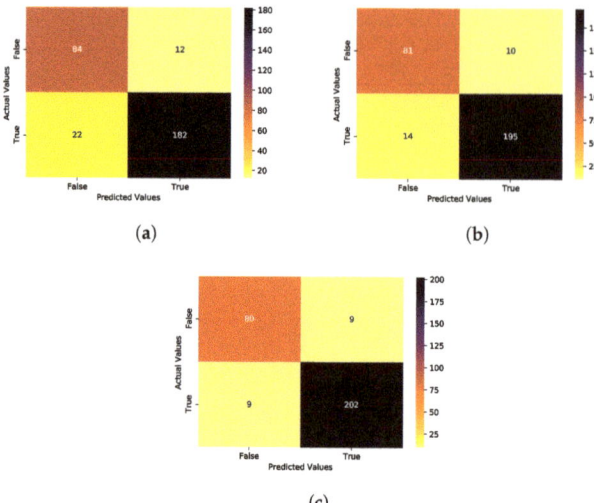

Figure 5. Parameters of confusion matrix for BreakHis40× dataset. (**a**) Confusion matrix for CNN with EfficientNetB0 as backbone. (**b**) Confusion matrix for CNN with ResNet50 as backbone. (**c**) Confusion matrix for CNN with Xception as backbone.

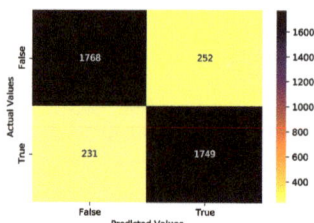

Figure 6. Parameters of confusion matrix for IDC dataset.

4.2. Receiver Operating Characteristics

The performance of the classification model can be evaluated by using the area under the receiver operating characteristics curve (AUC-ROC). There is a false positive rate on the X-axis of this curve, and on the Y-axis, there is a true positive rate. Using the AUC-ROC curve, the model's performance can be evaluated on all possible threshold values. The AUC-ROC curve for CNN using Xception as the pre-trained model is depicted in Figure 7. It is seen that the curve occupies a large area and is inclined toward the true positive rate parameter. The AUC value obtained for the proposed model is 0.9211.

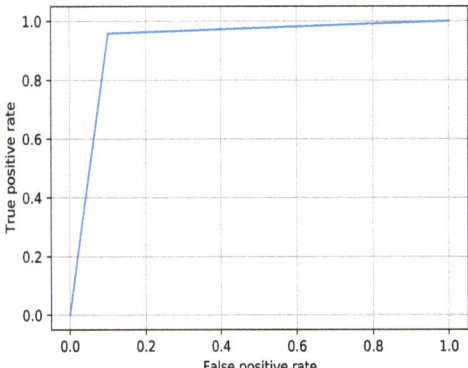

Figure 7. AUC-ROC curve for CNN using Xception model for BreakHis dataset

The AUC-ROC curve is also plotted for the IDC dataset in Figure 8. Again, CNN using the Xception model as a backbone is considered for classification on the IDC dataset. The AUC value obtained in this case is 88.07949.

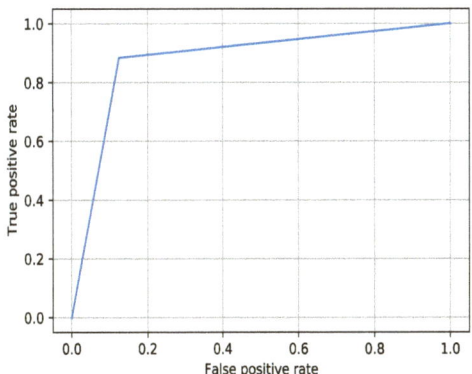

Figure 8. AUC-ROC curve for CNN using Xception model for IDC dataset.

4.3. Comparison with Other State of the Art Techniques

Other established methods are compared in Table 2 based on accuracy obtained on 40X magnification factor images of the BreakHis dataset. The study proposed in this article is magnification-dependent breast cancer detection, which applies to a zoom factor of 40X for breast histopathological images. Hence, comparative analysis involves comparing with other studies based on their accuracy obtained on the 40X magnification factor.

Table 2. Comparison with other state of the art methods for binary classification of BreakHis 40×.

Strategy Used for Binary Classification	Accuracy in Percentage
Multiple compact CNNs [54]	87.70%
Convolutional Neural network [55]	90.40%
Residual learning based CNN [56]	91.40%
Cubic SVM [57]	92.03%
Proposed Method of Xception with custom CNN layers	93.33%

4.4. Discussion

All the hyperparameters of the models are tweaked manually based on the accuracy criterion. The numbers of kernels in three custom convolutional layers are kept as 16, 32, and 64, respectively, with a filter size of 3 × 3. All convolutional custom layers use the Relu activation function. For all three CNN models, 50 epochs are used to fit the model with 50 steps per epoch on augmented training data. Adam is used as an optimizer with a learning rate of 0.0001. The image batch size is kept at 16. The three enhanced pre-trained models are compared based on accuracy for malignancy classification on the BreakHis dataset. The enhanced Xception pre-trained model obtained the highest accuracy of 93.33% as compared to the rest of the two models on the BreakHis classification task of malignancy detection. Further, the same model with BreakHis weights is used for binary classification on the IDC dataset. It achieved an accuracy of 88.08%.

5. Conclusions

In this study, an optimized CNN architectural framework for breast cancer histopathology slide binary classification is evaluated. The significant contribution of this work is the design, training, and testing of a custom CNN model, which uses the Xception pre-trained model as a backbone. The proposed model is used for the binary classification of breast malignancy. This deep learning approach consisting of stacking Xception architecture and custom CNN layers provided satisfactory accuracy of 93.33% for the BreakHis dataset of the 40× magnification factor. The data augmentation procedure is used to increase dataset size and ensure faithful training. The other two deep CNNs, one with EfficientNetB0 as the backbone and the other with ResNet50 as the backbone, were also evaluated for comparison of performances. Skipped connections in the form of drop-out layers are used to enhance the network's generalization capability. Batch normalization is used to improve and stabilize the overall training process. Another dataset called IDC was used to detect the presence of invasive carcinoma using an enhanced Xception model. By keeping weights, the same as tuned on the BreakHis dataset, the customized Xception CNN model is again fine-tuned for the IDC dataset. Furthermore, binary classification of IDC vs. non-IDC is achieved. This task achieved an accuracy of 88.08%. The use of a new dataset thus verifies the enhanced pre-trained model's generalization capability and transfer learning ability. Thus, such kind of work can help histopathologists in the primary diagnosis of breast cancer. Deep learning mechanisms developed for histopathological analysis work on patch-level data primarily. This is because of the huge size of whole slide images, which makes them unsuitable for deep convolutional neural networks. In the future, more sophisticated algorithms should be developed to obtain a proper diagnosis on the patient level and whole slide image (WSI) level histopathological image analysis.

Author Contributions: Conceptualization, S.A.J. and A.M.B.; methodology, S.A.J.; software, S.A.J.; validation, S.A.J., A.M.B. and P.O.O.; formal analysis, S.A.J., A.M.B., P.O.O. and S.U.; resources, S.A.J., A.M.B. and P.O.O.; data curation, S.A.J.; writing—original draft preparation, S.A.J. and A.M.B.; writing—review and editing, P.O.O., S.U., D.D. and A.B.; visualization, A.M.B.; supervision, A.M.B. All authors have read and agreed to the published version of the manuscript.

Funding: This research received no external funding.

Data Availability Statement: BreakHis dataset availability at https://web.inf.ufpr.br/vri/databases/breast-cancer-histopathological-database-breakhis/ accessed on 22 November 2022.

Conflicts of Interest: The authors declare no conflict of interest.

Abbreviations

The following abbreviations are used in this manuscript:

CNN	Convolutional Neural Network
WHO	World Health Organization
AI	Artificial Intelligence
GPU	Graphics Processing Unit
ICIAR	International Conference on Image Analysis and Recognition
BMI	Body Mass Index
VGG	Visual Geometry Group
RAM	Random Access Memory
PCA	Principal Component Analysis
AUC	Area under the curve
ROC	Receiver Operating Characteristics

References

1. Sung, H.; Ferlay, J.; Siegel, R.L.; Laversanne, M.; Soerjomataram, I.; Jemal, A.; Bray, F. Global cancer statistics 2020: GLOBOCAN estimates of incidence and mortality worldwide for 36 cancers in 185 countries. *CA Cancer J. Clin.* **2021**, *71*, 209–249. [CrossRef] [PubMed]
2. Barba, D.; León-Sosa, A.; Lugo, P.; Suquillo, D.; Torres, F.; Surre, F.; Trojman, L.; Caicedo, A. Breast cancer, screening and diagnostic tools: All you need to know. *Crit. Rev. Oncol.* **2021**, *157*, 103174. [CrossRef] [PubMed]
3. Gøtzsche, P.C.; Jørgensen, K.J. Screening for breast cancer with mammography. *Cochrane Database Syst. Rev.* **2013**, *2013*, 1–70. [CrossRef] [PubMed]
4. Geras, K.J.; Mann, R.M.; Moy, L. Artificial intelligence for mammography and digital breast tomosynthesis: Current concepts and future perspectives. *Radiology* **2019**, *293*, 246–259. [CrossRef]
5. Morrow, M.; Waters, J.; Morris, E. MRI for breast cancer screening, diagnosis, and treatment. *Lancet* **2011**, *378*, 1804–1811. [CrossRef]
6. Joshi, P.; Singh, N.; Raj, G.; Singh, R.; Malhotra, K.P.; Awasthi, N.P. Performance evaluation of digital mammography, digital breast tomosynthesis and ultrasound in the detection of breast cancer using pathology as gold standard: An institutional experience. *Egypt. J. Radiol. Nucl. Med.* **2022**, *53*, 1–11. [CrossRef]
7. Canino, F.; Piacentini, F.; Omarini, C.; Toss, A.; Barbolini, M.; Vici, P.; Dominici, M.; Moscetti, L. Role of Intrinsic Subtype Analysis with PAM50 in Hormone Receptors Positive HER2 Negative Metastatic Breast Cancer: A Systematic Review. *Int. J. Mol. Sci.* **2022**, *23*, 7079. [CrossRef]
8. Alom, M.Z.; Yakopcic, C.; Hasan, M.; Taha, T.M.; Asari, V.K. Recurrent residual U-Net for medical image segmentation. *J. Med. Imaging* **2019**, *6*, 014006. [CrossRef]
9. Hoorali, F.; Khosravi, H.; Moradi, B. IRUNet for medical image segmentation. *Expert Syst. Appl.* **2022**, *191*, 116399. [CrossRef]
10. Guo, S.; Xu, L.; Feng, C.; Xiong, H.; Gao, Z.; Zhang, H. Multi-level semantic adaptation for few-shot segmentation on cardiac image sequences. *Med. Image Anal.* **2021**, *73*, 102170. [CrossRef]
11. Zhao, M.; Cao, X.; Zhou, M.; Feng, J.; Xia, L.; Pogue, B.W.; Paulsen, K.D.; Jiang, S. MRI-guided near-infrared spectroscopic tomography (MRg-NIRST): System development for wearable, simultaneous NIRS and MRI imaging. In Proceedings of the Multimodal Biomedical Imaging XVII, SPIE, San Francisco, California, United States, 2 March 2022; Volume 11952, pp. 87–92.
12. Fusco, R.; Sansone, M.; Filice, S.; Carone, G.; Amato, D.M.; Sansone, C.; Petrillo, A. Pattern recognition approaches for breast cancer DCE-MRI classification: A systematic review. *J. Med. Biol. Eng.* **2016**, *36*, 449–459. [CrossRef]
13. Spanhol, F.A.; Oliveira, L.S.; Petitjean, C.; Heutte, L. A dataset for breast cancer histopathological image classification. *IEEE Trans. Biomed. Eng.* **2015**, *63*, 1455–1462. [CrossRef]
14. Kharya, S.; Soni, S. Weighted naive bayes classifier: A predictive model for breast cancer detection. *Int. J. Comput. Appl.* **2016**, *133*, 32–37. [CrossRef]
15. Shukla, K.; Tiwari, A.; Sharma, S. Classification of histopathological images of breast cancerous and non cancerous cells based on morphological features. *Biomed. Pharmacol. J.* **2017**, *10*, 353–366.
16. Valieris, R.; Amaro, L.; Osório, C.A.B.D.T.; Bueno, A.P.; Rosales Mitrowsky, R.A.; Carraro, D.M.; Nunes, D.N.; Dias-Neto, E.; Silva, I.T.D. Deep learning predicts underlying features on pathology images with therapeutic relevance for breast and gastric cancer. *Cancers* **2020**, *12*, 3687. [CrossRef] [PubMed]
17. Al Noumah, W.; Jafar, A.; Al Joumaa, K. Using parallel pre-trained types of DCNN model to predict breast cancer with color normalization. *BMC Res. Notes* **2022**, *15*, 14. [CrossRef] [PubMed]
18. Senan, E.M.; Alsaade, F.W.; Al-Mashhadani, M.I.A.; Theyazn, H.; Al-Adhaileh, M.H. Classification of histopathological images for early detection of breast cancer using deep learning. *J. Appl. Sci. Eng.* **2021**, *24*, 323–329.
19. Man, R.; Yang, P.; Xu, B. Classification of breast cancer histopathological images using discriminative patches screened by generative adversarial networks. *IEEE Access* **2020**, *8*, 155362–155377. [CrossRef]

20. Nahid, A.A.; Mehrabi, M.A.; Kong, Y. Histopathological breast cancer image classification by deep neural network techniques guided by local clustering. *BioMed Res. Int.* **2018**, *2018*, 2362108. [CrossRef]
21. Anupama, M.; Sowmya, V.; Soman, K. Breast cancer classification using capsule network with preprocessed histology images. In Proceedings of the 2019 International Conference on Communication and Signal Processing (ICCSP), Melmaruvathur, India, 4–6 April 2019; pp. 0143–0147.
22. Sharma, S.; Mehra, R.; Kumar, S. Optimised CNN in conjunction with efficient pooling strategy for the multi-classification of breast cancer. *IET Image Process.* **2021**, *15*, 936–946. [CrossRef]
23. Hameed, Z.; Zahia, S.; Garcia-Zapirain, B.; Javier Aguirre, J.; Maria Vanegas, A. Breast cancer histopathology image classification using an ensemble of deep learning models. *Sensors* **2020**, *20*, 4373. [CrossRef]
24. Zováthi, B.H.; Mohácsi, R.; Szász, A.M.; Cserey, G. Breast Tumor Tissue Segmentation with Area-Based Annotation Using Convolutional Neural Network. *Diagnostics* **2022**, *12*, 2161. [CrossRef]
25. Zheng, H.; Zhou, Y.; Huang, X. Spatiality Sensitive Learning for Cancer Metastasis Detection in Whole-Slide Images. *Mathematics* **2022**, *10*, 2657. [CrossRef]
26. Jin, Y.W.; Jia, S.; Ashraf, A.B.; Hu, P. Integrative data augmentation with U-Net segmentation masks improves detection of lymph node metastases in breast cancer patients. *Cancers* **2020**, *12*, 2934. [CrossRef]
27. Arooj, S.; Zubair, M.; Khan, M.F.; Alissa, K.; Khan, M.A.; Mosavi, A. Breast Cancer Detection and Classification Empowered With Transfer Learning. *Front. Public Health* **2022**, *10*, 1–18. [CrossRef] [PubMed]
28. Ukwuoma, C.C.; Hossain, M.A.; Jackson, J.K.; Nneji, G.U.; Monday, H.N.; Qin, Z. Multi-Classification of Breast Cancer Lesions in Histopathological Images Using DEEP_Pachi: Multiple Self-Attention Head. *Diagnostics* **2022**, *12*, 1152. [CrossRef]
29. Yao, H.; Zhang, X.; Zhou, X.; Liu, S. Parallel structure deep neural network using CNN and RNN with an attention mechanism for breast cancer histology image classification. *Cancers* **2019**, *11*, 1901. [CrossRef] [PubMed]
30. Guo, Y.; Dong, H.; Song, F.; Zhu, C.; Liu, J. Breast cancer histology image classification based on deep neural networks. In Proceedings of the International Conference Image Analysis and Recognition, Povoa de Varzim, Portugal, 27–29 June 2018; pp. 827–836.
31. Celik, Y.; Talo, M.; Yildirim, O.; Karabatak, M.; Acharya, U.R. Automated invasive ductal carcinoma detection based using deep transfer learning with whole-slide images. *Pattern Recognit. Lett.* **2020**, *133*, 232–239. [CrossRef]
32. Chen, Y.; Zhou, Y.; Chen, G.; Guo, Y.; Lv, Y.; Ma, M.; Pei, Z.; Sun, Z. Segmentation of Breast Tubules in H&E Images Based on a DKS-DoubleU-Net Model. *BioMed Res. Int.* **2022**, *2022*, 2961610.
33. Yari, Y.; Nguyen, T.V.; Nguyen, H.T. Deep learning applied for histological diagnosis of breast cancer. *IEEE Access* **2020**, *8*, 162432–162448. [CrossRef]
34. Anwar, F.; Attallah, O.; Ghanem, N.; Ismail, M.A. Automatic breast cancer classification from histopathological images. In Proceedings of the 2019 International Conference on Advances in the Emerging Computing Technologies (AECT), Al Madinah Al Munawwarah, Saudi Arabia, 10 February 2020; pp. 1–6.
35. Sudharshan, P.; Petitjean, C.; Spanhol, F.; Oliveira, L.E.; Heutte, L.; Honeine, P. Multiple instance learning for histopathological breast cancer image classification. *Expert Syst. Appl.* **2019**, *117*, 103–111. [CrossRef]
36. Attallah, O.; Anwar, F.; Ghanem, N.M.; Ismail, M.A. Histo-CADx: Duo cascaded fusion stages for breast cancer diagnosis from histopathological images. *PeerJ Comput. Sci.* **2021**, *7*, e493. [CrossRef] [PubMed]
37. Ghanem, N.M.; Attallah, O.; Anwar, F.; Ismail, M.A. AUTO-BREAST: A fully automated pipeline for breast cancer diagnosis using AI technology. In *Artificial Intelligence in Cancer Diagnosis and Prognosis, Volume 2: Breast and Bladder Cancer*; IOP Publishing: Bristol, UK, 2022.
38. Ahmad, N.; Asghar, S.; Gillani, S.A. Transfer learning-assisted multi-resolution breast cancer histopathological images classification. *Vis. Comput.* **2022**, *38*, 2751–2770. [CrossRef]
39. Wang, P.; Wang, J.; Li, Y.; Li, P.; Li, L.; Jiang, M. Automatic classification of breast cancer histopathological images based on deep feature fusion and enhanced routing. *Biomed. Signal Process. Control.* **2021**, *65*, 102341. [CrossRef]
40. Iesmantas, T.; Alzbutas, R. Convolutional capsule network for classification of breast cancer histology images. In Proceedings of the Image Analysis and Recognition: 15th International Conference, ICIAR 2018, Póvoa de Varzim, Portugal, 27–29 June 2018; Proceedings 15; Springer: Berlin/Heidelberg, Germany, 2018; pp. 853–860.
41. He, Z.; Lin, M.; Xu, Z.; Yao, Z.; Chen, H.; Alhudhaif, A.; Alenezi, F. Deconv-transformer (DecT): A histopathological image classification model for breast cancer based on color deconvolution and transformer architecture. *Inf. Sci.* **2022**, *608*, 1093–1112. [CrossRef]
42. Zou, Y.; Chen, S.; Sun, Q.; Liu, B.; Zhang, J. DCET-Net: Dual-stream convolution expanded transformer for breast cancer histopathological image classification. In Proceedings of the 2021 IEEE International Conference on Bioinformatics and Biomedicine (BIBM), Houston, TX, USA, 9–12 December 2021; pp. 1235–1240.
43. Wang, Y.; Acs, B.; Robertson, S.; Liu, B.; Solorzano, L.; Wählby, C.; Hartman, J.; Rantalainen, M. Improved breast cancer histological grading using deep learning. *Ann. Oncol.* **2022**, *33*, 89–98. [CrossRef]
44. Mahmood, T.; Arsalan, M.; Owais, M.; Lee, M.B.; Park, K.R. Artificial intelligence-based mitosis detection in breast cancer histopathology images using faster R-CNN and deep CNNs. *J. Clin. Med.* **2020**, *9*, 749. [CrossRef]
45. Naik, N.; Madani, A.; Esteva, A.; Keskar, N.S.; Press, M.F.; Ruderman, D.; Agus, D.B.; Socher, R. Deep learning-enabled breast cancer hormonal receptor status determination from base-level H&E stains. *Nat. Commun.* **2020**, *11*, 5727.

46. Rawat, R.R.; Ortega, I.; Roy, P.; Sha, F.; Shibata, D.; Ruderman, D.; Agus, D.B. Deep learned tissue "fingerprints" classify breast cancers by ER/PR/Her2 status from H&E images. *Sci. Rep.* **2020**, *10*, 1–13.
47. Farahmand, S.; Fernandez, A.I.; Ahmed, F.S.; Rimm, D.L.; Chuang, J.H.; Reisenbichler, E.; Zarringhalam, K. Deep learning trained on hematoxylin and eosin tumor region of Interest predicts HER2 status and trastuzumab treatment response in HER2+ breast cancer. *Mod. Pathol.* **2022**, *35*, 44–51. [CrossRef]
48. Chauhan, R.; Vinod, P.; Jawahar, C. Exploring Genetic-histologic Relationships in Breast Cancer. In Proceedings of the 2021 IEEE 18th International Symposium on Biomedical Imaging (ISBI), Nice, France, 13–16 April 2021; pp. 1187–1190.
49. Jaber, M.I.; Song, B.; Taylor, C.; Vaske, C.J.; Benz, S.C.; Rabizadeh, S.; Soon-Shiong, P.; Szeto, C.W. A deep learning image-based intrinsic molecular subtype classifier of breast tumors reveals tumor heterogeneity that may affect survival. *Breast Cancer Res.* **2020**, *22*, 12. [CrossRef]
50. Mooney, P. Breast Histopathology Images. Kaggle. 2020. Available online: https://www.kaggle.com/datasets/paultimothymooney/breast-histopathology-images (accessed on 828 November 2022).
51. Tan, M.; Le, Q. Efficientnet: Rethinking model scaling for convolutional neural networks. In Proceedings of the International Conference on Machine Learning, PMLR, Long Beach, CA, USA, 9–15 June 2019; pp. 6105–6114.
52. He, K.; Zhang, X.; Ren, S.; Sun, J. Deep residual learning for image recognition. In Proceedings of the IEEE Conference on Computer Vision and Pattern Recognition, Las Vegas, NV, USA, 27–30 June 2016; pp. 770–778.
53. Chollet, F. Xception: Deep learning with depthwise separable convolutions. In Proceedings of the IEEE Conference on Computer Vision and Pattern Recognition, Honolulu, HI, USA, 21–26 June 2017; pp. 1251–1258.
54. Zhu, C.; Song, F.; Wang, Y.; Dong, H.; Guo, Y.; Liu, J. Breast cancer histopathology image classification through assembling multiple compact CNNs. *BMC Med. Inform. Decis. Mak.* **2019**, *19*, 198. [CrossRef]
55. Spanhol, F.A.; Oliveira, L.S.; Petitjean, C.; Heutte, L. Breast cancer histopathological image classification using convolutional neural networks. In Proceedings of the 2016 International Joint Conference on Neural Networks (IJCNN), Vancouver, BC, Canada, 24–29 July 2016; pp. 2560–2567.
56. Gour, M.; Jain, S.; Sunil Kumar, T. Residual learning based CNN for breast cancer histopathological image classification. *Int. J. Imaging Syst. Technol.* **2020**, *30*, 621–635. [CrossRef]
57. Singh, S.; Kumar, R. Histopathological image analysis for breast cancer detection using cubic SVM. In Proceedings of the 2020 7th international conference on signal processing and integrated networks (SPIN), Noida, India, 27–28 February 2020; pp. 498–503.

Disclaimer/Publisher's Note: The statements, opinions and data contained in all publications are solely those of the individual author(s) and contributor(s) and not of MDPI and/or the editor(s). MDPI and/or the editor(s) disclaim responsibility for any injury to people or property resulting from any ideas, methods, instructions or products referred to in the content.

Article

Performance Investigation for Medical Image Evaluation and Diagnosis Using Machine-Learning and Deep-Learning Techniques

Baidaa Mutasher Rashed and Nirvana Popescu *

Computer Science Department, University POLITEHNICA of Bucharest, 060042 Bucharest, Romania; rashed.baidaa@stud.acs.upb.ro or baidaaalsafy@utq.edu.iq or baidaaalsafy@gmail.com
* Correspondence: nirvana.popescu@upb.ro

Abstract: Today, medical image-based diagnosis has advanced significantly in the world. The number of studies being conducted in this field is enormous, and they are producing findings with a significant impact on humanity. The number of databases created in this field is skyrocketing. Examining these data is crucial to find important underlying patterns. Classification is an effective method for identifying these patterns. This work proposes a deep investigation and analysis to evaluate and diagnose medical image data using various classification methods and to critically evaluate these methods' effectiveness. The classification methods utilized include machine-learning (ML) algorithms like artificial neural networks (ANN), support vector machine (SVM), k-nearest neighbor (KNN), decision tree (DT), random forest (RF), Naïve Bayes (NB), logistic regression (LR), random subspace (RS), fuzzy logic and a convolution neural network (CNN) model of deep learning (DL). We applied these methods to two types of datasets: chest X-ray datasets to classify lung images into normal and abnormal, and melanoma skin cancer dermoscopy datasets to classify skin lesions into benign and malignant. This work aims to present a model that aids in investigating and assessing the effectiveness of ML approaches and DL using CNN in classifying the medical databases and comparing these methods to identify the most robust ones that produce the best performance in diagnosis. Our results have shown that the used classification algorithms have good results in terms of performance measures.

Keywords: medical image dataset analysis; diagnosis; machine learning; deep learning

Citation: Rashed, B.M.; Popescu, N. Performance Investigation for Medical Image Evaluation and Diagnosis Using Machine-Learning and Deep-Learning Techniques. *Computation* **2023**, *11*, 63. https://doi.org/10.3390/computation11030063

Academic Editor: Anando Sen

Received: 17 February 2023
Revised: 12 March 2023
Accepted: 15 March 2023
Published: 20 March 2023

Copyright: © 2023 by the authors. Licensee MDPI, Basel, Switzerland. This article is an open access article distributed under the terms and conditions of the Creative Commons Attribution (CC BY) license (https://creativecommons.org/licenses/by/4.0/).

1. Introduction

The world is changing at such a rapid pace that the pressure on healthcare is increasing; adverse changes in climate, environment, and human lifestyle raise the degree of danger, as well as diseases in individuals. This work is focused on analyzing lung diseases and melanoma skin cancer, conditions which, if detected early, can be properly treated. One of the most seriously injured organs is the lung; people can develop a wide variety of lung diseases [1]. Skin cancer is a common type of cancer that affects people with fair skin, and melanoma is a particularly dangerous type of skin cancer; it may quickly transition between different body parts, and has the greatest fatality rate. However, if it is recognized and treated early, the chances of it being cured are higher, necessitating early detection [2].

The traditional methods of diagnosis are costly and time-consuming due to the involvement of trained experts, as well as the requirement of a well-equipped environment. Recent advances in computerized solutions for diagnosis are quite promising, showing increased accuracy and efficiency [3]. By applying medical image-processing techniques to chest X-ray and melanoma skin cancer dermoscopy images, we can assist in detecting diseases earlier and more accurately, which can save many humans. Lung and melanoma skin cancer are two diseases that can be detected earlier and more accurately thanks to the development of technology and computers [4].

Machine-learning and deep-learning techniques have recently attained impressive results in the image-processing field along with medical science. Several areas of healthcare have successfully used machine-learning algorithms [5]. In recent years, various researchers have suggested various artificial intelligence (AI)-based treatments for various medical issues. The DL using the CNN method has enabled researchers to achieve successful outcomes in a variety of medical applications, such as skin cancer classification from skin images and disease prediction from X-ray images. Due to this development, numerous studies have been conducted to determine how DL and ML may affect the healthcare field and medical-imaging diagnostics [6].

ML is the process of teaching a computer to use its prior expertise to solve a problem [7]. Because of the current availability of cheaper processing power and inexpensive memory, the concept of applying ML in several fields to resolve problems more quickly than humans has attracted substantial interest. This allows for the processing and analysis of vast amounts of data to identify insights and correlations within the data that would not be obvious to the human eye. Its intelligent behaviors are based on many algorithms that allow the computer to deliver salient conclusions [8]. In contrast, DL is a branch of ML offering a more advanced approach which allows computers to automatically extract, analyze, and interpret relevant information from raw data by mimicking how people learn and think. DL is a set of neural data-driven approaches based on autonomous feature engineering processes; its accuracy and performance are due to its automatic learning of features from inputs. CNN is regarded as one of the finest image-recognition and -classification models in DL [9].

In this work, we analyze medical images for two sets of medical databases: the first group is chest X-ray images to detect lung diseases and the second group is skin dermoscopy images to detect melanoma skin cancer. We focus on the use of the most common machine-learning techniques and convolutional neural networks for deep learning to classify the lungs in the first medical dataset into normal and abnormal, and skin lesions in the second medical dataset into malignant and benign to prove the efficiency and effectiveness of these methods in classification and medical diagnosis.

The major goal of this work is to investigate and assess the effectiveness of ML approaches and CNN for the classification of medical databases and compare these methods to identify the methods that have the best performance in diagnosis. We applied these methods to two different types of medical databases to diagnose two different diseases that are considered among the most dangerous, which threaten human life and can be treated if diagnosed early.

The paper is organized as follows: Section 2 presents the current advances in this research field, emphasizing our research in this context. Section 3 explains the components of the proposed system to diagnose lung diseases and melanoma skin cancer. Section 4 presents the chest X-ray and melanoma skin cancer dermoscopy image analysis in detail, as well as the databases resources. Here, the analysis methods are discussed, preprocessing and segmentation methods are presented, and the extraction-of-features methods are introduced, emphasizing the classification techniques employed for lung illness and melanoma skin cancer diagnosis and performance metrics. Section 5 introduces the performance of the results and the comparison between the two datasets. Section 6 provides the most important work contributions. Section 7 discusses the work results and future works. The paper concludes by drawing some conclusions.

2. Related Work

In recent years, several studies have been conducted on performance research for medical image evaluation and diagnosis using ML and DL techniques. In this research field, we did not find any research that incorporates most of the machine-learning methods (as we used) and makes a parallel comparison when using more than one medical database (we used two databases) to assess the effectiveness of ML and DL methods. The authors of [10] evaluated the SVM and RF machine-learning techniques, as well as CNN for detecting breast cancer in thermographic images; the images were preprocessed to improve them and then isolate the region of interest (ROI). The object-oriented image-segmentation method was used, which eliminated the salt-and-pepper noise from the image and increased its precision using spectral signatures. The forms of the items' border, thickness, and color of images were extracted as features. Classification algorithms were utilized and evaluated using a variety of metrics, such as validation accuracy, elapsed time, training error, and training precision. In terms of accuracy, precision, and the amount of data needed, CNN outperformed the SVM and RF approaches.

In [11], the effectiveness of deep learning and machine learning was assessed using skin cancer datasets. The proposed method used Laplacian and average filters to eliminate noise from images and dull-razor techniques to remove hair. For image segmentation, the region-growth method was used. Three techniques were utilized to extract hybrid features: GLCM, discrete wavelet transform (DWT), and local binary pattern (LBP). These features were then combined into a feature vector and categorized utilizing ANN and feed-forward neural network (FFNN) classifiers of machine-learning methods and CNN models (ResNet-50 and AlexNet). The approaches were evaluated using statistical measures (accuracy, precision, sensitivity, specificity, and AUC), where the FFNN and ANN classifiers outperformed the CNN models. The authors of [12] developed a system for predicting lung diseases such as pneumonia and COVID-19 from patients' chest X-ray images; they used median filtering and histogram equalization to improve the image quality. They developed a modified region-growing technique for extracting the ROI of the chest areas. They extracted a set of features represented by texture, shape, visual, and intensity features followed by normalization. ANN, SVM, KNN, ensemble classifiers, and deep-learning classifiers were utilized for classification. Deep-learning architecture based on recurrent neural networks (RNN) with long short-term memory was suggested for the accurate identification of lung illnesses (LSTM). The approaches were evaluated using metric measures (accuracy, specificity, precision, recall, and F-measure). The F-RNN-LSTM approach had the highest accuracy.

In [13], the findings and analysis of the UCI Heart Disease dataset were compared using various machine-learning and deep-learning methodologies. The dataset contained 14 major attributes which were used in the study. Several promising results were achieved and validated using accuracy, sensitivity, and specificity. Isolation forest was used to address certain uninteresting aspects of the dataset. Deep learning achieved higher accuracy compared with the ML methods that were used in the work. The authors of [14] created a model that aided in the diagnosis of chest X-ray medical images and classified the images into healthy and sick by employing six machine-learning techniques (DT, RF, KNN, AdaBoost, Gradient Boost, XGBboost) and a CNN model to improve efficiency and accuracy. The approach begins by reducing the size of chest X-ray images before identifying and classifying them using the conventional neural network framework, which extracts and classifies information from the images. The model's performance was estimated utilizing classification accuracy and cross-validation. Deep learning had the highest accuracy. The decision tree classifier, on the other hand, had the lowest performance.

The authors of [15] investigated the efficiency of various ML and DL algorithms for detecting Plasmodium on digital microscopy cell pictures. They used a publicly available dataset that included equal numbers of parasitized and uninfected cells. They used color constancy and spatially resampled all images to a specific size based on the classification architecture used, and they presented a swift CNN architecture. Additionally, they investigated and evaluated the effectiveness of transfer-learning algorithms built on well-known network topologies such as AlexNet, ResNet, VGG-16, and DenseNet. They also studied how well the bag-of-features model performed when used with an SVM for classification. Based on the average probabilities provided by all the developed CNN architectures, the probability of a cell image containing Plasmodium was calculated. All deep-learning- and transfer-learning-based techniques outperformed the bag-of-features and SVM-based classification models.

In comparison with previous studies, we noticed the following:

- The proposed system in this work dealt with two different types of databases for evaluating two different diseases and determining the performance of DL and ML methods, while all previous studies dealt with one database for diagnosing one type of disease for evaluating the performance of DL and ML methods. A continuous comparison between the application of the methods on both datasets has been made, emphasizing the fact that the type of evaluated disease matters.
- In comparison with [11], we note that the authors used a dull-razor tool to remove hair from skin images, while we suggested in our work an accurate algorithm to remove hair from skin images while preserving the shape of the lesion and the quality of the image.
- We noticed that most of the previous studies [10–12] used the object-oriented image method in the segmentation stage to extract the ROI from the images. In our study, we proposed methods for extracting the ROI (lung and skin lesions) that depend on the threshold techniques, binarization, negation, and morphological operations to segment the colored and gray level images. In addition to this, ref. [14] does not mention any image-segmentation method.
- Compared with previous studies, our study focused on extracting hybrid features from images, which included most types of features (texture, color, shape, geometry, and intensity), and different methods for extracting features were addressed.
- In the classification stage, we noticed that our study dealt with most of the methods of machine learning (nine methods) for an advanced comparison, while the rest of the studies dealt with a limited number of machine-learning methods, and may be limited to one [15], two [10,11], or three methods [12].

We concluded from the comparison that our work was more comprehensive and more in-depth in evaluating the performance of machine-learning and deep-learning methods in diagnosing diseases, and it provided good results in addition to offering a better overview of the ML methods. In this way, our work can be considered a good start for any researcher that has to choose an appropriate technique.

3. Workflow Design

This section describes the proposed framework, which is divided into three phases: the first phase is responsible for acquiring medical datasets to be analyzed in the next phase; the second phase is responsible for analyzing the input medical dataset that includes preprocessing medical images to improve them, extracting the region of interest (ROI), and then extracting significant features that help in classification. The third phase is responsible for the diagnosis and evaluation, where selected classification methods are applied to the selected datasets, and then these methods are evaluated. All the algorithms and processes of the proposed work were implemented using Matlab 2021. Figure 1 shows the whole workflow for the proposed framework.

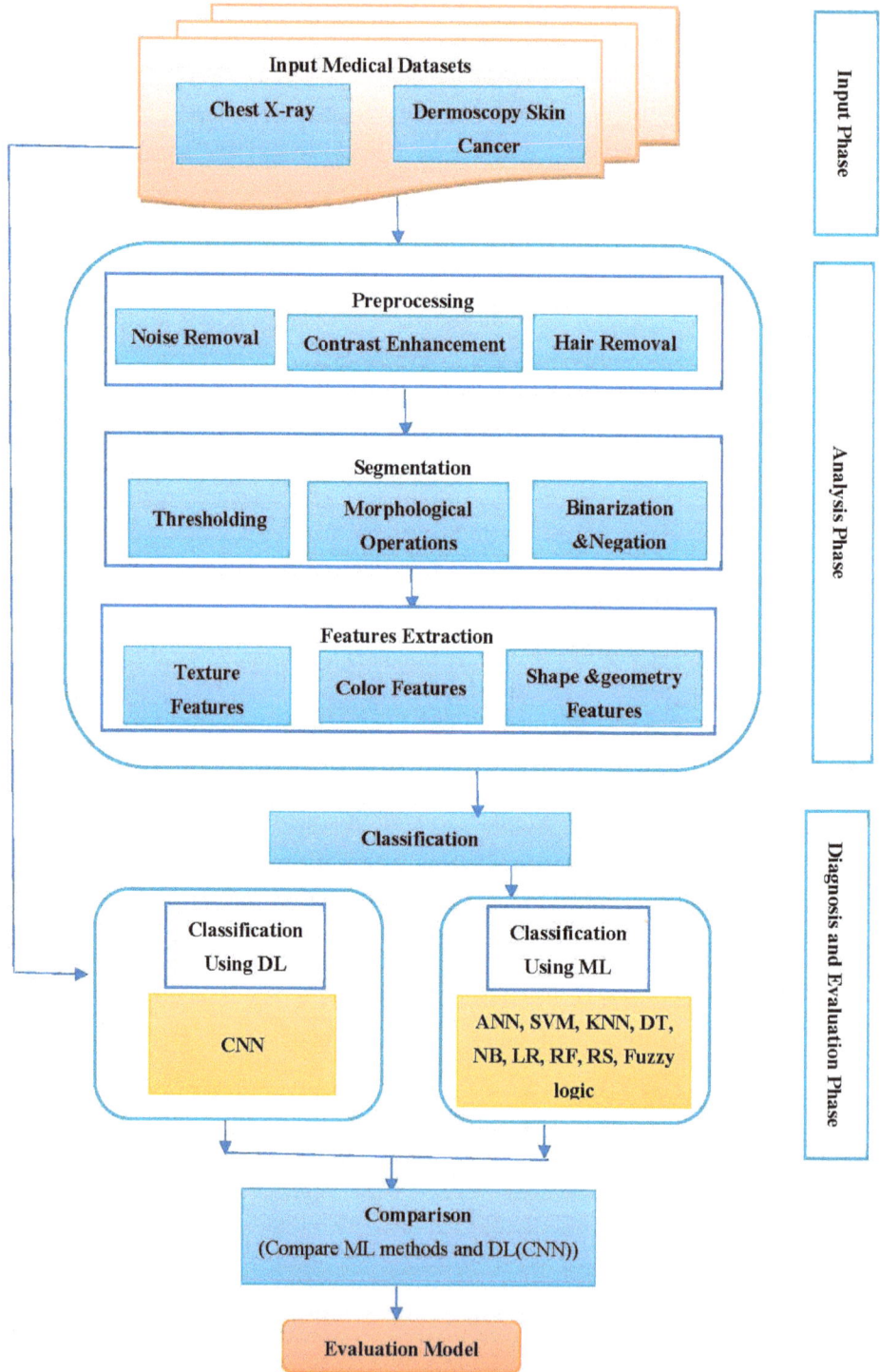

Figure 1. The workflow of the proposed framework.

4. Methods

This section is divided into three parts. The first part deals with the medical datasets used in this work. The second part deals with the analysis of the medical datasets, which involves the stages of preprocessing, segmentation, and feature extraction. The third part deals with the diagnosis and evaluation phase, which involves applying classification algorithms on the selected medical datasets and then evaluating these algorithms.

4.1. Medical Datasets

In this work, we used two sets of medical databases; the first group included images of chest X-rays, and the second group included dermoscopy images for melanoma skin cancer. These databases were divided into 70% for the training and 30% for the testing.

4.1.1. The Chest X-ray Dataset

The chest X-ray samples of normal and abnormal lung cases were obtained from Kaggle [16]. A dataset containing 612 images was used for the proposed methodology; among the total 612 images of lungs that were certified in this work, 288 images showed healthy lungs and 324 were images of lungs affected by different types of lung diseases like atelectasis, pneumonia, emphysema, fibrosis, lung opacity, COVID-19, and bacterial and viral diseases. Images were captured in the JPG format, with various resolution sizes; the sizes were standardized to 256 × 256 pixels. Figure 2 illustrates normal and abnormal lung images from the chest X-ray database.

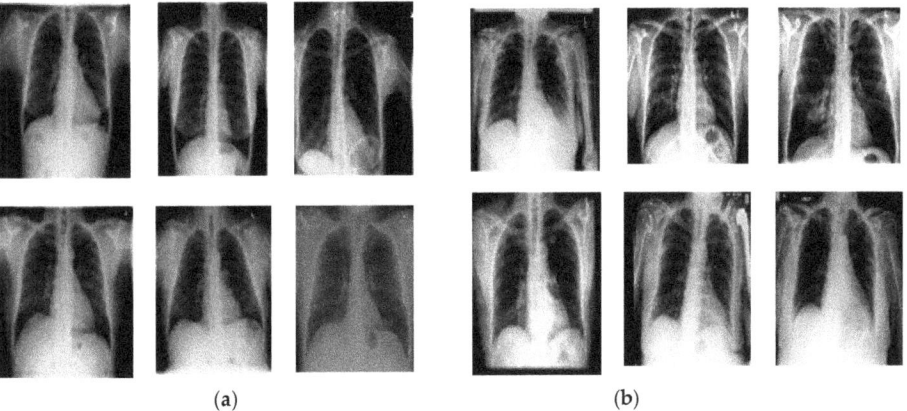

Figure 2. Samples of chest X-ray dataset: (**a**) Normal lung images; (**b**) Abnormal lung images.

4.1.2. The Dermoscopy Melanoma Skin Cancer Dataset

The dermoscopy samples of melanoma skin cancer were obtained from The Lloyd Dermatology and Laser Center [17], and the Dermatology Online Atlas [18]. A dataset containing 300 images was used to evaluate the proposed methodology; among the total of 300 images of melanoma skin cancer which were used in this work, 145 images represented benign conditions and 155 represented malignant ones, which include several types of malignant melanoma-like superficial spreading, nodular, lentigo, and acral malignant melanoma. The images were captured in the JPG format, as in the previous case, and the sizes were standardized to 256 × 256 pixels to extract accurate features that distinguish between benign and malignant melanoma skin cancer images. Figure 3 illustrates benign and malignant images from the melanoma skin cancer dermoscopy database.

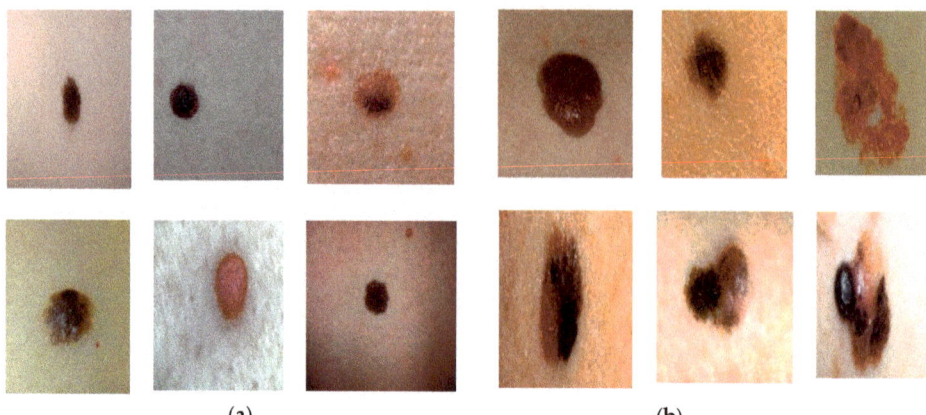

(a) (b)

Figure 3. Samples of melanoma skin cancer dermoscopy dataset: (**a**) Benign lesion images; (**b**) Malignant lesion images.

4.2. Datasets Analysis

In this section, we discuss the data analysis phase according to the suggested model (shown in Figure 1) employed in the study. This phase contains three stages: preprocessing, segmentation, and feature extraction.

4.2.1. Image Preprocessing

Preprocessing aims to enhance images and remove undesirable effects. Because the quality of the first medical dataset, which contains chest X-ray images, is low and contains noise, the suppression of lung regions affected by congestion or fluids may occur, and the X-ray scan also generates noise in the image. The suggested model for preprocessing the chest X-ray images involves applying three main processes: image cropping, noise removal, and contrast enhancement.

4.2.1.1. Image Cropping

Cropping is applied to the input original chest X-ray images to accentuate the ROI (lung) and remove all undesired artifacts. Image cropping is required to accelerate image processing. Manual cropping is used in this work, where the image is cropped into a square form consisting of the lung, as shown in Figure 4b.

4.2.1.2. Noise Removal

In the case of X-ray datasets, median filtering outperforms adaptive bilateral filtering, average filtering, and Wiener filtering [19]. Median filtering is a simple approach widely utilized in many image-processing applications because it is more successful in noise reduction and edge preservation and eliminates any additional noise present in the image [12]. In the proposed model, we used median filtering, which works by traversing across the image pixel by pixel and replacing every value with the median value of the adjacent pixel. The design of the neighbor is determined by the size of the window; a window size of a 3 × 3 neighborhood was utilized in this work. Figure 4c demonstrates the application of median filtering in a chest X-ray image.

4.2.1.3. Contrast Enhancement

In this process, we utilized the adaptive intensity values adjustment, which concentrates on adjusting the image intensity values for low-contrast X-ray images. In this way, the contrast is improved. Then, we applied histogram equalization for increased contrast to make the ROI clear. The histogram represents the distribution of image pixels; it is calculated by counting the number of times each pixel value appears, and is then mapped against the grayscale image's intensity [20]. Enhancing the contrast of an image can sharpen its border and increase segmentation accuracy because it creates a contrast between the object and the background. The applied contrast-enhancement process is shown in Figure 4d,e.

The result after applying the main processes of image preprocessing on an example of a chest X-ray image is shown in Figure 4.

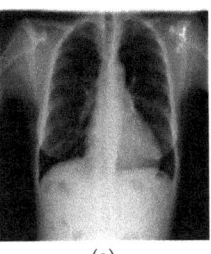 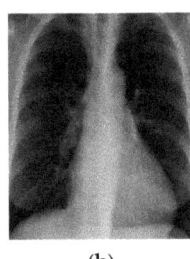 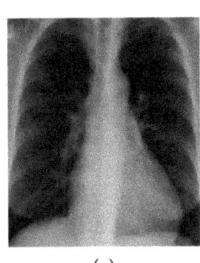 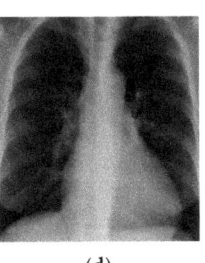 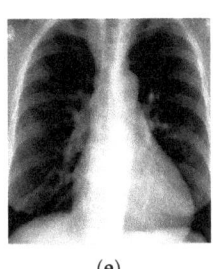

(a) (b) (c) (d) (e)

Figure 4. The main preprocessing process for chest X-ray images: (**a**) Original image; (**b**) Image cropping; (**c**) Applying median filter; (**d**) Applying contrast adjustment; (**e**) Applying histogram equalization.

Image preprocessing for the second group, which contains melanoma skin cancer dermoscopy images, involves presenting an algorithm for hair detection and removal. Some of the images of skin may contain hair, and this hair offers an inaccurate classification; as a result, it is preferable to remove the hair before moving on to the next stages. The proposed algorithm creates a clean dermoscopy image while maintaining the dermoscopy appearance by replacing the portions of the image containing hair structures with the neighboring pixels. The original RGB (red, green, blue) dermoscopy image is first transformed into a grayscale image, and then the resulting grayscale image is subjected to a morphological filter known as black top-hat [21,22].

The top-hat morphological filter fills in the image's minute gaps while preserving the original area sizes; thus, all background areas are removed for the pixel values that act as structuring elements. A thresholding technique is applied to the output of the used top-hat morphological filter to create a binary mask of the undesirable structures present on the dermoscopy image. After the creation of the binary mask of the hair structures, we replace the mask's pixels to remove undesired pixels while retaining the image's shape and extracting and restoring the clean skin lesion image [23]. The following four steps describe the hair-removal algorithm:

- The color image is converted into a grayscale image;
- Black top-hat transformation is utilized for the detection of dark and thick hairs and is represented as the following equation:

$$T_w(C_n) = C_n \bullet b - C_n$$

where $\bullet$ denotes the closing operation, C_n is the local contrasted image, and b is a grayscale structuring element.

- By filling the regions in the image that the mask specifies, we can use region fill to remove items from the image or to replace invalid pixel values with their neighbors. The mask's nonzero pixels specify the image pixels to be filled.
- The result is a fully preprocessed image maintained throughout the subsequent phases.

Figure 5 shows an example of applying steps of a hair-removal algorithm on a melanoma skin cancer image.

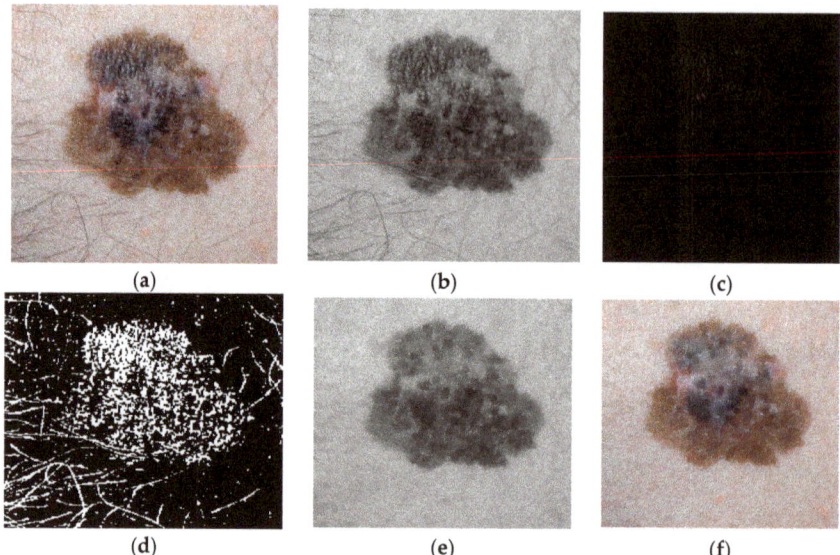

Figure 5. Example of the proposed hair-removal algorithm: (**a**) The original RGB image; (**b**) Converting to grayscale; (**c**) Black top-hat filter applied; (**d**) Thresholding applied; (**e**) Hair removal of a gray image; (**f**) Hair removal of a color image.

4.2.2. Image Segmentation

Image segmentation is the process of separating an object from an image according to criteria like the gray level of a pixel and the gray level of its nearby pixels. One approach for segmenting is the thresholding method, which divides the grayscale image into segments based on many classes according to the gray level [20]. In this work, image segmentation for the first group containing chest X-ray images involves applying thresholding and morphological operations to the preprocessed image to separate the ROI (lung) from the image. Here, global thresholding was utilized because the intensity distribution between the background and foreground of the image was considerably different. After that, we applied morphological operations, which are a large set of image-processing operations that process medical images according to shapes and facilitate object segmentation from images [24]. In the proposed algorithm, the fill operation was used, which smooths the contour and closes small holes as the inner lining of the shapes fills inward; the restoring fill skips the closed holes, thus making this operation effective for closing holes. Thus, morphological operation helps smooth and simplify the borders of objects without changing their size and improves the specific region for accurate segmentation. Figure 6 shows the steps of the suggested segmentation algorithm for chest X-ray images.

Image segmentation for the second group, which contains melanoma skin cancer dermoscopy images, involves applying Otsu thresholding, binarization, and image negation to the preprocessed image to separate the object (skin lesion) from the image. Here, Otsu thresholding was utilized, which is a thresholding approach that automatically finds the threshold point that splits the gray-level image histogram into two distinct sections. The image's gray level is expressed as I to L, where I is 0 pixels and L is 255 pixels. The Otsu approach was utilized to automatically determine the threshold according to the input images [20]. Following that, the processes of binarization and image negation were carried out. Binarization via thresholding is the process of converting a grayscale image to a binary image; here, every pixel in the improved image's gray-level value was computed, and if the value was larger than the global threshold, the pixel value was set to one; otherwise, was is set to zero [25]. During the image negation, an image with white pixels is replaced with black pixels. Meanwhile, white pixels replace dark pixels [20]. After applying the binary mask, the mask is then multiplied by the three color channels (red, green, and blue) to extract the region of interest. Figure 7 shows the steps of the suggested segmentation algorithm for melanoma skin cancer images.

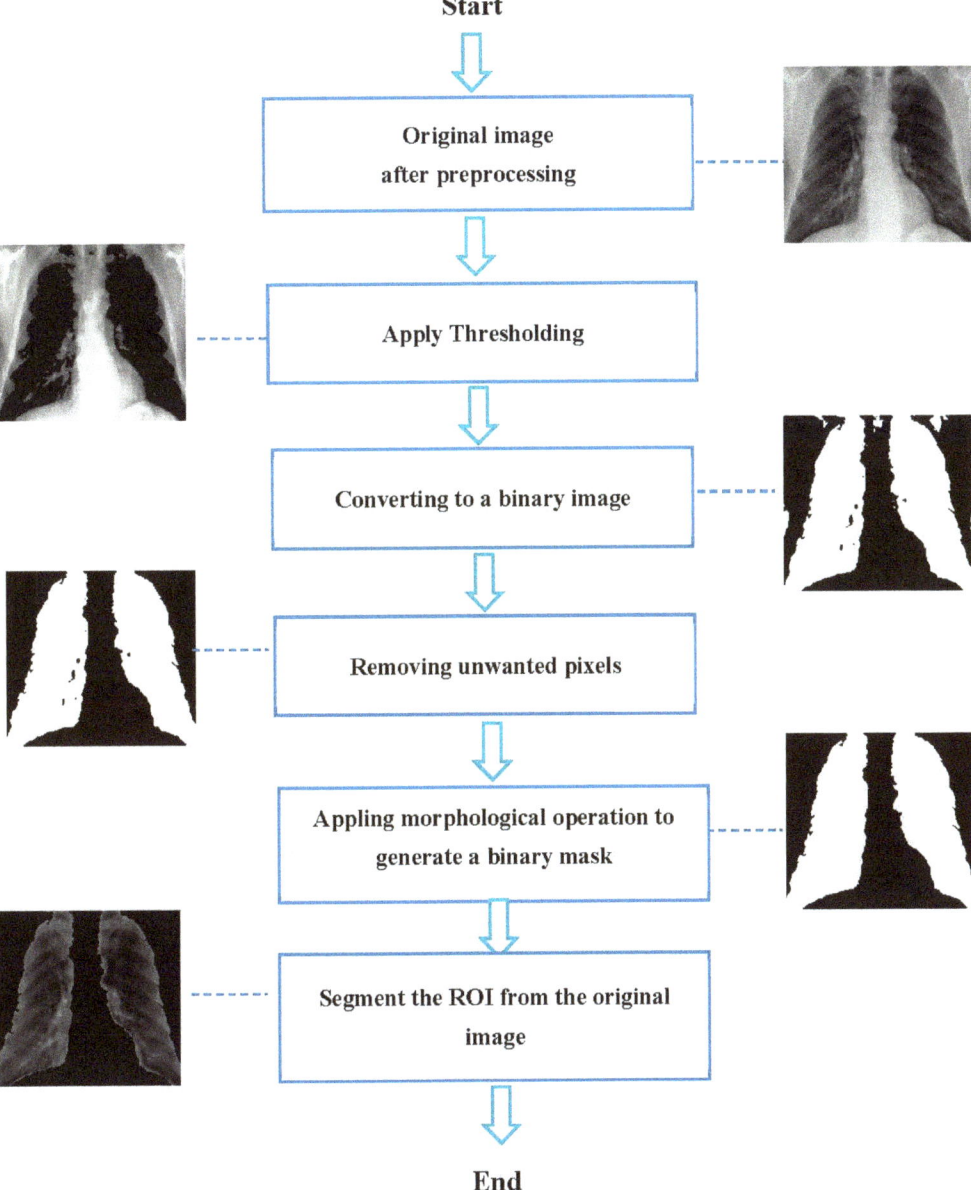

Figure 6. Flowchart of the proposed segmentation algorithm for chest X-ray images.

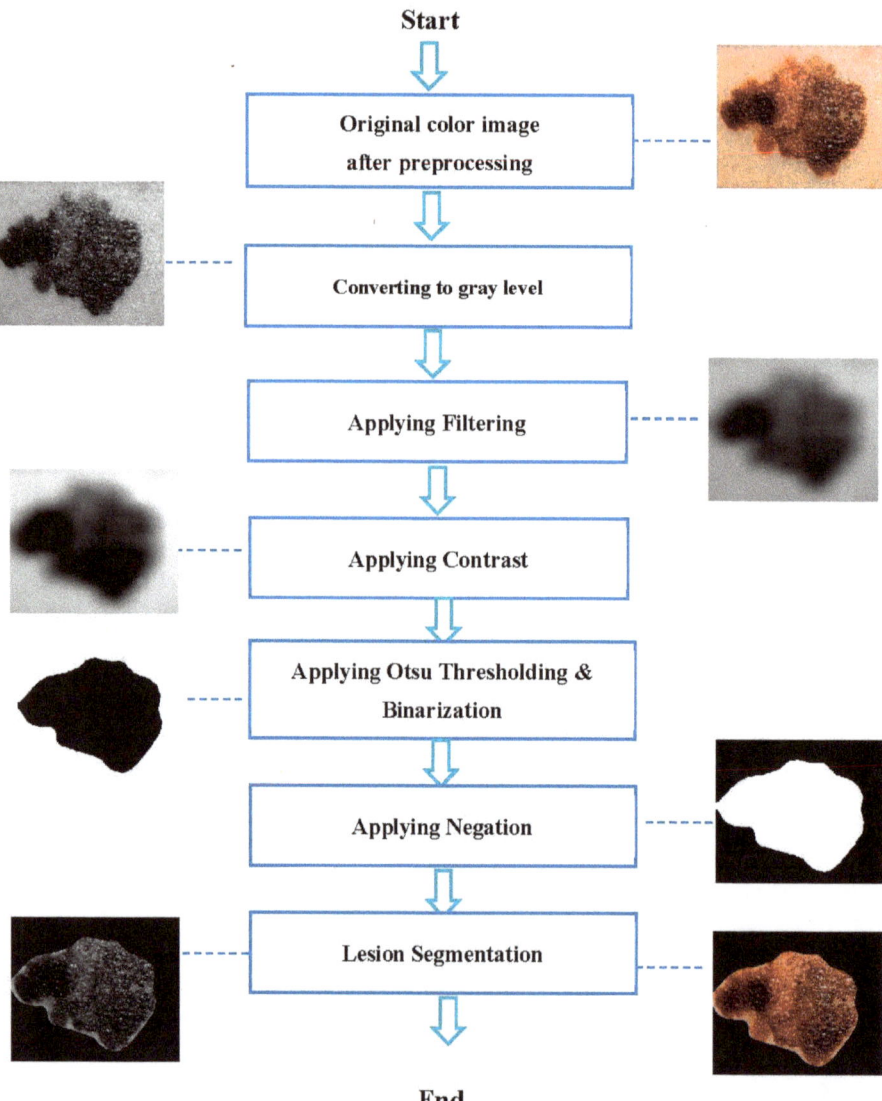

Figure 7. Flowchart of the proposed segmentation algorithm for melanoma skin cancer images.

4.2.3. Feature Extraction

In the feature-extraction stage, we extracted a set of features from images that represent meaningful information fundamental for classification and diagnosis. Several methods are utilized for extracting features such as texture, shape, color, etc. In this work, the color, texture, shape, and geometry features were extracted. To detect normal and abnormal lungs for the first medical dataset (chest X-ray), we combined two types of features, the texture and shape features. The texture features were extracted from the lung in a gray-level image and represented by computing contrast, correlation, energy, and homogeneity from the Gray-Level Co-Occurrence Matrix (GLCM) in four directions (0, 45, 90, and 135), in addition to computing Short-Run Emphasis (SRE), Long-Run Emphasis (LRE), Run Percentage (RP), Low Gray-level Run Emphasis (LGRE) from the Gray-Level Run-Length

Matrix (GLRLM) in four directions (0, 45, 90, and 135), and the shape features represented by seven features for moment invariants (MI). Thus, we obtained 16 features from the GLCM method, 16 features from the GLRLM method for texture features, and 7 features from the MI method for shape features. These features were been combined to create a single feature descriptor with 39 features to achieve an accurate output for good classification. Figure 8 shows the extracted features from the lung images in the first medical database.

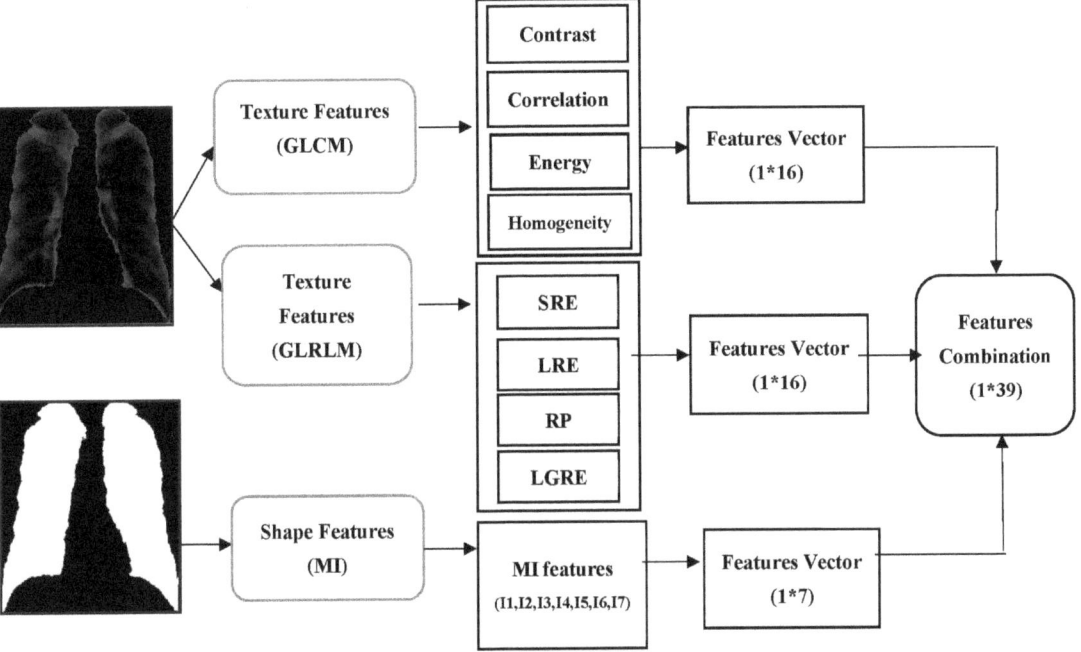

Figure 8. Diagram for the features extracted from lung images.

A. *Texture features set*

To extract texture features from lung images in the gray level, two suitable methods were used:

1. Gray-Level Co-occurrence Matrix (GLCM)

GLCM is a well-known statistical method for obtaining texture information from gray-level images [26]. It is the representation of the spatial distribution and the interdependence of the gray levels within a local area. The location of a very gray-level pixel can be found by the GLCM method [27].

Let Ng be the total number of gray levels, $g(i,j)$ be the entry (i,j) in the GLCM, μ be the mean of the GLCM, and σ^2 be the variance of the GLCM. Table 1 shows the GLCM features with descriptions and equations.

2. Gray-Level Run-Length Matrix (GLRLM)

GLRLM is a type of two-dimensional (2D) histogram-like matrix that records the occurrence of all conceivable gray-level values and gray-level run combinations in an ROI for a given direction. Gray-level values and runs are generally denoted as row and column keys in the matrix; hence, the (i,j)-th entry in the matrix identifies the number of pairings whose gray-level value is i and whose run length is j [28,29].

Let P denote a GLRLM, then, Pij is the (i, j)-th entry of the GLRLM, Nr denotes the set of various run lengths, Ng is the set of various gray levels, Np is the number of voxels in the image, and lastly, N represents the number of total pixels. Table 2 shows the GLRLM features with descriptions and equations.

Table 1. GLCM features with descriptions and equations.

GLCM Feature	Description	Equation
Contrast	It measures the extreme difference in grayscale between adjacent pixels.	$\sum_{i=0}^{Ng-1}\sum_{j=0}^{Ng-1}(i-j)^2 \cdot g^2(i-j)$
Correlation	It examines the linear dependency between the gray levels of adjacent pixels.	$\sum_{i=0}^{Ng-1}\sum_{j=0}^{Ng-1}(i-\mu)\cdot(j-\mu)\cdot g(i,j)/\sigma^2$
Energy	It measures texture uniformity or pixel-pair repetitions.	$\sqrt{\sum_{i=0}^{Ng-1}\sum_{j=0}^{Ng-1}g^2(i,j)}$
Homogeneity	It measures the homogeneity of the image and the degree of local uniformity that is present in the image.	$\sum_{i=0}^{Ng-1}\sum_{j=0}^{Ng-1}\frac{1}{1+(i-j)^2}\cdot g(i,j)$

Table 2. GLRLM features with descriptions and equations.

GLRLM Feature	Description	Equation
SRE	It measures the distribution of small run lengths, with a higher value indicating shorter run lengths and finer textures.	$\frac{\sum_{i=1}^{Ng}\sum_{j=1}^{Nr}\frac{P(i,j|\theta)}{j^2}}{Nr(\theta)}$
LRE	It measures the distribution of lengthy run lengths, with higher values indicating longer run lengths and coarser structural textures.	$\frac{\sum_{i=1}^{Ng}\sum_{j=1}^{Nr}P(i,j|\theta)j^2}{Nr(\theta)}$
RP	It measures the coarseness of the texture by comparing the number of runs to the number of voxels in the ROI.	$\frac{Nr(\theta)}{Np}$
LGRE	It measures the distribution of low grayscale values in an image, with a larger value denoting a higher concentration of low grayscale values.	$\frac{\sum_{i=1}^{Ng}\sum_{j=1}^{Nr}\frac{P(i,j|\theta)}{i^2}}{Nr(\theta)}$

B. *Shape features set*

MI is characteristic of connected regions in binary images which are invariant to translation, rotation, and scaling. The moments can be used to illustrate the shape of objects. Invariance recognizes the multidimensional moment-invariant-features space. The seven shape attributes are derived from the central moments and are not affected by the object scale, orientation, or translation. The translation of invariant moments at the origins of the central moments is calculated using the center of gravity [30].

The moments are represented by the image of the ij plane with nonzero elements. For a 2D ROI image, the moment invariance of order (p, q) is calculated as [12]:

$$m_{pq} = \sum_{i=1}^{m}\sum_{j=1}^{n} i^p j^q I^3(i,j)$$

Lower-order geometric moments have intuitive meaning: *m00* is the ROI's "mass," and *m10/m00* and *m01/m00* determine the ROI image's centroid. In the case of moments invariance, we have central moments of order (p, q) [31]:

$$\mu_{pq} = \sum_{i=1}^{m}\sum_{j=1}^{n} (i-\bar{i})^q I^3(i,j)$$

where $i = m10/m00$ and $j = m01/m00$ are the coordinates of the object centroid. In this way, we calculated seven moments as outlined in Table 3; this table shows the MI features with equations.

Table 3. MI features with equations.

MI Feature	Equation
I1	$m_{00} = \sum_{i=1}^{m} \sum_{j=1}^{n} I^3(i,j)$
I2	$m_{10} = \sum_{i=1}^{m} \sum_{j=1}^{n} i I^3(i,j)$
I3	$m_{01} = \sum_{i=1}^{m} \sum_{j=1}^{n} j I^3(i,j)$
I4	$\mu_{11} = \sum_{i=1}^{m} \sum_{j=1}^{n} (i-\bar{i})(j-\bar{j}) I^3(i,j)$
I5	$\mu_{12} = \sum_{i=1}^{m} \sum_{j=1}^{n} (i-\bar{i})(j-\bar{j})^2 I^3(i,j)$
I6	$\mu_{21} = \sum_{i=1}^{m} \sum_{j=1}^{n} (i-\bar{i})^2 (j-\bar{j}) I^3(i,j)$
I7	$\mu_{30} = \sum_{i=1}^{m} \sum_{j=1}^{n} ((i-\bar{i}))^2 ((j-\bar{j}))^2 I^3(i,j)$

To detect benign and malignant skin lesions in melanoma skin cancer images for the second medical dataset (melanoma skin cancer dermoscopy), we combined three kinds of features: color, texture, and geometry features.

The color features were extracted from the skin lesion in color images in the HSV (hue, saturation, value) system, represented by computing the mean, the standard deviation (STD), and skewness for each H, S, and V channel by applying the color-moments (CM) method. The texture features were extracted from the skin lesion in a gray-level image and represented by computing coarseness, contrast, and directionality by applying the Tamura method. The geometry features were extracted from the skin lesion in binary images, represented by computing area, perimeter, diameter, and eccentricity. Thus, we obtained nine features from the CM method for color features, three features from the Tamura method for texture features, and four shape features. These features were combined to create a single feature descriptor with 16 features to achieve an accurate output for good classification. Figure 9 shows the extracted features from the skin lesion for the second medical database images.

1. Color features set

CMs are one of the simplest and most active features compared with other color features; the features of common moments are mean, standard deviation, and skewness [32].

Let f_{ij} be the color value of the i-th color component of the j-th image pixel and N be the image's total number of pixels. $\mu_i, \sigma_i, \gamma_i$ ($i = 1,2,3$) represent the mean, standard deviation, and skewness of every channel of an image, respectively. Table 4 shows the CM features with equations.

2. Texture features set

Tamura is a method for devising texture features based on human visual perception. It identified six textural features (coarseness, contrast, directionality, regularity, roughness, and line-likeness). The first three are very effective outcomes [33,34]. Let n denote the image size, k be the value that maximizes the differences in the moving averages, μ_4 represent the fourth moment of the image, σ represent the image standard deviation, H_D denote the local direction histogram, n_p be the peak number of H_D, $\varnothing_p$ be the p-th peak position of H_D, w_p

be the *p*-th peak range between valleys, r be a normalizing factor, and ∅ be the quantized direction code. Table 5 shows the Tamura features with descriptions and equations.

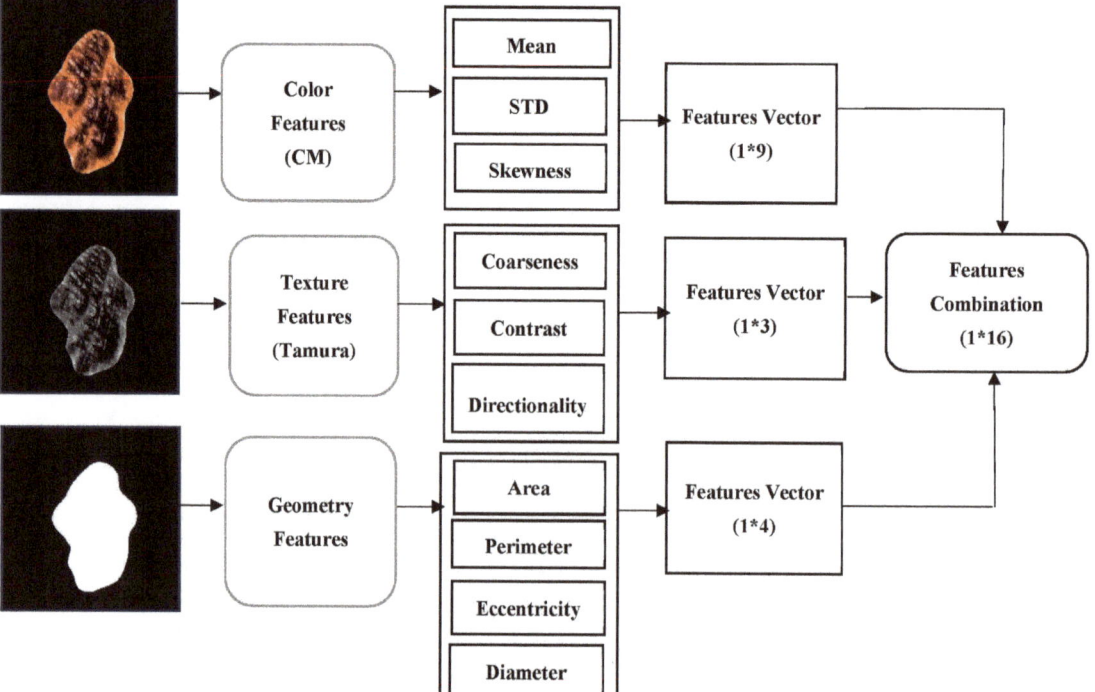

Figure 9. Diagram for the features extracted from skin lesion images.

Table 4. CM features with equations.

CM Feature	Equation
Mean	$\mu_i = \frac{1}{N} \sum_{j=1}^{N} f_{ij}$
STD	$\sigma_i = \left(\frac{1}{N} \sum_{j=1}^{N} (f_{ij} - \mu_i)^2 \right)^{\frac{1}{2}}$
Skewness	$Y_i = \left(\frac{1}{N} \sum_{j=1}^{N} (f_{ij} - \mu_i)^3 \right)^{\frac{1}{3}}$

3. Geometry features set

The geometry features extracted from a skin lesion in binary images are represented by computing the lesion area, lesion perimeter, eccentricity, and diameter of the lesion. Variable "A" is the lesion area and is a segmented image of x rows and y columns, (x_1), (x_1, y_1), and (x_2, y_2) are endpoints on the major axis, $z_1, \ldots, z_n$ is a boundary list, and di is the distance [35]. Table 6 shows the geometry features with descriptions and equations.

Table 5. Tamura features with description and equations.

Tamura Features	Description	Equation
Coarseness	It represents the size and number of textures primitives. It seeks to find the maximum size at which a texture exists.	$\frac{1}{n^2}\sum_{i}^{n}\sum_{j}^{n} 2^k\, p(i,j)$
Contrast	It indicates the difference in intensity between adjacent pixels.	$\dfrac{\sigma}{\left(\frac{\mu_4}{\sigma^4}\right)^{\frac{1}{4}}}$
Directionality	It is used to calculate directionality. The frequency distribution of oriented local edges against their directional angles is used to calculate an image's directionality.	$1 - r \cdot n_p \cdot \sum\limits_{p}^{n_p} \sum\limits_{\emptyset \in w_p} (\emptyset - \emptyset_p)^2 \cdot H_D(\emptyset)$

Table 6. Geometry features with descriptions and equations.

Geometry Feature	Description	Equation
Area (A)	It is the real number of pixels in the region which is returned as a scalar. The lesion area can be represented by the region of the lesion containing the total number of pixels.	$A = \sum_{x=1}^{n} \cdot \sum_{y=1}^{m} B(x,y)$
Perimeter (P)	It is a distance around the boundary of a region which is returned as a scalar by computing the distance between every contiguous pair of pixels around the border of the region.	$P = \sum\limits_{i=1}^{N-1} d_i = \sum\limits_{i=1}^{N-1} |z_i - z_{i+1}|$
Eccentricity (Ecc)	It is the ratio of the length of the short (minor) axis to the length of an object's long (major) axis; it is defined as the proportion of eigenvalues of the covariance matrix that matches a binary image of the shape.	$Ecc = \dfrac{axislength_{short}}{axislength_{long}}$
Diameter (D)	The diameter is identified by calculating the distance between every pair of points in a binary image and taking the maximum of these distances.	$D = \sqrt{(x_1 - x_2)^2 + (y_1 - y_2)^2}$

4.3. Diagnosis and Evaluation

In this section, we discuss the diagnosis and evaluation phase according to the suggested model (shown in Figure 1) employed in the study.

4.3.1. Classification

Classification is the most significant part of the diagnosis of the medical databases. In this work, we used most of the known ML algorithms such as ANN, SVM, KNN, DT, RF, NB, LR, RS, and fuzzy logic to classify the lungs as normal or abnormal in the first database and the skin lesions as benign or malignant in the second database. Most of these algorithms performed well. In contrast, CNN is regarded as one of the finest image-recognition and classification models in DL [36]. As a result, we used the CNN model to classify the chest X-ray and melanoma skin cancer dermoscopy medical datasets. The primary distinction between ML and DL is the method utilized to extract the features on which the classifier operates. DL extracts the features from numerous nonlinear hidden layers, while ML classification relies on extracting features manually [37].

In this section, we describe the classification methods used in this work, in addition to making a comparison among the methods used in terms of the advantages and disadvantages of each one.

1. Artificial Neural Network (ANN) Classifier

ANN is a data-processing system made up of many simple, interconnected processing components called neurons that are interconnected so that each neuron's output functions as one or more other neurons' input. The neurons are organized into layers in a parallel architecture inspired by the cerebral cortex of the brain, which has a superior ability to interpret and analyze complicated data and create clear and explanatory patterns to solve complicated issues [11]. There are two main kinds of layers: hidden and output layers; the data are fed into the hidden layer and then processed and delivered to the output-layer neurons, where they are compared to the required output, and the network error is calculated [38]. In our work, we utilized a backpropagation artificial neural network to classify the medical images according to the computed features for distinguishing the normal and abnormal lungs in the first medical dataset and the benign and malignant skin lesions in the second medical dataset. The neural network performance relies on the network architecture [39]. Figure 10 shows the network structure for training the first and the second medical datasets. The network input layer has 39 inputs for the first dataset and 16 inputs for the second dataset, two hidden layers, and two output layers.

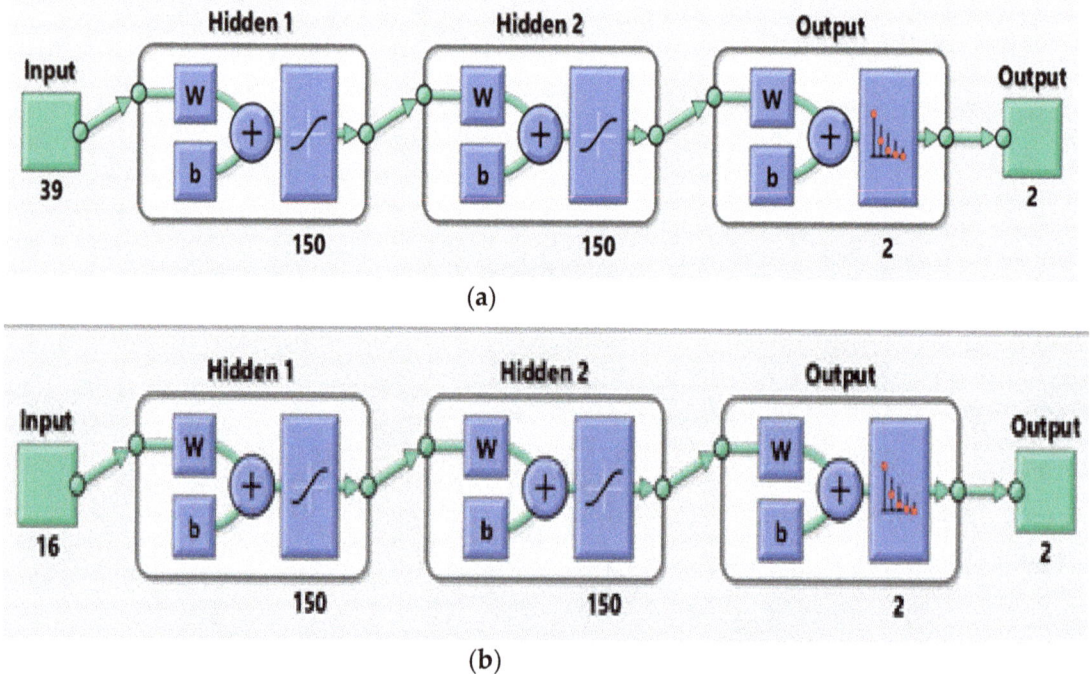

Figure 10. The structure of a neural network model: (a) ANN for the first medical dataset (chest X-ray); (b) ANN for the second medical dataset (dermoscopy melanoma skin cancer).

To construct, train, and test the neural network for disease (skin cancer and lung disease) diagnostics, the ANN architecture mentioned above and the feed-forward backpropagation-learning algorithm were utilized. Datasets were divided into two sets, training and testing. The important parameters were determined, the learning rate was set to 1, the maximum number of epochs was set to 1000, the training time was infinity, the data-division function was used (divide rand), the transfer function of the *i*-th layer hyperbolic tangent sigmoid transfer function was utilized (tansig), the activation function was chosen for the output layer (softmax) which is considered a good function to assign the input image's probability distribution to each of the classes in which the network was trained, the performance function was set to (mse) to minimize the error between actual and predicted probabilities, and the training function was a backpropagation function; weights were created randomly.

2. K Nearest Neighbor (K-NN) Classifier

K-NN is a one-of-a-kind instance-based prediction model; the testing sample's class label was decided by the bigger part class of its k-nearest neighbors according to their Euclidean distance [40]. For the K-NN algorithm, a data sample was compared to other data samples using a distance metric [41]. There were two phases to the algorithm: the training phase and the testing phase.

Training phase: The classifier fed the patterns of features and class labels of normal and abnormal images for lung images and of benign and malignant skin lesion images using the feature characteristics that were extracted during feature extraction.

Testing phase: An unidentified test pattern was provided and, using the knowledge learned through the training phase, the unidentified pattern was classified and plotted once more in the feature space, each sample image and its properties were represented as a point in an n-dimensional space known as a feature space, and the number of features that were employed to characterize the patterns determined their dimension [42].

3. Support Vector Machine (SVM) Classifier

SVM is a popular algorithm that is widely utilized in disease diagnosis. The main idea of SVM is to utilize hyperplanes to discriminate between different groups. This classifier tries to identify the hyperplane (decision boundaries) that aids in building the effective separation of classes according to statistical-learning theory [43]. The fundamental algorithm is based on the idea of "margin," or either side of a hyperplane that divides two classes of data. The fundamental goal of the SVM classification system is to find an overview that differentiates positive from negative data with the least amount of error [44].

4. Naïve Bayes (NB) Classifier

NB is a statistical classifier that utilizes the Bayes theorem as an underlying concept. A supervised learning-based Bayesian approach contains two phases: the learning phase and the testing phase. During the learning phase, an estimation is created based on the applied attributes, which keeps track of these attributes and categorizes their features; in the testing phase, predictions are produced based on the learning phase, and the likelihood of the desired outcome is calculated when new test data are tested. These characteristics offer a self-sufficient benefit in the development of the result [45].

5. Decision Tree (DT) Classifier

DT is a tree-structured classifier that has two nodes: decision node and leaf node. Decision nodes make decisions and have numerous branches, whereas leaf nodes display the outcomes of those decisions and do not have any additional branches. The features of the given dataset are used to perform the test or make the decision. It is called a decision tree because, like a tree, it starts with the root node and then extends to form more branches and a tree-like structure [46]. The most extensive DT algorithm for classification is C4.5. The C4.5 algorithm was employed in this work to classify lung images and melanoma skin cancer images.

6. Random Forest (RF) Classifier

RF is a supervised machine-learning technique that utilizes several decision trees on different subsets of a given dataset and takes the average to enhance the expected accuracy of that dataset. Rather than relying on a single decision tree, RF accumulates forecasts from each tree and estimates the ultimate output according to the majority vote of the predictions [47]. Each decision tree is constructed using randomly sampled training data and splitting nodes with subsets of features. The input is provided at the top of a decision tree, and as it passes down, the data are bucketed into smaller and smaller sets. RF is generated in two stages: the first is the combining of N decision trees to generate the random forest, and the second is making predictions for each tree produced in the first stage [48].

7. Random Subspace (RS) Classifier

The RS approach is a methodology for ensemble learning. The idea is to promote variety among ensemble members by limiting classifiers to working on distinct random subsets of the complete feature space. Every classifier learns with a subset of size n drawn at random from the entire collection of size N [49]. Because it uses subspaces of the real data size, this method is extremely advantageous because smaller parts can be better trained. Researchers are interested in this method because it reduces overlearning, introduces a broad model, requires less training time, and has an easier-to-understand and more straightforward structure than other classical models [50].

8. Logistic Regression (LR) Classifier

LR is a machine-learning method utilized to process classification problems. An LR model is built on a probabilistic foundation, with projected values ranging from 0 to 1. There are three kinds of LR: binary logistic regression, multinomial logistic regression, and ordinal logistic regression, and the most popular utilized case is binary logistic regression, where the result is binary (yes or no). LR employs the cost function, sometimes known as the sigmoid function. Each real number between 0 and 1 is transformed by the sigmoid function. LR can be used in the medical field to determine whether the ROI is normal or abnormal and whether a tumor is either benign or malignant [51,52].

9. Fuzzy logic Classifier

Fuzzy logic is a mathematical approach to computing and inference that uses the concept of a fuzzy set to generalize classical logic and set theory. Fuzzy inference involves all the components outlined in membership functions, logical processes, and if–then rules. FIS (Fuzzy Inference System) is a system that maps inputs (features) to outputs (classes) using fuzzy set theory [53]. To build an FIS, first, we chose the input numerical variables, which should be precise, and determined their ranges for each term. The correspondence between the input values and each fuzzy set were then defined during the fuzzification stage; this was accomplished through the use of membership functions, which reflect the degree to which a parameter value belongs to each class. The set of fuzzy rules that characterize FIS rules using logical operators, as well as the method for merging fuzzy outputs from each rule, were then described, finally extracting the output distribution from a mixture of fuzzy rules, followed by defuzzification to obtain the crisp classification result [54].

10. CNN of DL Classifier

CNNs are the best type of deep-learning model for image analysis. CNN is made up of numerous layers that use convolution filters to convert the input. The performance of CNN relies on the network architecture [55]. CNN is made up of a series of layers that form its architecture, in addition to the input layer, which is commonly an image with width and height. There are three primary layers: (1) The convolutional layer: this layer is made up of several filters (kernels) that can be learned via training. The kernels are tiny matrices with real values that can be interpreted as weights. (2) The pooling layer: this layer is employed after a convolution layer to minimize the spatial size of the generated convolution matrices; as a result, this strategy decreases the number of parameters to be learned in the network, which contributes to overfitting control. (3) The fully connected layer: this layer connects each element of the convolution output matrices to an input neuron. The output of the convolutional and pooling layers represents the features extracted from the input image [56,57].

In this work, we suggested and assessed a deep convolutional neural network structure for diagnosing lung diseases and melanoma skin cancer. We performed an ablation analysis of the proposed structure and tested other topologies to compare their results with the results of the proposed CNN structure. We also manually tested extracted features in this work on the proposed CNN structure and compared the results.

The proposed CNN architecture used to resolve our classification problem is shown in Figure 11. The model contains three convolution layers, three max-pooling layers, three batch-normalization layers, and a fully connected layer. When the image was input into the CNN structure, the image was represented as image height × image width. The image size was standardized in the system to obtain robust outcomes. After the image passed through the convolutional layers, the feature map included the feature depth, represented as image height × image width × image depth. Filter size, stride, and padding zero were the most important parameters of the convolutional layers that impacted the performance of the convolutional layers. Convolutional layers wrapped with the filter size (in this case, we used a 3 × 3 matrix to achieve more precision when traversing the matrix containing the images) around the image, learned the weights through the training phase, processed the input, and passed it to the next layer. Zero padding was the process of filling neurons with zeros to maintain the size of the resulting neurons. When zero padding was one, the neurons were padded with a row and a column around the edges. Rectified linear unit (ReLU) layers were also utilized after convolutional layers for image processing. The objective of ReLU was to pass the positive output and repress the negative output. The dimensions were decreased by the pooling layer, as the dimensions of the image were decreased by grouping numerous neurons and representing them in one neuron based on the maximum or average method, which is named the max-pooling layer. The maximum value of the groups of neurons was chosen utilizing the maximum method, and the average value of the neurons was selected utilizing the average method. In the fully connected layers, the last layer of the convolutional neural networks, each neuron was connected to all neurons. Feature maps were transformed into flat representations (unidirectional). Softmax is the activation function utilized in the last phase of the convolutional neural network model; it is nonlinear and is utilized in multiple classes.

Figure 12 describes the number of layers, the size of each filter, and the parameter of the CNN structure that was used in diagnosing the two medical datasets (chest X-ray and melanoma skin cancer).

Figure 13 illustrates the results of the datasets (lung X-ray and melanoma skin cancer dermoscopy) classification by the proposed CNN structure.

As can be seen from the confusion matrices for the first and second medical image datasets in Figure 14, for the lung dataset, we used 612 samples that were classified as 288 normal and 324 abnormal. The samples were divided by CNN into 70% for training and 30% for testing. The number of samples for training was 429 (227 samples for abnormal and 202 for normal). The number of samples for testing was 183 (97 for abnormal and 86 for normal), and for the skin cancer dataset, we used 300 samples that have been classified (145 benign and 155 malignant). The samples were divided by CNN into 70% for training and 30% for testing. The number of samples for training was 211 (102 samples for benign and 109 for malignant). The number of samples for testing was 89 (43 for benign and 46 for malignant); the accuracy of the classification for the first medical dataset was 95% in the last epoch at 25 epochs, 75 iterations, and the learning rate was equal to 0.01; the accuracy of the classification for the second medical dataset is 93% in the last epoch at 25 epochs, 25 iterations, with a learning rate equal to 0.01.

Ablation analyses of the proposed CNN: Some ablation tests were conducted through the three crucial parameters: the number of layers, the kernel size, and the network parameter. The number of layers is the first critical network parameter, and it is directly proportional to the network's description capabilities. Nevertheless, having more layers means having more variables to optimize, which necessitates even more training data, without which overfitting occurs. Given the restricted amount of samples available and the insignificant impact of additional layers (up to 25 were evaluated), we determined that three convolution layers, three max-pooling layers, three batch-normalization layers, and a fully linked layer were sufficient for our architecture to obtain good results, but when trying to remove any layer from the suggested architecture, the result started to deteriorate. Table 7 shows the results of the ablation test for the first parameter (the number of layers) for the two medical datasets.

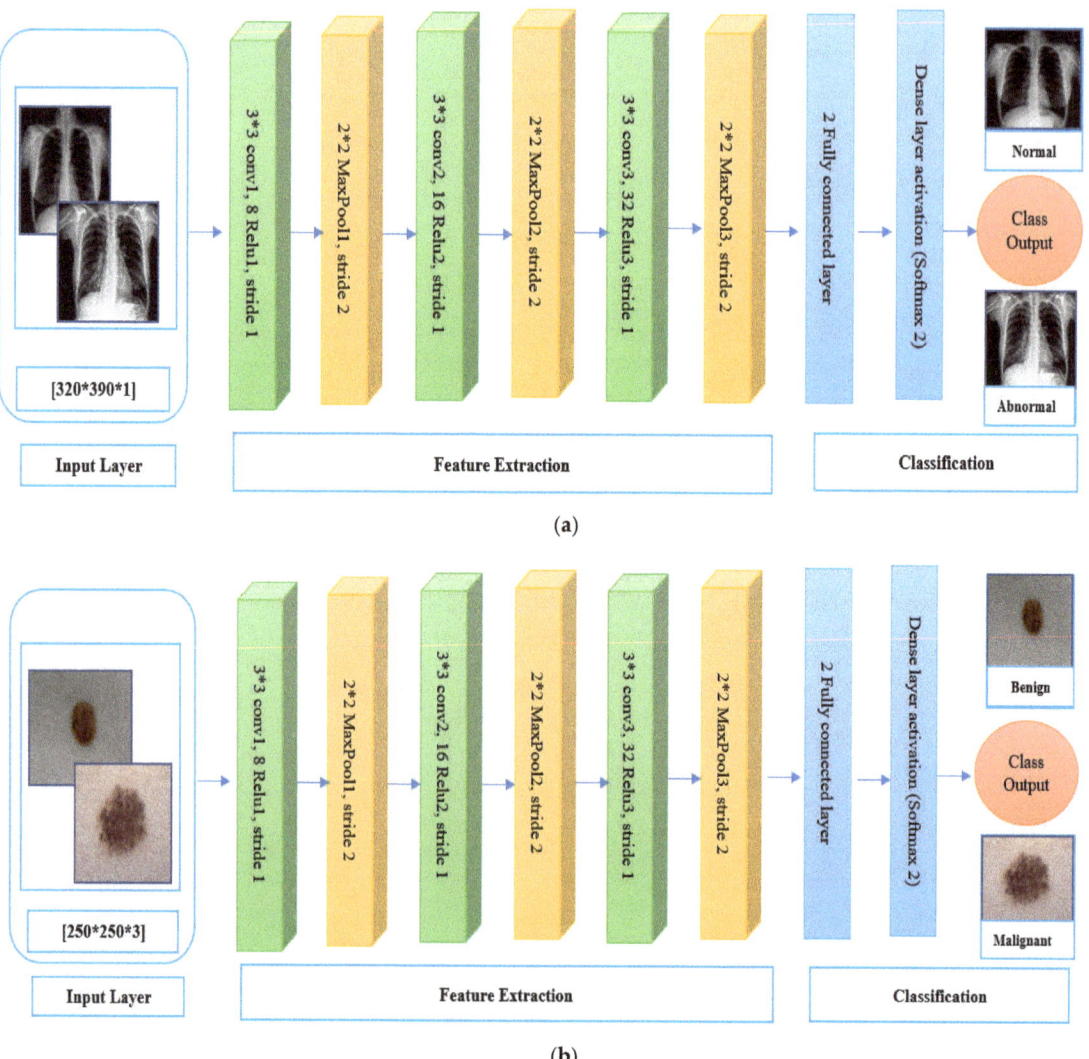

Figure 11. The architecture of the employed CNN model: (**a**) CNN for the first medical dataset (chest X-ray); (**b**) CNN for the second medical dataset (dermoscopy melanoma skin cancer).

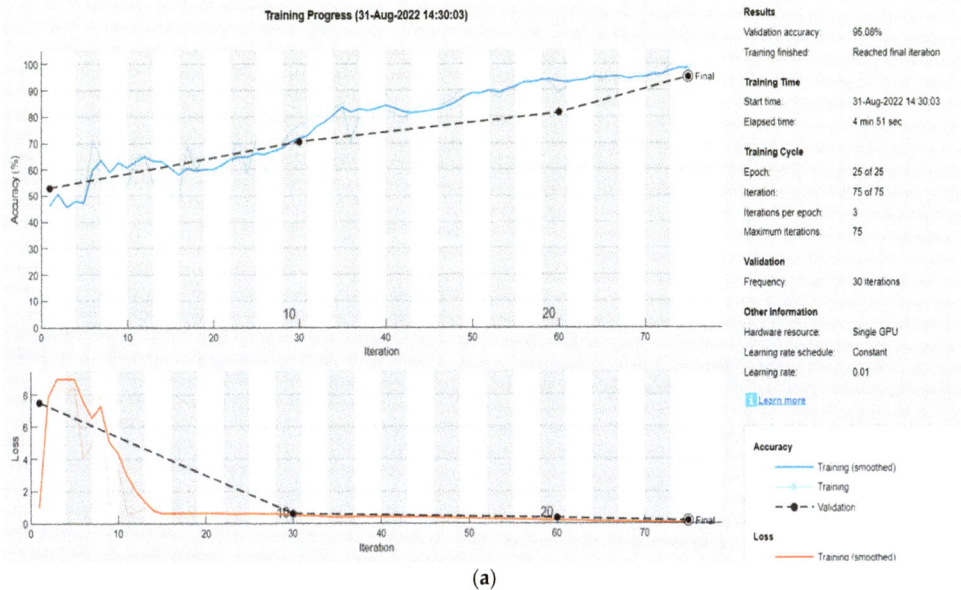

Figure 12. The detailed structure of the proposed CNN model.

Figure 13. *Cont.*

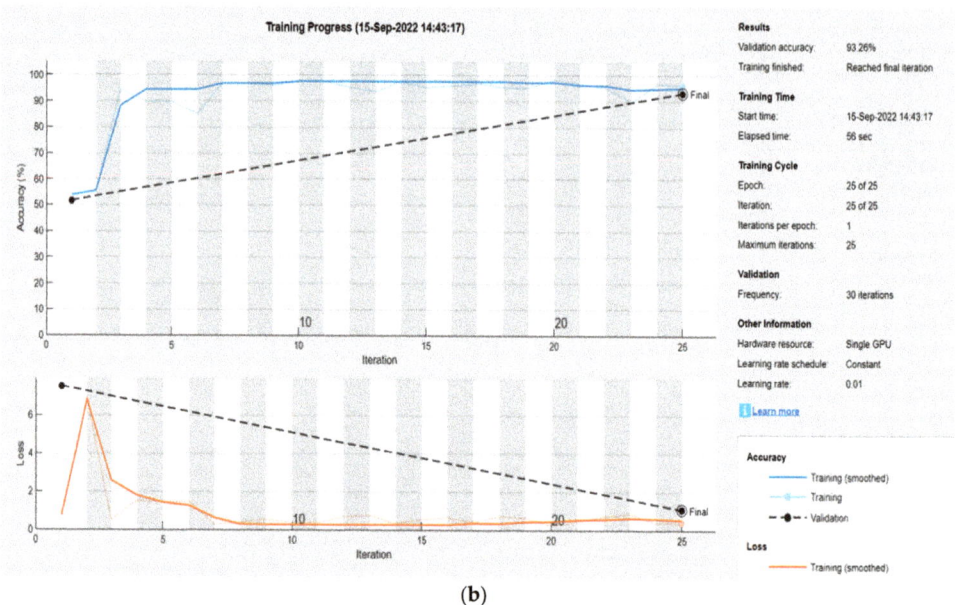

(b)

Figure 13. Result of the dataset classification by the proposed CNN structure (**a**) The first dataset (lung X-ray); (**b**) The second dataset (melanoma skin cancer dermoscopy).

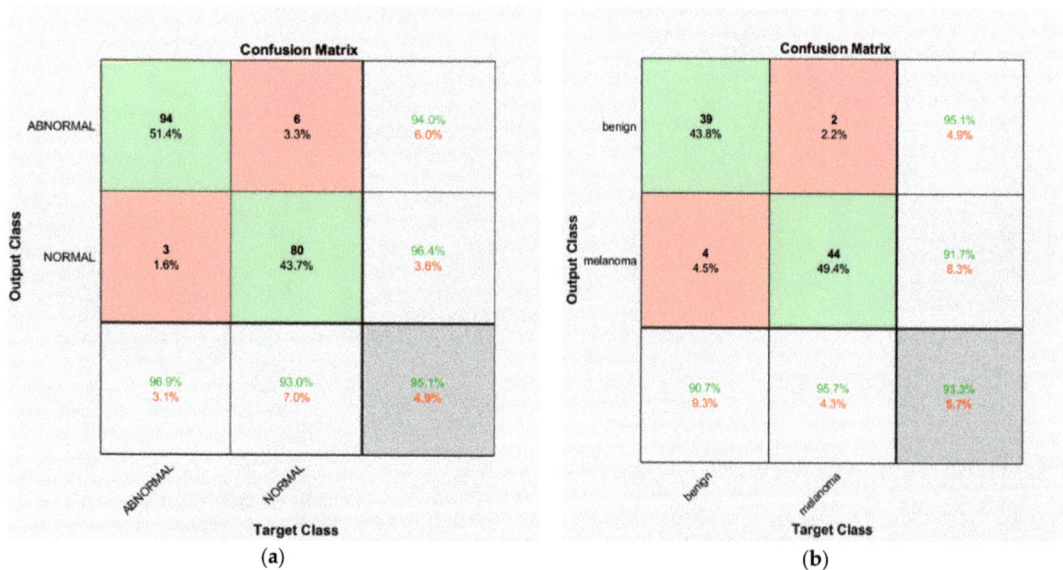

Figure 14. CNN confusion matrices: (**a**) CNN confusion matrices for the first dataset; (**b**) CNN confusion matrices for the second dataset.

Table 7. The results of the ablation test for the first parameter (number of layers) for the two medical datasets.

Test No.	Number of Layers	Result	
		Lung Dataset	Skin Cancer Dataset
1	Two convolution layers, two max-pooling layers, two batch-normalization layers	81.5%	80.6%
2	One convolution layer, one max-pooling layer, one batch-normalization layer	73.5%	71.5%

As far as the kernel size is concerned, experiments were conducted from 1×1 up to 9×9 pixels. The 3×3 size in each convolutional layer had a strong positive impact, which, nevertheless, did not increase with a larger spatial radius; regardless, with the size of 3×3, we achieved an accuracy of more than 90% for the two sets of selected medical databases. Figure 15 shows the relative results of ablation test for the second parameter (kernel size) for the two medical datasets.

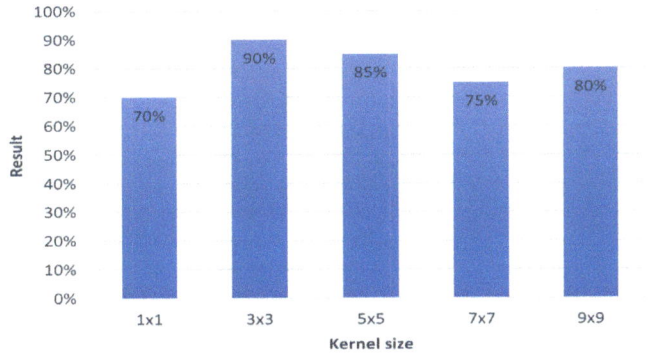

Figure 15. The relative results of the ablation test for the second parameter (kernel size) for the two medical datasets.

As far as the crucial network parameters, we conducted experiments on the most important network information:

Learning rate (LR): The network's learning rate is inversely proportional to convergence speed. We experimented with a large spectrum of values; however, their effect on overall performance was insignificant and the model did not exhibit pathological behavior. The training of the proposed CNN network was realized with a learning rate of 1×10^{-2}.

Epochs: We selected 25 epochs for training the network because training over many epochs is common in applications and often results in greater potential for overfitting, 25 epochs were enough to train the datasets and obtain good accuracy.

Activation function: We conducted several experiments on choosing the activation functions and changing the function in each experiment (we used several activation functions commonly used like the sigmoid function, and the hyperbolic tangent tanh(x)), but we did not obtain satisfactory results for the proposed network except in the case of activation functions for the ReLU and softmax, because it is considered the most effective activation function. In comparison to sigmoid and tanh, ReLU is more trustworthy and speeds up convergence by six times.

Through experiments with regard to these main parameters in the network, it is not possible to remove any one of them because removing them from the network architecture may lead to a deterioration in the result, or it may not work.

Test different topologies: Some advanced CNNs have more complicated topologies and network architecture for different tasks, for example, GoogLeNet, ResNet, AlexNet, VGGNet, and inception modules. In this work, we tested the ResNet18 model to compare the result with the proposed CNN structure result. The ResNet model's architecture is shown in Figure 16.

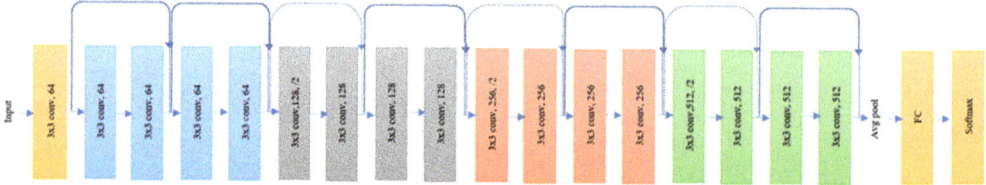

Figure 16. Representing original ResNet-18 architecture [58].

ResNet18 has 18 layers, the first of which is a 7×7 kernel. It has four identical layers of ConvNets. Each layer is made up of two residual blocks. Each block is made up of two weight layers connected by a skip connection to the output of the second weight layer through a ReLU. If the result equals the ConvNet layer's input, the identity connection is used. If the input and output are not similar, convolutional pooling is performed on the skip connection. The ResNet18 input size is (224, 224, 3), which is achieved through preprocessing augmentation with the AugStatic package. In (224, 224, 3), 224 denotes the width and height. The RBG channel is number three. The result is an FC layer that feeds data to the sequential layer [59,60]. We tested the two selected medical databases in this work on the ResNet18 model and changed the size to $224 \times 224 \times 3$ to fit with the network input to compare the result of the network ResNet18 with the proposed CNN network. The table shows the comparison results for the two medical datasets.

Figure 17 describes the detailed structure of the part of the ResNet18 model (the number of layers, the size of each filter, and the parameter of the ResNet18 model).

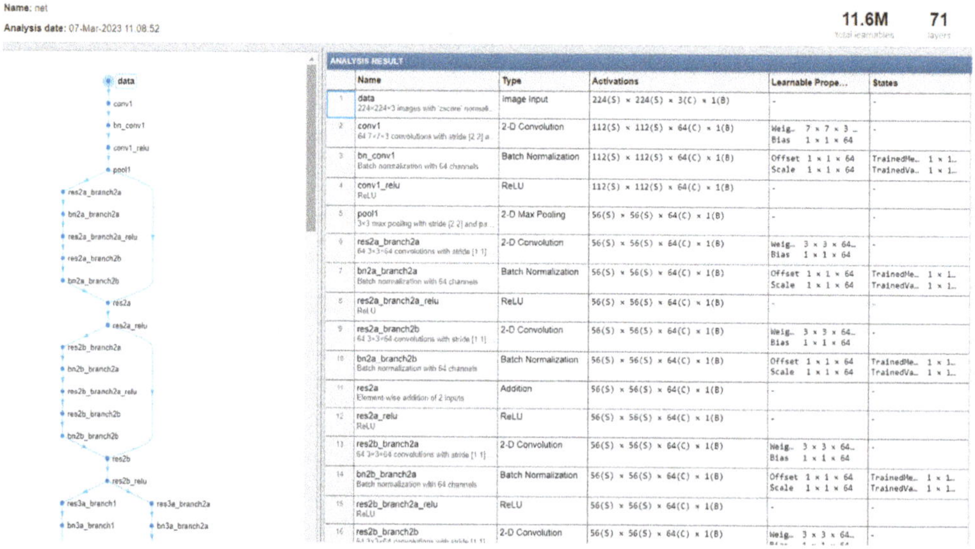

Figure 17. The detailed structure of the ResNet18 model.

Figure 18 illustrates the results of the dataset (lung X-ray and melanoma skin cancer dermoscopy) classification by the ResNet18 model.

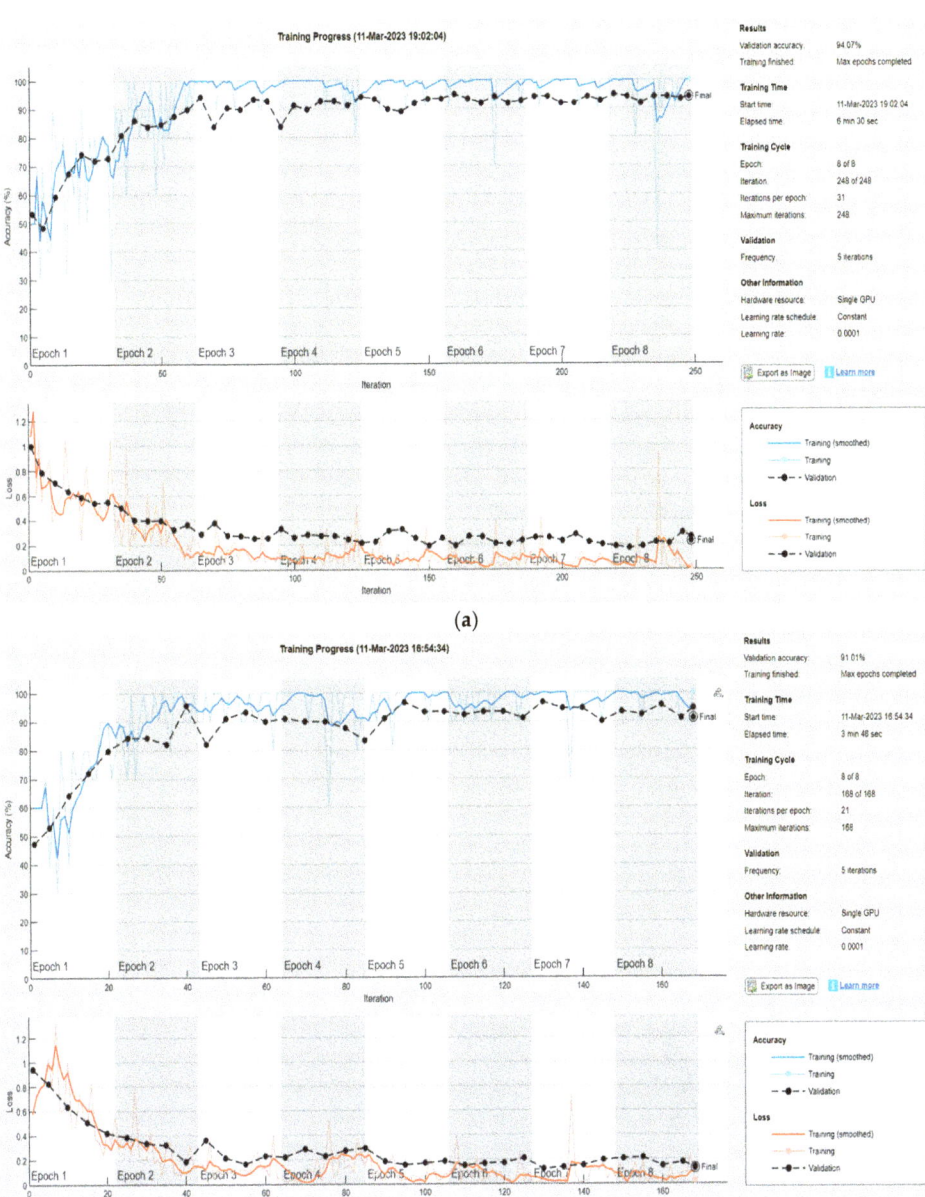

Figure 18. Result of the dataset classification by the ResNet18 model: (**a**) The first dataset (lung X-ray); (**b**) The second dataset (melanoma skin cancer dermoscopy).

As can be seen from the confusion matrices for the first and second medical image datasets in Figure 19, the accuracy of the classification for the first medical dataset is 94%, and 91% for the second medical dataset.

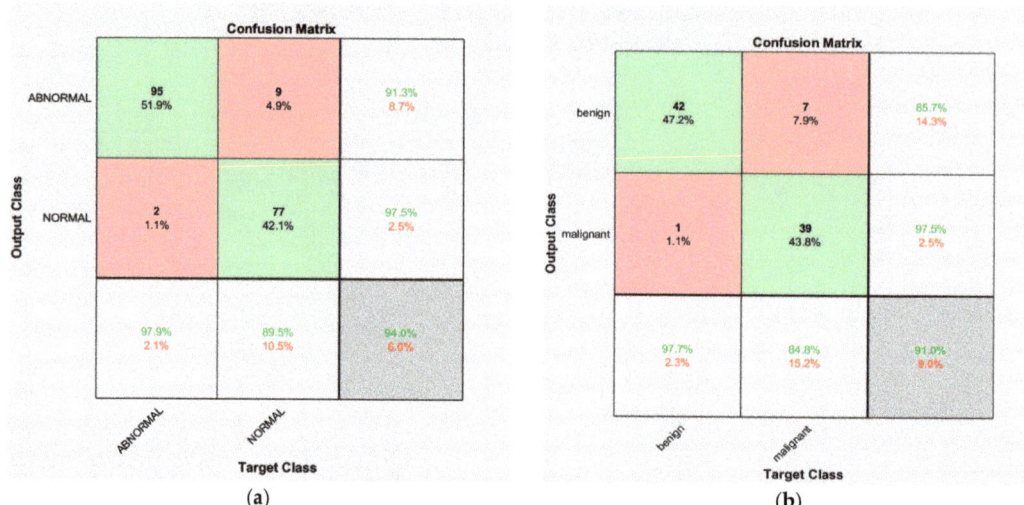

Figure 19. ResNet18 confusion matrices: (**a**) Confusion matrices for the first dataset; (**b**) Confusion matrices for the second dataset.

From Table 8, it can be seen that the proposed CNN structure outperformed ResNet18. It has been noticed that as the number of added layers increases, training neural networks becomes more difficult, and in some cases, accuracy reduces. Through experiments, we found that the proposed CNN structure is sufficient to classify the selected medical databases and obtain good results. In addition, the proposed network takes less time to train, and this is one of the important advantages that made us adopt the proposed structure in the diagnosis of the selected medical databases.

Table 8. The results of the performance of ResNet18 and the proposed CNN structure for the two medical datasets.

The Type of Architecture	Accuracy	
	Lung Dataset	Skin Cancer Dataset
The proposed CNN	95.1%	93.3%
ResNet18	94%	91%

Additionally, we tested the performance of the proposed CNN structure in the case of the inputs of extracted features for the other ML methods concatenated to the final layer. As can be seen from the confusion matrices for the first and second medical image datasets in Figure 20, the accuracy of the classification for the first medical dataset is 83.1%, and 87.6% for the second medical dataset, in the case of training the network using the extracted features manually. Therefore, the performance of the network using the features extracted manually is less efficient; we noticed through experiments that the features extracted from the network are more robust and effective, and suit the performance of the network and provide better results. Therefore, the proposed CNN network structure was adopted in the diagnosis of the selected medical dataset in this work because it achieved high accuracy in diagnosis.

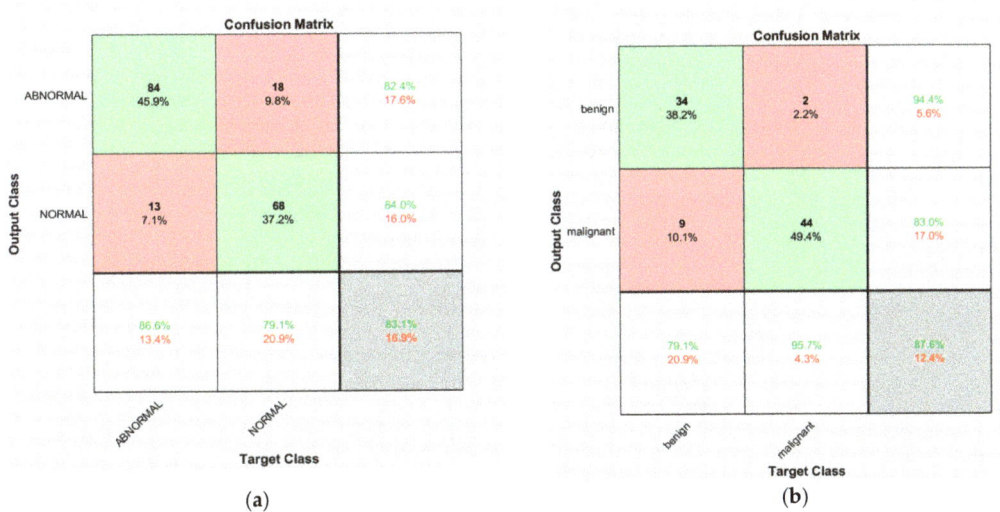

Figure 20. CNN confusion matrices: (**a**) Confusion matrices for the first dataset; (**b**) Confusion matrices for the second dataset.

Table 9 shows a comparison of the classification methods used in this work in terms of the most important advantages and disadvantages of each one.

Table 9. Comparison of the classification methods used in this work.

Method	Advantage	Disadvantage
1. ANN	Advanced predictive ability Parallel processing ability	Computationally costly Long time to process massive amounts of data
2. SVM	The ability to handle structured and semistructured data Appropriate for nonlinear problems and those with little samples and high dimensions	Decreased performance with large amounts of data Imperfect work with noisy data
3. KNN	Flexibility Easy to implement	Sensitive to k-value selection Requires well-classified training data
4. DT	Ease and speed in implementation The ability to generate rules easily	Difficulty controlling tree size Can suffer from overfitting
5. NB	Speed in predicting the dataset category Simplicity in implementation	Accuracy decreases with a small amount of data Necessitates a vast number of records
6. LR	Speed in training Ease in implementation and application	Not suited for predicting the value of a binary variable, only accepts Boolean values Unable to solve nonlinear problems
7. RF	Flexibility There is no need to normalize data because it employs a rule-based approach	Takes a long time to train Takes a lot of resources and computational effort to build multiple trees and integrate their outputs
8. RS	Precise and reliable predictions Implements a random subset of features to a combined group of foundation classifiers	Takes a long time to train Risk of overfitting
9. Fuzzy Logic	Flexibility Active system for nonlinear problems	Necessitates a large amount of data Rules need to be updated frequently
10. CNN	Effective with large amounts of data Extremely good at image identification and classification	Requires sufficient data and time for training High computational cost

4.3.2. Model Evaluation and Validation

This stage is the primary metric for assessing the performance of the classification models. Lung diseases and melanoma skin cancer can be classified as true positive (TP) or true negative (TN) if correctly diagnosed, or false positive (FP) or false negative (FN) if incorrectly diagnosed. Accuracy, sensitivity, specificity, precision, recall, F-measure, and AUC (area under curve) are the most popular assessment metrics used for lung diseases and melanoma skin cancer classification. We describe these metrics briefly below [11,41]:

Accuracy (Acc): This metric measures the number of correctly classified cases; it can be represented utilizing the following equation:

$$Acc = \frac{(TP + TN)}{(TP + TN + FN + FP)}$$

Sensitivity (Sn): This metric reveals the number of correctly estimated total positive cases; it can be represented utilizing the following equation:

$$Sn = \frac{TP}{(TP + FN)}$$

Specificity (Sp): This metric reveals the number of correctly estimated total negative cases; it can be represented by the equation:

$$Sp = \frac{TN}{(TN + FP)}$$

Precision (Pr): This metric indicates how accurate the overall positive forecasts are; it can be represented by the equation:

$$Pr = \frac{TP}{(TP + FP)}$$

Recall: This metric reveals how well the total number of positive instances is predicted; it can be represented by the equation:

$$Recall = \frac{TP}{(TP + FN)}$$

F-Measure: This metric is a way of checking how accurately the model operates by distinguishing the right true positives from the expected ones; the following equation can be used to express it:

$$F1\ Score = \frac{(2TP)}{(2TP + FN + FP)}$$

AUC: This metric is one of the most important evaluation metrics for measuring the performance of any classification model and is used to summarize the ROC curve, which represents the area under the ROC curve. The greater the AUC, the better the model distinguishes between positive and negative classifications, and it can be represented by the equation [58]:

$$AUC = \frac{TPR - FPR + 1}{2}$$

5. Results and Comparison

This section introduces the outcomes of all the methods utilized in the classification, which were implemented using Matlab 2021, as well as a comparison of the results.

The results of the feature extraction for some samples of the first medical dataset (chest X-ray dataset) and the second medical dataset (melanoma skin cancer dermoscopy dataset) for the feature extraction methods used in this work are introduced in Appendix A.

The results of various classification algorithms are described in Tables 10 and 11, where the performance metrics for all ML algorithms and deep-learning CNN that were applied to the chest X-ray dataset and melanoma skin cancer dermoscopy dataset are described.

Table 10. Results of classification of lung disease dataset.

Algorithm	Acc%	Sn%	Sp%	Pr	Recall	F-Measure	AUC
ANN	91.1	94.7	88.4	0.916	0.911	0.912	0.945
SVM	84.4	84.2	84.6	0.846	0.844	0.845	0.844
KNN	86.6	84.2	88.4	0.867	0.867	0.867	0.800
DT	74.4	73.6	75.5	0.747	0.747	0.747	0.743
NB	81.1	76.3	84.6	0.811	0.811	0.811	0.887
LR	92	92.3	91.6	0.920	0.920	0.920	0.947
RF	93.3	94.7	92.3	0.935	0.933	0.934	0.992
RS	84.4	92.1	78.8	0.860	0.844	0.845	0.948
Fuzzy Logic	81.1	71	88.4	0.821	0.811	0.809	0.798
CNN	95.1	94	96.3	0.969	0.94	0.954	0.994

Table 11. Results of classification of melanoma skin cancer dataset.

Algorithm	Acc%	Sn%	Sp%	Pr	Recall	F-Measure	AUC
ANN	96.6	95.4	97.8	0.967	0.967	0.967	0.974
SVM	84.4	97.7	71.7	0.871	0.844	0.842	0.847
KNN	95.5	95.4	95.6	0.956	0.956	0.956	0.930
DT	84.4	100	69.5	0.882	0.844	0.841	0.848
NB	80	84	76	0.803	0.800	0.800	0.874
LR	87.7	93.1	82.6	0.883	0.878	0.878	0.949
RF	94.6	94.8	94.4	0.947	0.947	0.947	0.984
RS	93.3	94.8	91.6	0.934	0.933	0.933	0.986
Fuzzy Logic	90	100	80.4	0.917	0.900	0.899	0.902
CNN	93.3	95.1	91.6	0.906	0.915	0.928	0.919

From Table 10, it can be noted that classification based on a random forest (RF) algorithm performs better than other machine-learning techniques applied to lung datasets in terms of performance metrics, and CNN for deep-learning classification provides better outcomes than all other algorithms when applied to X-ray lung diseases. LR, ANN, KNN, SVM, and RS are also effective algorithms for classification issues, and they perform well in the issue at hand.

From Table 11, it can be seen that classification based on the ANN algorithm outperforms all the other ML algorithms that were applied to melanoma skin cancer datasets in terms of performance metrics. Other algorithms such as KNN, RF, RS, Fuzzy Logic, and CNN are also effective algorithms for classification issues, and they perform well in the issue at hand.

Table 12 shows the comparison of the test accuracy of the classification algorithms applied to the two medical image datasets used in our work.

Table 12. Comparison of the accuracy of classification algorithms for the two medical datasets.

Algorithm	Accuracy in the First Database (Chest X-ray)	Accuracy in the Second Database (Melanoma Skin Cancer Dermoscopy)
ANN	91.1%	96.6%
SVM	84.4%	84.4%
KNN	86.6%	95.5%
DT	74.4%	84.4%
NB	81.1%	80%
LR	92%	87.7%
RF	93.3%	94.6%
RS	84.4%	93.3%
Fuzzy Logic	81.1%	90%
CNN	95.1%	93.3%

From Table 12, we notice that CNN and ANN were the highest-performing classifiers for the lung disease and skin cancer disease classification, respectively. This is due to the characteristics of the CNN network (the number of layers and the number of filters that were utilized in addition to the efficiency of the activation functions) and the superior ability of DL networks in classification tasks, especially CNNs, which are considered to be the state-of-the-art systems for image classification, especially in the classification of medical databases.

For an ANN network, the classification accuracy was high if the extracted features were strong and good for classification, the network structure was good, and the number of layers was large (we used many layers and 150 complex neurons for each hidden layer), in addition to the efficiency of the activation functions utilized in the network (the activation function was chosen for the output layer (softmax) which is considered a good function to assign the input images' probability distribution to each of the classes in which the network was trained).

To check the accuracy of the final optimized model, a new set of 100 images from the medical datasets used in this work was used for validation. Table 13 shows the results of the validation accuracy of the classification algorithms applied to the two medical image datasets used in our work. We note that the results of the validation are very close to the test results of the classification algorithms applied to the two medical image datasets, shown in Table 12.

Table 13. The results of the validation accuracy of classification algorithms for the two medical datasets.

Algorithm	Accuracy in the First Database (Chest X-ray)	Accuracy in the Second Database (Melanoma Skin Cancer Dermoscopy)
ANN	92.%	95.2%
SVM	88.8%	84.5%
KNN	86.2%	95.8%
DT	75%	83.8%
NB	80.9%	80.4%
LR	92.8%	88.3%
RF	92.9%	93.7%
RS	85.7%	94.8%
Fuzzy Logic	80.9%	90.6%
CNN	95.08%	92.1%

Figure 21 shows the comparison of the accuracy of the classification algorithms for the lung dataset.

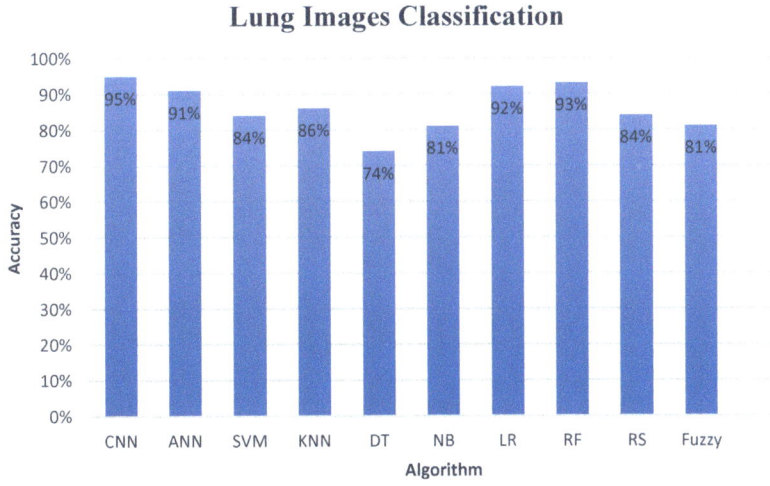

Figure 21. Comparison of the accuracy of classification algorithms for the lung dataset.

Figure 22 shows the comparison of the accuracy of the classification algorithms for the melanoma skin cancer dataset.

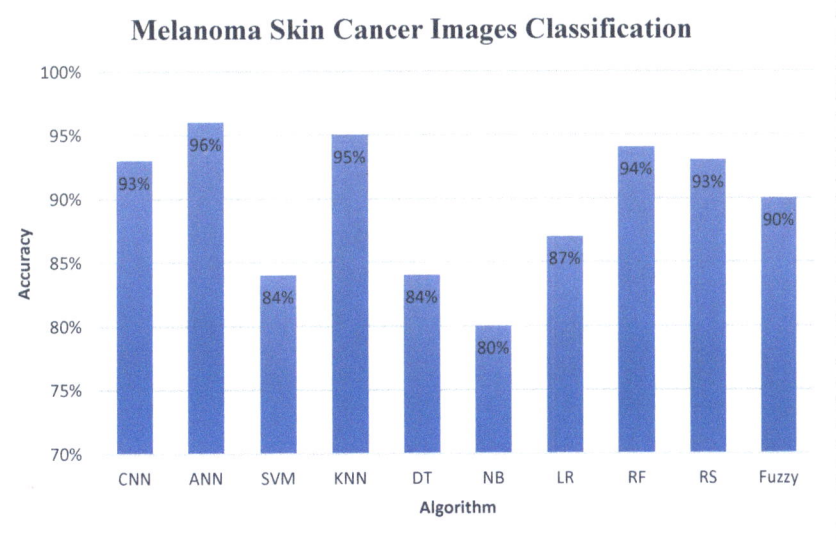

Figure 22. Comparison of the accuracy of classification algorithms for the skin dataset.

As seen in Figure 21, the highest accuracy was 95% for the CNN algorithm. For the machine-learning algorithms, the highest accuracy was 93% for the RF algorithm, followed by the LR and ANN algorithms with an accuracy of 92% and 91%, respectively. On the other hand, the DT classifier attained the lowest performance of 74%.

As seen in Figure 22, the highest accuracy was 96% for ANN, followed by the KNN, RF, and RS algorithms, with an accuracy of 95%, 94%, and 93%, respectively, for ML algorithms. Additionally, the CNN algorithm attained a high accuracy of 93%. On the other hand, the NB classifier attained the lowest performance of 80%.

Figures 23 and 24 show the graphical representation of the performance measures for the algorithms applied to the lung dataset and the melanoma skin cancer dataset, which shows a comparison of the performance measures in terms of accuracy, sensitivity, specificity, precision, recall, F-measure, and AUC.

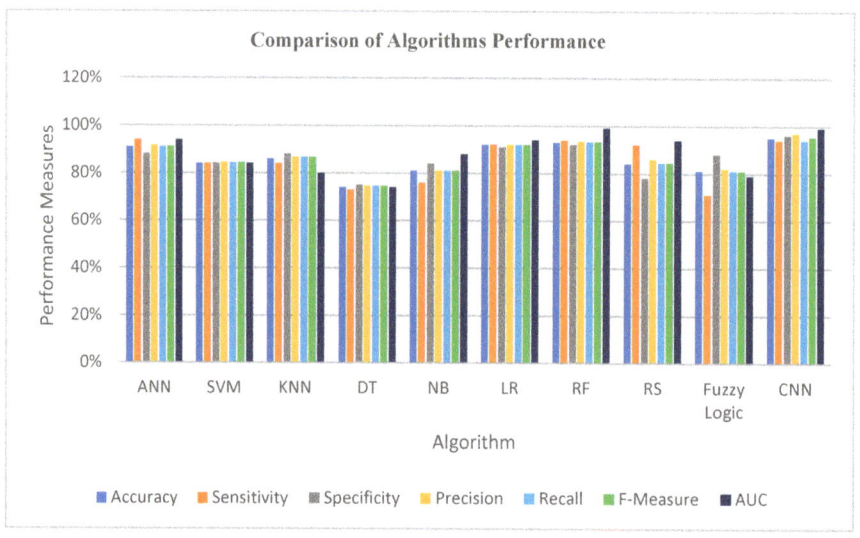

Figure 23. Graphical representation of the classifier performance-evaluation comparison for the lung dataset.

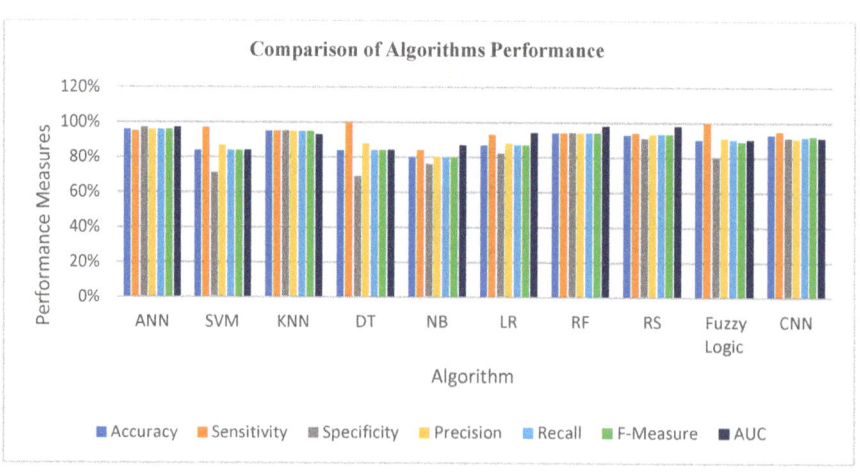

Figure 24. Graphical representation of the classifier performance-evaluation comparison for the melanoma skin cancer dataset.

Figures 25 and 26 illustrate the receiver operating characteristic (ROC), which is a system performance curve used in medical testing for diagnosing medical datasets. A ROC curve is constructed by plotting the true-positive rate (TPR) against the false-positive rate

(FPR). The true-positive rate is the fraction of all positive observations that were correctly expected to be positive. The false-positive rate is the proportion of negative observations that were wrongly anticipated to be positive. Each figure represents an assessment curve for ten methods for each database.

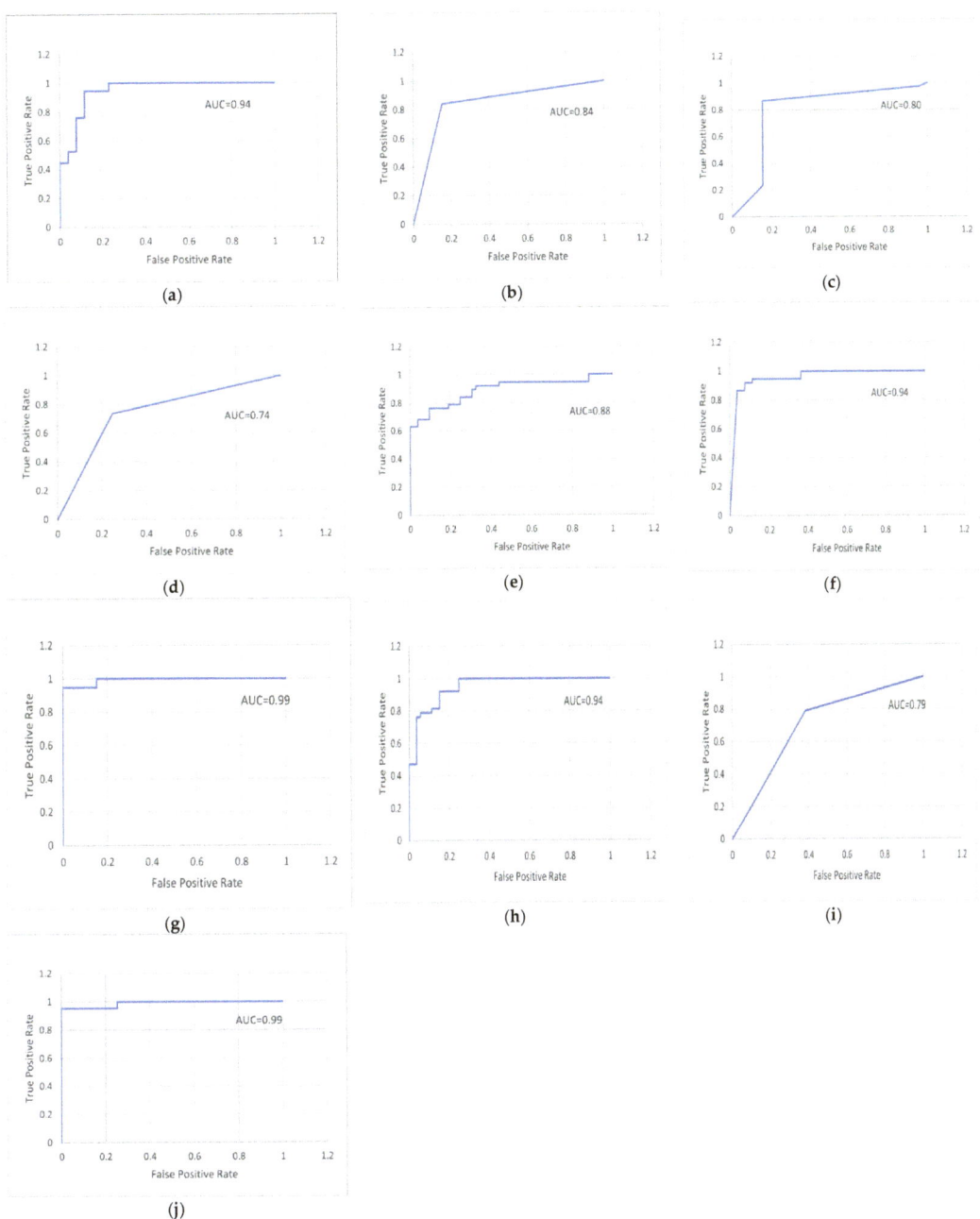

Figure 25. ROC curves of the classifier performance evaluation for the lung dataset: (**a**) ANN; (**b**) SVM; (**c**) KNN; (**d**) DT; (**e**) NB; (**f**) LR; (**g**) RF; (**h**) RS; (**i**) Fuzzy; (**j**) CNN.

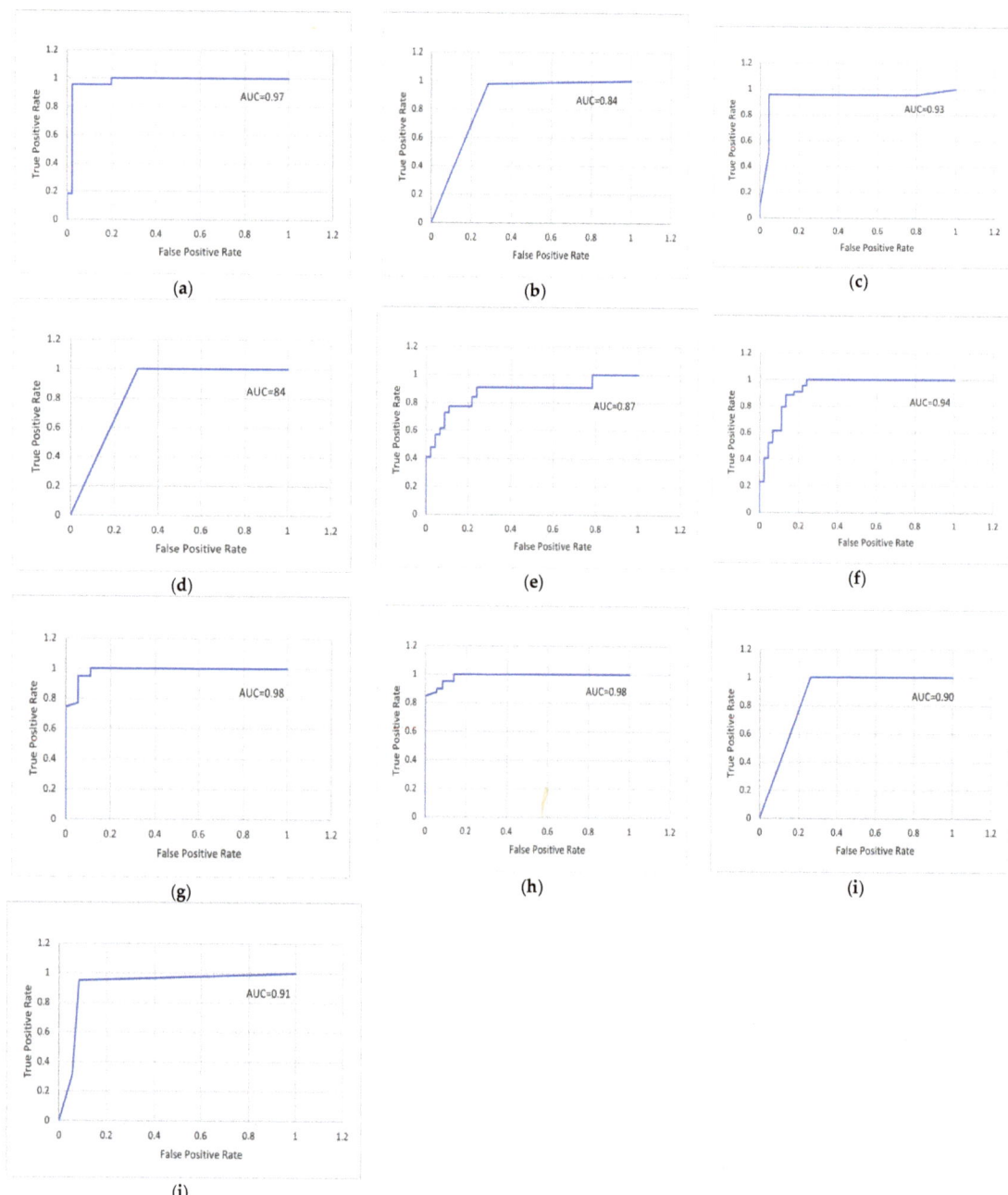

Figure 26. ROC curves of the classifier performance evaluation for the melanoma skin cancer dataset: (**a**) ANN; (**b**) SVM; (**c**) KNN; (**d**) DT; (**e**) NB; (**f**) LR; (**g**) RF; (**h**) RS; (**i**) Fuzzy; (**j**) CNN.

6. Contributions

The essential contributions of this work are summarized as follows:

- We exploited ML and DL to find the most precise techniques for diagnosis to provide directions for future research.
- We analyzed more than one medical image database to evaluate more than one disease using the proposed system.
- By improving the raw images, finding the ROI (lung and lesion), extracting ROI-specific features, and applying ML and DL algorithms for automatic classification, we present an integrated framework for identifying lung disease utilizing chest X-ray scans and melanoma skin cancer using skin dermoscopy.
- We suggest an algorithm for image preprocessing, where the raw X-ray images were processed and their quality was improved. Additionally, an algorithm was proposed to remove hair from dermoscopy skin images to enhance them and obtain a precise diagnosis. The proposed preprocessing algorithms provided good results in the work.
- We suggest an algorithm for image segmentation to separate the ROI from the image to extract only lung regions from chest X-ray images and lesion regions from dermoscopy skin images. The proposed segmentation algorithms achieved good results in the work.
- We extracted a robust collection of features from ROI (lung and skin lesion) images, including color, texture, shape, and geometry features to help us achieve satisfactory results in the classification.
- Good results were obtained for the proposed system, utilizing two scalable datasets and an appropriate training-to-testing ratio of 70% to 30%. The CNN model and machine-learning techniques such as SVM, KNN, ANN, NB, LR, RF, RS, and fuzzy logic were trained for assessment. In the end, the results of the suggested model methods were compared.

7. Concluded Discussion and Future Directions

In this section, we discuss the techniques utilized in this work, and the medical dataset utilized to diagnose diseases through discussion and future directions.

7.1. Discussion

In this work, chest X-ray and melanoma skin cancer dermoscopy datasets were used for classification purposes. The data were divided into 70% for training and 30% for testing. The data were analyzed by applying the main analysis processes, where the medical data images were preprocessed to remove noise and improve contrast. Some filters were applied to improve the images; the chest X-ray images were preprocessed using the median filter and applying the contrast adjustment with the histogram equalization enhancement to improve the images and remove noise. Hair was removed from the melanoma skin cancer images using the algorithm proposed in the work to improve the images and prepare them well for the next stage of the analysis, which was to separate the object of interest from the rest of the image.

The suggested segmentation algorithms were applied to separate the lung the from chest X-ray images by using thresholding and morphological operations, and to separate the lesion from skin cancer images using Otsu thresholding with binarization and negation processes. The proposed segmentation algorithms provided excellent results in separating the object from the image to move to the other stage, which is the extraction of features. Here, the best methods were applied to obtain the most relevant features from the images such as texture and shape, extracting features from lung images and lesion images, such as color, texture, and some geometry features, and to move to the most important stage, which is the classification.

At this point, we applied a set of the most important and best common classification methods based on machine learning such as ANN, SVM, KNN, DT, NB, LR, RF, RS, and fuzzy logic in addition to CNN for deep learning to identify the performance of these algorithms and the accuracy of their classification. All these were realized by training the selected databases and obtaining results, where most of the methods proved effective in diagnosing lung diseases and melanoma skin cancer and showed good results.

The model was evaluated using accuracy, sensitivity, specificity, precision, recall, and F-measure. However, the analysis of the results indicates that there is enough space to improve the performance obtained for some methods by applying other techniques that may be hybrid or improved methods to increase the speed of performance and reduce the time, cost, and effort in diagnosing diseases. The comparison in the application of diagnostic and evaluation methods to both sets of medical image datasets revealed that the accuracy and performance of the algorithm depend on the type of the medical dataset, the amount of data, the type of disease, and the performance of the methods applied in the preprocessing, segmentation, and feature-extraction stages.

In particular, the results of the applied classification methods are generalizable to the diagnosis of the selected databases, even if this work focused on classifying melanoma skin cancer and lung diseases.

In general, the results of the classification algorithms utilized in this work can be generalized to all medical data if appropriate preprocessing methods are applied to treat databases because each database differs in its processing according to the conditions of taking pictures and the noise in the images. However, through what we have seen in this work, the step of extracting the most relevant features from the images differs depending on the type of dataset. For this reason, we conducted our experiments and tested two different specific types of medical databases to diagnose two different diseases and, in this way, create a model that shows its capacity for generalization.

7.2. Future Directions

In future work, we suggest the following:
- Increase the number of diseases that are diagnosed and employ other classifiers.
- We also plan to work with more sophisticated medical image data.
- Employ new sets of features for more medical images, to improve performance.
- Although good findings were produced in this work, more research should be conducted by merging the algorithms employed in classification or by adding optimization tools.
- There is a need to develop or create a new classification system for the diagnosis of diseases based on medical image databases.
- Further extensive studies or experiments with vast datasets and hybrid or optimized classification approaches are necessary.

In the future, we plan to work on a new classification model architecture by building a hybrid classifier by merging two or more of the classification methods that we used in this work or by improving one of the classifiers used by adding tools for improvement to increase the speed of performance and reduce the time, cost, and effort in diagnosing diseases, and applying the new system to more than one database that includes types of diseases, to prove its effectiveness and increase the percentage of generalizing the system on the most diseases that pose a threat to human life and detect them early so that we can treat them.

8. Conclusions

This work examines the effectiveness of several machine-learning- and deep-learning-based classification algorithms for medical data diagnosis, and describes a machine-learning- and deep-learning-based diagnosis system for two of the world's most serious diseases. Experiments were performed on two different medical datasets (chest X-ray dataset and dermoscopy melanoma skin cancer dataset). The following conclusions can be drawn based on the experimental results:
- Most of the classification algorithms based on machine learning that were applied to the two selected databases provided good results in terms of various classification per-

- formance metrics such as accuracy, sensitivity, specificity, precision, recall, F-measure, and AUC.
- The deep-learning-based convolutional neural network algorithm outperformed in others when applied to the two selected medical databases, as it provided high classification accuracy, reaching 95% in classifying the lung dataset into normal and abnormal, and 93% in classifying the melanoma skin cancer dataset into benign and malignant.
- Additionally, the outcomes varied from one dataset to another, according to the type of medical dataset, the type of medical imaging, and the efficiency of the methods applied in the preprocessing, segmentation, and feature extraction to classify the medical dataset; whenever the methods that were applied to a dataset to train the model were accurate and worked well, the performance of the classification model was better.
- The work provides some crucial insights into modern ML/DL methodologies in the medical field that are applied in disease research nowadays.
- Better outcomes are anticipated with the usage of hybrid algorithms and combined ML and DL techniques. Even minor adjustments can sometimes yield good results. We found that training data quality is an important consideration when creating ML- and DL-based systems.

As an outline of what we have achieved in this work, we analyzed two medical datasets (chest X-ray and melanoma skin cancer dermoscopy) by applying the main analysis processes, where preprocessing of the medical data images was conducted to remove noise and improve contrast. Hair was removed from the melanoma skin cancer images using the algorithm proposed in the work to improve the images. The suggested segmentation algorithms were applied to separate the lung from the chest X-ray images by using thresholding and morphological operations, and to separate the lesion from skin cancer images using Otsu thresholding with binarization and negation processes. Relevant features were extracted from the images for use in the classification such as color, texture, shape, and geometry features. In the classification stage, we applied a set of the most important and best common classification methods based on machine learning such as ANN, SVM, KNN, DT, NB, LR, RF, RS, and fuzzy logic in addition to CNN for deep learning to investigate the performance of these algorithms and the accuracy of their classification. The model was evaluated utilizing accuracy, sensitivity, specificity, precision, recall, F-measure, and AUC. Most of the methods proved effective in diagnosing lung diseases and melanoma skin cancer and provided good results.

Author Contributions: The concept of the article was proposed by N.P., the data resources and validation were contributed by B.M.R., and the formal analysis, investigation, and draft preparation were performed by B.M.R. The supervision and review of the study were headed by N.P. The final writing was critically revised by N.P. and finally approved by the authors. All authors have read and agreed to the published version of the manuscript.

Funding: This research received no external funding.

Institutional Review Board Statement: Not applicable.

Informed Consent Statement: Not applicable.

Data Availability Statement: The study does not report any data.

Conflicts of Interest: The authors declare no conflict of interest.

Abbreviations

ML—Machine Learning; DL—Deep Learning; ANN—Artificial Neural Network; SVM—Support Vector Machine; KNN—K nearest Neighbor; DT—Decision Tree; RF—Random Forest; NB—Naïve Bayes; RS—Random Subspace; LR—Logistic Regression; CNN—Convolutional Neural Network; ROI—Region Of Interest; RGB—Red–Green–Blue; HSV—Hue–Saturation–Value; 2-D—Two-Dimensional; RNN—

Recurrent Neural Network; GLCM—Gray-Level Co-Occurrence Matrix; GLRLM—Gray-Level Run-Length Matrix; SRE—Short-Run Emphasis; LRE—Long-Run Emphasis; RP—Run Percentage; LGRE—Low Gray-level Run Emphasis; MI—Moment Invariant; CM—Color Moment; A—Area; Pr—Perimeter; Ecc—Eccentricity; D—Diameter; FIS—Fuzzy Inference System; TP—True Positive; TN—True Negative; FP—False Positive; FN—False Negative; Acc—Accuracy; Sn—Sensitivity; Sp—Specificity; Pr—Precision; AUC—Area Under Curve.

Appendix A

This section introduces the results of the feature extraction for some samples of the first medical dataset (chest X-ray dataset) and the second medical dataset (melanoma skin cancer dermoscopy dataset) for the feature-extraction methods used in this work.

Tables A1 and A2 show the results of different random samples of the normal and abnormal lungs for the first medical dataset (chest X-ray dataset) using the GLCM texture-features method.

Table A1. Results of normal sample lungs using GLCM features.

Samples	GLCM Features															
	Contrast				Correlation				Energy				Homogeneity			
	0°	45°	90°	135°	0°	45°	90°	135°	0°	45°	90°	135°	0°	45°	90°	135°
Image 1	0.23	0.31	0.15	0.28	0.931	0.933	0.96	0.951	0.294	0.241	0.25	0.23	0.68	0.65	0.69	0.61
Image 22	0.18	0.34	0.23	0.18	0.933	0.924	0.94	0.942	0.259	0.243	0.24	0.24	0.69	0.66	0.67	0.62
Image 69	0.25	0.23	0.17	0.27	0.94	0.931	0.95	0.953	0.281	0.252	0.23	0.22	0.66	0.64	0.63	0.65
Image 80	0.26	0.33	0.21	0.19	0.921	0.913	0.961	0.939	0.278	0.251	0.25	0.24	0.62	0.63	0.68	0.64
Image 187	0.14	0.208	0.15	0.24	0.924	0.915	0.952	0.941	0.284	0.247	0.24	0.23	0.67	0.62	0.69	0.66

Table A2. Results of abnormal sample lungs using GLCM features.

Samples	GLCM Features															
	Contrast				Correlation				Energy				Homogeneity			
	0°	45°	90°	135°	0°	45°	90°	135°	0°	45°	90°	135°	0°	45°	90°	135°
Image 1	0.504	0.61	0.37	0.54	0.809	0.87	0.82	0.86	0.196	0.168	0.15	0.14	0.77	0.701	0.75	0.701
Image 22	0.47	0.53	0.35	0.48	0.806	0.89	0.83	0.85	0.201	0.147	0.14	0.16	0.76	0.703	0.76	0.702
Image 69	0.44	0.49	0.38	0.55	0.807	0.84	0.86	0.88	0.188	0.156	0.12	0.15	0.74	0.702	0.73	0.703
Image 80	0.43	0.63	0.31	0.49	0.804	0.86	0.84	0.87	0.202	0.138	0.13	0.13	0.72	0.711	0.72	0.711
Image 187	0.502	0.54	0.43	0.51	0.803	0.88	0.81	0.89	0.191	0.127	0.16	0.14	0.71	0.712	0.76	0.721

Tables A3 and A4 show the results of different random samples of the normal and abnormal lungs for the first medical dataset (chest X-ray dataset) using the GLRLM texture features method.

Table A3. Results of normal sample lungs using GLRLM features.

Samples	GLRLM Features															
	SRE				LRE				RP				LGRE			
	0°	45°	90°	135°	0°	45°	90°	135°	0°	45°	90°	135°	0°	45°	90°	135°
Image 1	0.17	0.33	0.31	0.32	221.9	191.2	313.2	173.2	0.15	0.105	0.12	0.104	91.4	86.9	85.2	88.1
Image 22	0.23	0.28	0.28	0.36	189.2	188.6	298.3	177.3	0.13	0.104	0.14	0.112	87.6	86.5	85.6	77.8
Image 69	0.28	0.31	0.26	0.29	196.9	196.5	322.5	191.5	0.17	0.106	0.16	0.115	79.1	88.5	77.8	76.9
Image 80	0.15	0.27	0.33	0.38	203.7	189.6	389.7	188.4	0.14	0.103	0.13	0.108	94.1	79.8	79.3	85.8
Image 187	0.29	0.25	0.32	0.28	206.9	186.9	299.5	169.9	0.11	0.102	0.11	0.103	89.5	83.8	83.9	79.2

Table A4. Results of abnormal sample lungs using GLRLM features.

Samples	GLRLM Features															
	SRE				LRE				RP				LGRE			
	0°	45°	90°	135°	0°	45°	90°	135°	0°	45°	90°	135°	0°	45°	90°	135°
Image 1	0.48	0.51	0.501	0.51	396.3	288.5	675.7	357.1	0.25	0.216	0.21	0.214	60.8	65.8	53.4	59.1
Image 22	0.52	0.48	0.46	0.46	323.8	292.5	586.8	287.5	0.23	0.218	0.19	0.215	59.8	58.7	58.8	57.2
Image 69	0.46	0.46	0.45	0.54	391.5	253.7	564.9	267.6	0.24	0.215	0.23	0.206	61.8	44.8	61.1	61.2
Image 80	0.39	0.45	0.44	0.48	378.6	304.5	621.3	311.5	0.27	0.211	0.201	0.209	58.8	49.7	59.4	49.8
Image 187	0.45	0.52	0.503	0.52	369.4	312.2	584.7	312.4	0.21	0.214	0.212	0.211	60.8	52.1	66.3	55.8

Table A5 shows the results of different random samples of the normal and abnormal lungs for the first medical dataset (chest X-ray dataset) using the MI shape features method.

Table A5. Results of normal and abnormal samples of the lungs using MI features.

Samples	MI Features													
	Normal							Abnormal						
	I1	I2	I3	I4	I5	I6	I7	I1	I2	I3	I4	I5	I6	I7
Image 1	2.08	7.02	9.08	8.56	−15.78	−13.65	18.11	2.74	4.5	7.6	9.9	−18.5	−9.9	15.4
Image 22	2.04	7.01	8.99	8.44	−16.65	−12.88	18.15	2.66	4.9	6.9	10.1	−18.6	−10.9	14.8
Image 69	1.81	6.98	9.06	7.64	−14.12	−13.81	17.8	2.82	5.9	7.4	9.8	−19.1	−10.2	14.9
Image 80	1.99	6.44	8.87	8.88	−16.76	−12.76	17.5	2.65	4.8	6.8	10.07	−18.7	−9.7	15.2
Image 187	1.92	7.04	9.11	7.89	−16.82	−13.32	18.2	2.52	5.8	7.1	9.08	−19.3	−11.1	13.9

Table A6 shows the results of different random samples of the benign and malignant skin lesions for the second medical dataset (melanoma skin cancer dermoscopy dataset) using the CM color features method.

Table A6. Results of benign and malignant samples of skin lesions using CM features.

Samples	CM Features																	
	Benign									Malignant								
	Mean (H)	Mean (S)	Mean (V)	STD (H)	STD (S)	STD (V)	Skewness (H)	Skewness (S)	Skewness (V)	Mean (H)	Mean (S)	Mean (V)	STD (H)	STD (S)	STD (V)	Skewness (H)	Skewness (S)	Skewness (V)
Image 1	0.11	0.16	0.62	0.02	0.03	0.04	3.09	1.26	1.28	0.36	0.38	0.78	0.16	0.107	0.102	1.52	0.49	0.88
Image 33	0.13	0.12	0.59	0.01	0.02	0.03	2.08	1.37	1.25	0.49	0.4	0.76	0.13	0.104	0.101	1.44	0.39	0.87
Image 88	0.12	0.15	0.55	0.01	0.02	0.05	3.07	1.4	1.32	0.32	0.28	0.8	0.14	0.103	0.101	1.48	0.45	0.78
Image 101	0.13	0.14	0.66	0.02	0.03	0.03	3.08	1.32	1.31	0.45	0.35	0.69	0.17	0.108	0.103	1.58	0.51	0.83
Image 203	0.12	0.13	0.57	0.01	0.02	0.06	3.06	1.41	1.35	0.5	0.27	0.79	0.15	0.104	0.101	1.44	0.42	0.79

Table A7 shows the results of different random samples of the benign and malignant skin lesions for the second medical dataset (melanoma skin cancer dermoscopy dataset) using the Tamura texture features method.

Table A7. Results of benign and malignant samples of skin lesions using Tamura features.

Samples	Tamura Features					
	Benign			Malignant		
	Coarseness	Contrast	Directionality	Coarseness	Contrast	Directionality
Image 1	14.1	14.2	0.06	23.2	31.8	0.02
Image 33	12.2	10.9	0.05	22.6	25.1	0.01
Image 88	12.8	16.8	0.05	21.8	32.8	0.03
Image 101	13.5	11.5	0.04	24.4	27.3	0.02
Image 203	11.9	12.1	0.06	20.5	30.9	0.03

Table A8 shows the results of different random samples of the benign and malignant skin lesions for the second medical dataset (melanoma skin cancer dermoscopy dataset) using the geometry features method.

Table A8. Results of benign and malignant samples of skin lesions using geometry features.

Sample	Geometry Features							
	Benign				Malignant			
	Area	Perimeter	Eccentricity	Diameter	Area	Perimeter	Eccentricity	Diameter
Image 1	421	166.09	0.51	28.8	842	289.2	0.82	49.8
Image 33	511	107.7	0.46	33.6	721	301.1	0.71	44.9
Image 88	399	133.09	0.43	22.9	711	302.5	0.85	48.6
Image 101	451	196.02	0.54	30.5	802	299.4	0.79	56.1
Image 203	411	145.02	0.49	34.1	741	284.2	0.74	53.08

References

1. Tripathi, S.; Shetty, S.; Jain, S.; Sharma, V. Lung disease detection using deep learning. *Int. J. Innov. Technol. Explor. Eng.* **2021**, *10*, 154–159.
2. Saba, T.; Javed, R.; Rahim, M.; Rehman, A.; Bahaj, S. IoMT Enabled Melanoma Detection Using Improved Region Growing Lesion Boundary Extraction. *Comput. Mater. Contin.* **2022**, *71*, 6219–6237. [CrossRef]
3. Usama, M.; Naeem, M.A.; Mirza, F. Multi-Class Skin Lesions Classification Using Deep Features. *Sensors* **2022**, *22*, 8311. [CrossRef] [PubMed]
4. Chola, C.; Mallikarjuna, P.; Muaad, A.Y.; Bibal Benifa, J.; Hanumanthappa, J.; Al-antari, M.A. A hybrid deep learning approach for COVID-19 diagnosis via CT and X-ray medical images. *Comput. Sci. Math. Forum* **2022**, *2*, 13.
5. Canayaz, M.; Şehribanoğlu, S.; Özdağ, R.; Demir, M. COVID-19 diagnosis on CT images with Bayes optimization-based deep neural networks and machine learning algorithms. *Neural Comput. Appl.* **2022**, *34*, 5349–5365. [CrossRef] [PubMed]
6. Clement, J.C.; Ponnusamy, V.; Sriharipriya, K.; Nandakumar, R. A survey on mathematical, machine learning and deep learning models for COVID-19 transmission and diagnosis. *IEEE Rev. Biomed. Eng.* **2021**, *15*, 325–340.
7. Varoquaux, G.; Cheplygina, V. Machine learning for medical imaging: Methodological failures and recommendations for the future. *Npj Digit. Med.* **2022**, *5*, 48. [CrossRef]
8. Sujatha, R.; Chatterjee, J.M.; Jhanjhi, N.Z.; Brohi, S.N. Performance of deep learning vs machine learning in plant leaf disease detection. *Microprocess. Microsyst.* **2021**, *80*, 103615. [CrossRef]
9. Sarker, I.H. Deep Learning: A Comprehensive Overview on Techniques, Taxonomy, Applications and Research Directions. *SN Comput. Sci.* **2021**, *2*, 420. [CrossRef]
10. Allugunti, V.R. Breast cancer detection based on thermographic images using machine learning and deep learning algorithms. *Int. J. Eng. Comput. Sci.* **2022**, *4*, 49–56.
11. Abunadi, I.; Senan, E.M. Deep learning and machine learning techniques of diagnosis dermoscopy images for early detection of skin diseases. *Electronics* **2021**, *10*, 3158. [CrossRef]

12. Goyal, S.; Singh, R. Detection and classification of lung diseases for pneumonia and COVID-19 using machine and deep learning techniques. *J. Ambient. Intell. Humaniz. Comput.* **2021**, *12*, 1–21. [CrossRef]
13. Bharti, R.; Khamparia, A.; Shabaz, M.; Dhiman, G.; Pande, S.; Singh, P. Prediction of heart disease using a combination of machine learning and deep learning. *Comput. Intell. Neurosci.* **2021**, *2021*, 8387680. [CrossRef]
14. Mamlook, R.E.A.; Chen, S.; Bzizi, H.F. Investigation of the performance of Machine Learning Classifiers for Pneumonia Detection in Chest X-ray Images. In Proceedings of the 2020 IEEE International Conference on Electro Information Technology (EIT), Chicago, IL, USA, 31 July–1 August 2020; pp. 98–104.
15. Narayanan, B.N.; Ali, R.; Hardie, R.C. Performance analysis of machine learning and deep learning architectures for malaria detection on cell images. *Appl. Mach. Learn.* **2019**, *11139*, 240–247.
16. Data Availability: Data Available for Free at the Kaggle Repository. Available online: www.kaggle.com/amanullahasraf/covid19-pneumonia-normal-chest-xray-pa-dataset; https://www.kaggle.com/datasets/paultimothymooney/chest-xray-pneumonia (accessed on 10 August 2022).
17. The Lloyd Dermatology and Laser Center. Available online: https://lloyd-derm.com/searchresults.php?search=images&sort=score (accessed on 9 August 2022).
18. Dermatology Online Atlas. Available online: http://homepages.inf.ed.ac.uk/rbf/DERMOFIT/ (accessed on 9 August 2022).
19. Roy, A.; Maity, P. A Comparative Analysis of Various Filters to Denoise Medical X-ray Images. In Proceedings of the 2020 4th International Conference on Electronics, Materials Engineering & Nano-Technology (IEMENTech), Kolkata, India, 2–4 October 2020; pp. 1–5.
20. Pitoya, P.A.; Suputraa, I.P.G.H. Dermoscopy image segmentation in melanoma skin cancer using Otsu thresholding method. *J. Elektron. Ilmu Komput. Udayana* **2021**, *2301*, 5373. [CrossRef]
21. Ashraf, H.; Waris, A.; Ghafoor, M.F.; Gilani, S.O.; Niazi, I.K. Melanoma segmentation using deep learning with test-time augmentations and conditional random fields. *Sci. Rep.* **2022**, *12*, 3948. [CrossRef] [PubMed]
22. Sivaraj, S.; Malmathanraj, R. Detection and Classification of Skin Lesions using Probability Map based Region Growing with BA-KNN Classifier. *JMIR Publications* **2021**. [CrossRef]
23. Zafar, K.; Gilani, S.O.; Waris, A.; Ahmed, A.; Jamil, M.; Khan, M.N.; Sohail Kashif, A. Skin lesion segmentation from dermoscopic images using convolutional neural network. *Sensors* **2020**, *20*, 1601. [CrossRef]
24. Naqvi, S.; Tauqeer, A.; Bhatti, R.; Ali, S.B. Improved lung segmentation based on U-Net architecture and morphological operations. *arXiv* **2022**, arXiv:2210.10545.
25. Khairnar, S.; Thepade, S.D.; Gite, S. Effect of image binarization thresholds on breast cancer identification in mammography images using OTSU, Niblack, Burnsen, Thepade's SBTC. *Intell. Syst. Appl.* **2021**, *10–11*, 200046. [CrossRef]
26. Park, Y.; Guldmann, J.-M. Measuring continuous landscape patterns with Gray-Level Co-Occurrence Matrix (GLCM) indices: An alternative to patch metrics? *Ecol. Indic.* **2020**, *109*, 105802. [CrossRef]
27. Zhou, J.; Yang, M. Bone region segmentation in medical images based on improved watershed algorithm. *Comput. Intell. Neurosci.* **2022**, *2022*, 3975853. [CrossRef]
28. Venkatesh, U.; Balachander, B. Analysis of Textural Variations in Cerebellum in Brain to Identify Alzheimers by using Haralicks in Comparison with Gray Level Co-occurrence Matrix (GLRLM). In Proceedings of the 2022 2nd International Conference on Innovative Practices in Technology and Management (ICIPTM), Gautam Buddha Nagar, India, 23–25 February 2022; pp. 549–556.
29. Chandraprabha, K.; Akila, S. Texture Feature Extraction for Batik Images Using GLCM and GLRLM with Neural Network Classification. *Int. J. Sci. Res. Comput. Sci. Eng. Inf. Technol.* **2019**, *5*, 6–15. [CrossRef]
30. Khan, S.; Kaklis, P.; Serani, A.; Diez, M.; Kostas, K. Shape-supervised Dimension Reduction: Extracting Geometry and Physics Associated Features with Geometric Moments. *Comput. Aided Des.* **2022**, *150*, 103327. [CrossRef]
31. Zhang, H.; Hung, C.-L.; Min, G.; Guo, J.-P.; Liu, M.; Hu, X. GPU-accelerated GLRLM algorithm for feature extraction of MRI. *Sci. Rep.* **2019**, *9*, 1–13.
32. Vishnoi, V.K.; Kumar, K.; Kumar, B. A comprehensive study of feature extraction techniques for plant leaf disease detection. *Multimed. Tools Appl.* **2022**, *81*, 367–419. [CrossRef]
33. Hammad, B.T.; Jamil, N.; Ahmed, I.T.; Zain, Z.M.; Basheer, S. Robust Malware Family Classification Using Effective Features and Classifiers. *Appl. Sci.* **2022**, *12*, 7877. [CrossRef]
34. Khan, P.; Kader, M.F.; Islam, S.R.; Rahman, A.B.; Kamal, M.S.; Toha, M.U.; Kwak, K.-S. Machine learning and deep learning approaches for brain disease diagnosis: Principles and recent advances. *IEEE Access* **2021**, *9*, 37622–37655. [CrossRef]
35. Hariraj, V.; Khairunizam, W.; Vikneswaran, V.; Ibrahim, Z.; Shahriman, A.; Zuradzman, M.; Rajendran, T.; Sathiyasheelan, R. Fuzzy multi-layer SVM classification of breast cancer mammogram images. *Int. J. Mech. Eng. Tech.* **2018**, *9*, 1281–1299.
36. Tripathi, M. Analysis of Convolutional Neural Network based Image Classification Techniques. *J. Innov. Image Process.* **2021**, *3*, 100–117. [CrossRef]
37. Janiesch, C.; Zschech, P.; Heinrich, K. Machine learning and deep learning. *Electron. Mark.* **2021**, *31*, 685–695.
38. Sarwar, A.; Ali, M.; Manhas, J.; Sharma, V. Diagnosis of diabetes type-II using hybrid machine learning based ensemble model. *Int. J. Inf. Technol.* **2020**, *12*, 419–428.
39. Suri, J.S.; Puvvula, A.; Biswas, M.; Majhail, M.; Saba, L.; Faa, G.; Singh, I.M.; Oberleitner, R.; Turk, M.; Chadha, P.S.; et al. COVID-19 pathways for brain and heart injury in comorbidity patients: A role of medical imaging and artificial intelligence-based COVID severity classification: A review. *Comput. Biol. Med.* **2020**, *124*, 103960. [CrossRef] [PubMed]

40. Singh, H.; Sharma, V.; Singh, D. Comparative analysis of proficiencies of various textures and geometric features in breast mass classification using k-nearest neighbor. *Vis. Comput. Ind. Biomed. Art* **2022**, *5*, 3. [CrossRef] [PubMed]
41. Houssein, E.H.; Emam, M.M.; Ali, A.A.; Suganthan, P.N. Deep and machine learning techniques for medical imaging-based breast cancer: A comprehensive review. *Expert Syst. Appl.* **2021**, *167*, 114161.
42. Rezaei, K.; Agahi, H.; Wyld, D.C. Segmentation and Classification of Brain Tumor CT Images Using SVM with Weighted Kernel Width. *Comput. Sci. Inf. Technol.* **2017**, *7*, 39–50.
43. Ahsan, M.M.; Luna, S.A.; Siddique, Z. Machine-Learning-Based Disease Diagnosis: A Comprehensive Review. *Healthcare* **2022**, *10*, 541.
44. Arumugam, K.; Naved, M.; Shinde, P.P.; Leiva-Chauca, O.; Huaman-Osorio, A.; Gonzales-Yanac, T. Multiple disease prediction using Machine learning algorithms. *Mater. Today Proc.* **2021**, in press. [CrossRef]
45. Balaji, V.R.; Suganthi, S.T.; Rajadevi, R.; Krishna Kumar, V.; Saravana Balaji, B.; Pandiyan, S. Skin disease detection and segmentation using dynamic graph cut algorithm and classification through Naive Bayes classifier. *Measurement* **2020**, *163*, 107922. [CrossRef]
46. Hazra, R.; Banerjee, M.; Badia, L. Machine Learning for Breast Cancer Classification with ANN and Decision Tree. In Proceedings of the 2020 11th IEEE Annual Information Technology, Electronics and Mobile Communication Conference (IEMCON), Vancouver, BC, Canada, 4–7 November 2020; pp. 0522–0527.
47. Subudhi, A.; Dash, M.; Sabut, S. Automated segmentation and classification of brain stroke using expectation-maximization and random forest classifier. *Biocybern. Biomed. Eng.* **2020**, *40*, 277–289. [CrossRef]
48. Amini, N.; Shalbaf, A. Automatic classification of severity of COVID-19 patients using texture feature and random forest based on computed tomography images. *Int. J. Imaging Syst. Technol.* **2022**, *32*, 102–110. [PubMed]
49. Assam, M.; Kanwal, H.; Farooq, U.; Shah, S.K.; Mehmood, A.; Choi, G.S. An efficient classification of MRI brain images. *IEEE Access* **2021**, *9*, 33313–33322. [CrossRef]
50. Deegalla, S.; Walgama, K.; Papapetrou, P.; Boström, H. Random subspace and random projection nearest neighbor ensembles for high dimensional data. *Expert Syst. Appl.* **2022**, *191*, 116078. [CrossRef]
51. Almeida, M.A.; Santos, I.A. Classification models for skin tumor detection using texture analysis in medical images. *J. Imaging* **2020**, *6*, 51. [CrossRef]
52. Ali, N.M.; Aziz, N.; Besar, R. Comparison of microarray breast cancer classification using support vector machine and logistic regression with LASSO and boruta feature selection. *Indones. J. Electr. Eng. Comput. Sci.* **2020**, *20*, 712–719.
53. Roy, S.; Chandra, A. On the detection of Alzheimer's disease using fuzzy logic based majority voter classifier. *Multimed. Tools Appl.* **2022**, *81*, 43145–43161. [CrossRef]
54. Maqsood, S.; Damasevicius, R.; Shah, F.M. An efficient approach for the detection of brain tumor using fuzzy logic and U-NET CNN classification. In Proceedings of the Computational Science and Its Applications–ICCSA 2021: 21st International Conference, Cagliari, Italy, 13–16 September 2021; Part V 21. pp. 105–118.
55. Sarvamangala, D.; Kulkarni, R.V. Convolutional neural networks in medical image understanding: A survey. *Evol. Intell.* **2022**, *15*, 1–22.
56. Mijwil, M.M.; Al-Zubaidi, E.A. Medical image classification for coronavirus disease (COVID-19) using convolutional neural networks. *Iraqi J. Sci.* **2021**, *62*, 2740–2747.
57. Ashraf, R.; Habib, M.A.; Akram, M.; Latif, M.A.; Malik, M.S.A.; Awais, M.; Dar, S.H.; Mahmood, T.; Yasir, M.; Abbas, Z. Deep convolution neural network for big data medical image classification. *IEEE Access* **2020**, *8*, 105659–105670.
58. Ramzan, F.; Khan, M.U.G.; Rehmat, A.; Iqbal, S.; Saba, T.; Rehman, A.; Mehmood, Z. A deep learning approach for automated diagnosis and multi-class classification of Alzheimer's disease stages using resting-state fMRI and residual neural networks. *J. Med. Syst.* **2020**, *44*, 1–16.
59. Sai Abhishek, A.V. Resnet18 Model with Sequential Layer for Computing Accuracy on Image Classification Dataset. *Int. J. Creat. Res. Thoughts* **2022**, *10*, 2320–2882.
60. Sarwinda, D.; Paradisa, R.H.; Bustamam, A.; Anggia, P. Deep learning in image classification using residual network (ResNet) variants for detection of colorectal cancer. *Procedia Comput. Sci.* **2021**, *179*, 423–431.

Disclaimer/Publisher's Note: The statements, opinions and data contained in all publications are solely those of the individual author(s) and contributor(s) and not of MDPI and/or the editor(s). MDPI and/or the editor(s) disclaim responsibility for any injury to people or property resulting from any ideas, methods, instructions or products referred to in the content.

Article

Feet Segmentation for Regional Analgesia Monitoring Using Convolutional RFF and Layer-Wise Weighted CAM Interpretability

Juan Carlos Aguirre-Arango *, Andrés Marino Álvarez-Meza and German Castellanos-Dominguez

Signal Processing and Recognition Group, Universidad Nacional de Colombia, Manizales 170003, Colombia; amalvarezme@unal.edu.co (A.M.Á.-M.); cgcastellanosd@unal.edu.co (G.C.-D.)
* Correspondence: jucaguirrear@unal.edu.co

Abstract: Regional neuraxial analgesia for pain relief during labor is a universally accepted, safe, and effective procedure involving administering medication into the epidural. Still, an adequate assessment requires continuous patient monitoring after catheter placement. This research introduces a cutting-edge semantic thermal image segmentation method emphasizing superior interpretability for regional neuraxial analgesia monitoring. Namely, we propose a novel Convolutional Random Fourier Features-based approach, termed CRFFg, and custom-designed layer-wise weighted class-activation maps created explicitly for foot segmentation. Our method aims to enhance three well-known semantic segmentation (FCN, UNet, and ResUNet). We have rigorously evaluated our methodology on a challenging dataset of foot thermal images from pregnant women who underwent epidural anesthesia. Its limited size and significant variability distinguish this dataset. Furthermore, our validation results indicate that our proposed methodology not only delivers competitive results in foot segmentation but also significantly improves the explainability of the process.

Keywords: infrared thermal segmentation; regional neuraxial analgesia; deep learning; random fourier features; class activation maps

1. Introduction

The use of regional neuraxial analgesia for pain relief during labor is widely acknowledged as a safe method [1]. It involves the administration of medication into the epidural or subarachnoid space in the lower back. This procedure blocks pain signals from the uterus and cervix to the brain. This method is considered safe and effective for most women and is associated with lower rates of complications than other forms of pain relief [1,2]. Electrophysiological testing measures nerve fiber reactions to painful stimuli with electromyography, excitatory or inhibitory reflexes, evoked potentials, electroencephalography, and magnetoencephalography [3]. In addition, imaging techniques objectively measure relevant bodily function patterns (such as blood flow, oxygen use, and sugar metabolism) using positron emission tomography (PET), single-photon emission computed tomography (SPECT), and functional magnetic resonance imaging (fMRI) [4].

Nonetheless, imaging techniques can be costly and are generally prohibited in obstetric patients, limiting their use. A cost-effective alternative approach is utilizing thermographic skin images to measure body temperature and predict the distribution and efficacy of epidural anesthesia [5]. This approach is achieved by identifying areas of cold sensation [5]. The use of thermal imaging provides an objective and non-invasive solution to assess warm modifications resulting from blood flow redistribution after catheter placement [6]. However, an adequate assessment requires temperature measurements from the patient's foot soles at various times after catheter placement to accurately characterize early thermal modifications [7,8]. Regarding this, semantic segmentation of feet in infrared thermal images in obstetric environments is challenging due to various factors. Firstly, thermal images possess inherent characteristics such as low contrast, blurred edges, and uneven intensity distribution, making it difficult to identify objects accurately [9,10]. The second challenge is the high variability of foot position in clinical settings. Additionally, the specialized equipment required for collecting these images and the limited willingness of mothers to participate in research studies resulted in a need for more available samples and the challenge of acquiring annotated data, which is crucial for developing effective segmentation techniques.

Semantic segmentation is crucial in medical image analysis, with deep learning widely used. Fully Convolutional Networks (FCN) [11] is a popular approach that uses Convolutional layers for pixel-wise classification but produces coarse Region of Interest (ROI) and poor boundary definitions for medical images [12]. Likewise, U-Net [13] consists of encoders and decoders that handle objects of varying scales but have difficulty dealing with opaque or unclear goal masks [14]. U-Net++ [15] extends U-Net with nested skip connections for highly accurate segmentation but with increased complexity and overfitting risk. Besides, SegNet [16] is an encoder–decoder architecture that handles objects of different scales but cannot handle fine details. Mask R-CNN [17] extends Faster R-CNN [18] for instance segmentation with high accuracy but requires a large amount of training data and has high computational complexity. On the other hand, PSPNet uses a pyramid pooling module for multi-scale contextual information and increased accuracy but with high computational complexity and a tendency to produce fragmented segmentation maps for small objects [19].

Specifically for semantic segmentation of feet from infrared thermal images, most works were developed in the context of diabetic foot disorders. In [20], the authors combine RGB, infrared, and depth images to perform plantar foot segmentation based on a U-Net architecture together with RANdom SAmple Consensus (RANSAC) [21], which relies too much on depth information. The authors in [22] use a similar approach to integrating thermal and RGB images to be fed into a U-Net model. Their experiments show that RGB images help in more complex cases. In [23], the authors compare multiple models on thermal images, including U-Net, Segnet, FCN, and prior shape active contour-based methodology, proving Segnet outperforms them all. Similarly, in [24], the authors compare multiple infrared thermographic feet segmentation models using transfer learning and removal algorithms based on morphological operations on U-Net, FCN, and Segnet, showing that Segnet outperforms the rest of the models but with high computational cost.

On the other hand, Visual Transformers (VIT) [25] have revolutionized self-attention mechanisms to identify long-range image dependencies. Several recent works have leveraged VIT capabilities to enhance global image representation. For instance, in [26], a U-Net architecture fused with a VIT-based transformer significantly improves model performance. However, this approach requires a pre-trained model and many iterations. Similarly, in [27], a pure U-Net-like transformer is proposed to capture long-range dependencies. Another recent work [28] suggests parallel branches, one based on transformers to capture long-range dependencies and the other on CNN to conserve high resolution. The authors of [29] propose a squeeze-and-expansion transformer that combines local and global information to handle diverse representations effectively. This method has unlimited practical receptive fields, even at high feature resolutions. However, it relies on a large dataset and has higher computational costs than conventional methods. To address the data-hungry nature of transformer-based models, the work in [30] proposes a semi-supervised cross-teaching approach between CNN and Transformers. The most recent work in this field, Meta Segment Anything [31], relies on an extensive natural database (around 1B images) for general segmentation. However, medical and natural images have noticeable differences, including color and blurriness. It is also pertinent to note that accepting ambiguity can incorporate regions that may not be part of the regions of interest. Specifically, while transformers excel at capturing long-range dependencies, they still face challenges in scenarios where data is scarce [32].

Likewise, transfer learning-based strategies in medical image segmentation is a powerful technique that utilizes pre-trained models to enhance performance, minimize data requirements, and optimize computational resources [33]. Nevertheless, choosing an appropriate and representative pre-trained model is crucial to avoid suboptimal results and potential bias [34,35]. Nevertheless, in our study, we aim to assess the effectiveness of our proposal independently, thus excluding the use of transfer learning.

Here, we present a cutting-edge Convolutional Random Fourier Features (CRFFg) technique for foot segmentation in thermal images, leveraging layer-wise weighted class activation maps. Our proposed data-driven method is twofold. First, it integrates Random Fourier Features within a convolutional framework, enabling weight updates through gradient descent. To assess the efficacy of our approach, we benchmark it against three widely-used architectures: U-Net [13], FCN [11], and ResUNet [36]. We enhance these architectures by incorporating CRFFg at the skip connections, bolsters representation, and facilitate the fusion of low-level semantics from the decoder to the encoder. Second, we introduce a layer-wise strategy for quantitatively analyzing Class Activation Maps (CAMs) for semantic segmentation tasks [37]. Our experimental findings showcase the competitive performance of our models and the accurate quantitative assessment of CAMs. The proposed CRFFg method offers a promising solution for foot segmentation in thermal images, tailored explicitly for regional analgesia monitoring. Additionally, layer-wise weighted class activation maps contribute to a more comprehensive understanding of feature representations within neural networks.

The paper is organized as follows: Section 2 describes the materials and methods used in the study. Sections 3 and 4 present the experimental setup and results, respectively, followed by Section 5, which provides the concluding remarks.

2. Material and Methods

2.1. Deep Learning for Semantic Segmentation

Provided an image set, $\{I_n \in \mathbb{R}^{H \times \tilde{W} \times C} : n \in N\}$, we will call a label mask the corresponding matrix M_n that encodes the membership of each n-th image pixel to a particular class, where H is height, $\tilde{W}$ is width, and C holds the color channels of the image set. For simplicity, $C = 1$ is assumed. As regards the semantic segmentation task under consideration, each mask is binary, $M \in \{0,1\}^{H \times \tilde{W}}$, representing either the background or the foreground.

An estimate for matrix mask $\hat{M} \in [0,1]^{H \times \tilde{W}}$ can be obtained through deep learning models for semantic segmentation, stacking convolutional layers as follows:

$$\hat{M} = (\varphi_L \circ \cdots \circ \varphi_L)(I) \tag{1}$$

where $\varphi_l : \mathbb{R}^{H_{l-1} \times \tilde{W}_{l-1} \times D_{l-1}} \to \mathbb{R}^{H_l \times \tilde{W}_l \times D_l}$ denotes a function composition for the l-th layer ($l \in L$), which comprises learnable parameters represented by $W_l \in \mathbb{R}^{\tilde{k}_l \times \tilde{k}_l \times D_{l-1} \times D_l}$ and $b_l \in \mathbb{R}^{D_l}$ ($\tilde{k}_l$ holds the l-th convolutional kernel size). Of note, the feature map $F_l = \varphi_l(F_{l-1}) = \varsigma_l(W_l \otimes F_{l-1} + b_l) \in \mathbb{R}^{H_l \times \tilde{W}_l \times D_l}$, is comprised of D_l distinct features extracted, $\varsigma_l(\cdot)$ is a nonlinear activation function, and $\otimes$ stands for image-based convolution. Essentially, the function composition in Equation (1) transforms the input feature map from the previous layer, $(l-1)$, into the output feature map for the current layer, l, by employing the learnable parameters W_l and b_l. The resulting F_l captures the salient information within the l-th network layer.

The parameter set $\Theta = \{W_l, b_l : l \in L\}$ is estimated within the following optimizing framework [38]:

$$\Theta^* = \arg\min_{\Theta} \mathbb{E}\{\mathcal{L}\{M_n, \hat{M}_n | \Theta\} : \forall n \in N\}, \tag{2}$$

where $\mathcal{L} : \{0,1\}^{H \times \tilde{W}} \times [0,1]^{H \times \tilde{W}} \to \mathbb{R}$ in Equation (2) is a given loss function and notation $\mathbb{E}\{\cdot\}$ stands for the expectation operator.

2.2. Convolutional Random Fourier Features Gradient—CRFFg

Random Fourier Features establish a finite-dimensional, explicit mapping that approximates shift-invariant kernels $k(\cdot)$ as described in Rahimi et al. (2009) [39]. This explicit mapping, denoted by $z : \mathbb{R}^{\tilde{Q}} \to \mathbb{R}^Q$, serves to transform the input space into a finite-dimensional space $\mathcal{H} \subset \mathbb{R}^Q$, where the inner product can be obtained as:

$$k(x - x') = \langle \phi(x), \phi(x') \rangle_{\mathcal{H}} \approx z(x)^\top z(x'). \tag{3}$$

The mapping z in Equation (3) is defined through Bochner's theorem [40]:

$$k(x - x') = \int_{\mathbb{R}^{\tilde{Q}}} p(\omega) \exp(i\omega^\top (x - x')) d\omega = \mathbb{E}_\omega \{ \exp(i\omega^\top (x - x')) \}, \tag{4}$$

where $x, x' \in \mathbb{R}^{\tilde{Q}}$, $p(w)$ is the probability density function of $w \in \mathbb{R}^{\tilde{Q}}$ that defines the type of kernel. Specifically, the Gaussian kernel, favored for its universal approximating properties and mathematical tractability [41], is achieved from Equation (4) by setting $p(w) = \mathcal{N}(0, \sigma^2 \hat{I}); \sigma \in \mathbb{R}^+$ is a length-scale and $\hat{I}$ is an identity matrix of proper size.

As both the kernel and the probability are real values, the imaginary component can be disregarded by employing the Euler equation. This leads to the use of a cosine function rather than an exponential, ensuring the following relationship:

$$z(x) = \sqrt{\frac{2}{Q}} \left[\cos(\omega_1^\top x + b_1), \ldots, \cos(\omega_Q^\top x + b_Q) \right]^\top, \tag{5}$$

where $\omega_q \in \mathbb{R}^{\tilde{Q}}$, $b_q \in \mathbb{R}$, and $q \in Q$.

We aim to extend the kernel-based mapping depicted in Equation (5) for application to spatial data, such as images, by utilizing the power of convolutional operations. These operations have garnered significant attention for their efficacy in processing grid data [42]. Convolutional operations exhibit two crucial properties—translation equivariance and locality—that render them particularly suitable for handling spatial data [42]. In order to integrate these properties into the Random Fourier Features framework, we adapt the z mapping to operate within local regions of the grid input space. This results in the computation of the feature map $F_l \in \mathbb{R}^{H_l \times \tilde{W}_l \times Q_l}$, where the mapping is defined as $z : \mathbb{R}^{H_{l-1} \times \tilde{W}_{l-1} \times D_{l-1}} \to \mathbb{R}^{H_l \times \tilde{W}_l \times Q_l}$, yielding:

$$F_l = z(F_{l-1}) = \cos\left(\frac{W_l}{\Delta_l} \otimes F_{l-1} + b_l \right), \tag{6}$$

where $\Delta_l \in \mathbb{R}^+$ is a scale parameter. The parameters $W_l \in \mathbb{R}^{\tilde{k}_l \times \tilde{k}_l \times D_{l-1} \times Q_l}$ and $b_l \in \mathbb{R}^{Q_l}$ are initialized as in Equations (4) and (5), and updated through gradient descent under a back-propagation-based optimization of Equation (2) [38]. Consequently, we refer to the layers in Equation (6) as Convolutional Random Fourier Features Gradient (CRFFg).

The conceptual depiction of the proposed CRFFg layer is shown in Figure 1. Using this approach, we aim to integrate the advantageous attributes of kernel methods into a deep learning-based feature representation enhancement. In addition, using convolutions for local and equivariant representation of spatial data provides a robust and efficient strategy for image processing.

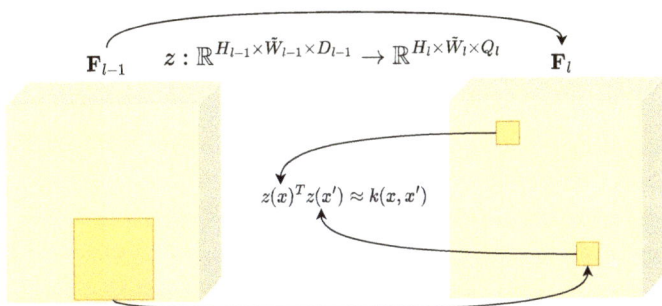

Figure 1. The Convolutional Random Fourier Features Gradient (CRFFg) mapping, grounded in kernel methods, is employed for image-based data examination within deep learning frameworks.

2.3. Layer-Wise Weighted Class Activation Maps for Semantic Segmentation

Class Activation Maps (CAMs) are a powerful tool to enhance the interpretability of outcomes derived from deep learning models. They achieve this by emphasizing the critical image regions in determining the model's predicted output. To evaluate the contribution of these regions to a specific class $r \in \{0,1\}$, a linear combination of feature maps from a designated convolutional neural network layer l can be employed [37]. Here, given an input image I and a target class r, the salient input spatial information coded by the l-th layer into a trained deep learning semantic segmentation model with parameter set Θ^*, as in Equation (2), is gathered through the Layer-CAM algorithm, yielding [43]:

$$S_l^r = (\Lambda \circ ReLU) \left(\sum_{d \in D_l} \alpha_l^{rd} \odot F_l^{rd} \right) \quad (7)$$

where $S_l^r \in \mathbb{R}^{H \times \tilde{W}}$ holds the Layer-CAM for class r at layer l, $\Lambda : \mathbb{R}^{H_l \times \tilde{W}_l} \to \mathbb{R}^{H \times \tilde{W}}$ is the up-sampling operator, $ReLU(x) = \max(0, x)$ is the Rectified Linear activation function, and $\odot$ stands for Hadamard product. Besides, $F_l^{rd} \in \mathbb{R}^{H_l \times \tilde{W}_l}$ collects the d-th feature map and $\alpha_l^{rd} \in \mathbb{R}^{H_l \times \tilde{W}_l}$ is a weighting matrix holding elements:

$$\alpha_l^{rd}[i,j] = ReLU\left(\partial y^r / \partial F_l^{rd}[i,j]\right), \quad (8)$$

with $\alpha_l^{rd}[i,j] \in \alpha_l^{rd}$ and $F_l^{rd}[i,j] \in F_l^{rd}$. y^r is the score for class r that is computed using the approach in [44] adopted for the semantic segmentation tasks, as follows:

$$y^r = \mathbb{E}\{\tilde{F}_L[i,j] : \forall i,j | M[i,j] = r\} \quad (9)$$

where $\tilde{F}_L[i,j] \in \tilde{F}_L$ holds the feature map elements for layer L in Equation (1) fixing a linear activation function.

As previously mentioned, the use of CAM-based representations enhances the explainability of deep learning models for segmentation tasks. To evaluate the interpretability of CAMs for a given model, we propose the following semantic segmentation measures, where higher scores indicate better interpretability:

- CAM-based Cumulative Relevance (ρ_r) : It involves computing the cumulative contribution from each CAM representation to detect class r within the segmented region of interest. This can be expressed as follows:

$$\rho_r = \mathbb{E}_l \left\{ \mathbb{E}_n \left\{ \frac{\mathbf{1}^\top (\tilde{M}_n^r \odot S_{nl}^r) \mathbf{1}}{\mathbf{1}^\top S_{nl}^r \mathbf{1}} : \forall n \in N \right\} \forall l \in L \right\}, \quad \rho_r \in [0,1], \quad (10)$$

where $\tilde{M}_n^r \in \{0,1\}^{H \times W}$ collects a binary mask that identifies the pixel locations associated with the class r, and S_{nl}^r holds the Layer-CAM for image n with respect to layer l (see Equation (7)).

- Mask-based Cumulative Relevance (ϱ_r): It assesses the relevance averaged across the class pixel set related to the target mask of interest. Then, each class-based cumulative relevance is computed as follows:

$$\varrho_r = \mathbb{E}_l \left\{ \mathbb{E}_n \left\{ \frac{\mathbf{1}^\top (\tilde{M}_n^r \odot S_{nl}^r) \mathbf{1}}{\mathbf{1}^\top \tilde{M}_n^r \mathbf{1}} : \forall n \in N \right\} \forall l \in L \right\}, \varrho_r \in \mathbb{R}^+. \quad (11)$$

The normalized Mask-based Cumulative Relevance can be computed as:

$$\rho'_r = \frac{\rho'_r}{\max\limits_{r' \in \{0,1\}} \rho_{r'}}, \quad \rho'_r \in [0,1]. \quad (12)$$

- CAM-Dice (D'): A version of the Dice measure that quantifies mask thickness and how the extracted CAM is densely filled:

$$D'_r = \mathbb{E}_l \left\{ \mathbb{E}_n \left\{ 2 \frac{\mathbf{1}^\top (\tilde{M}_n^r \odot S_{nl}^r) \mathbf{1}}{\mathbf{1}^\top \tilde{M}_n^r \mathbf{1} + \mathbf{1}^\top S_{nl}^r \mathbf{1}} : \forall n \in N \right\} : \forall l \in L \right\}, \quad D'_r \in [0,1]. \quad (13)$$

The proposed measures enable the weighting of each layer's contribution to a given class across the model by adjusting the normalization term related to the target mask, the estimated CAM, or both pixel-based salient activations. Figure 2 depicts a graphical representation of the proposed measures. The green circle represents the CAM generated for a specific region, as indicated by the white circle. These measures are designed to capture the relationship between the CAMs and the regions of interest. Furthermore, Figure 3 presents some exemplary scenarios. For instance, the ρ measure is associated with the proportion of the CAM inside the region of interest. On the other hand, ρ is based on the proportion of CAMs that, on average, belong to each pixel of the region of interest while maintaining the relationship between the classes (in this case, green for the foreground and red for the background). Additionally, D'_r follows a similar concept as the Dice coefficient used in segmentation, assessing the homogeneity of the intersection of the regions. In this case, we want to determine if the CAM is uniformly distributed.

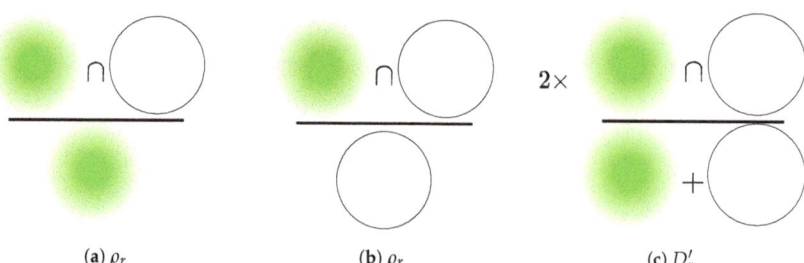

(a) ρ_r (b) ϱ_r (c) D'_r

Figure 2. Graphic depiction of the proposed relevance measures for Layer-Wise Class Activation Maps used in semantic segmentation tasks.

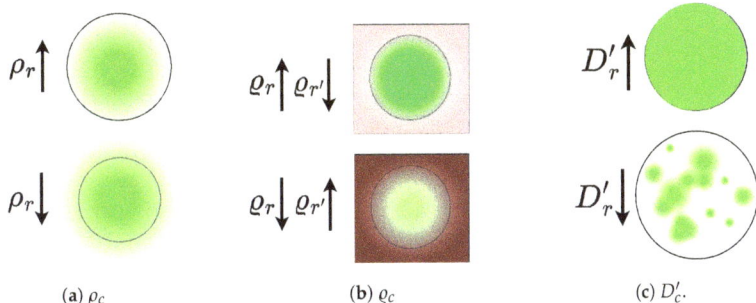

(a) ρ_c (b) ϱ_c (c) D'_c.

Figure 3. Illustrative scenarios regarding our novel Layer-Wise Class Activation Maps for semantic segmentation.

2.4. Feet Segmentation Pipeline from Thermal Images

In a nutshell, the proposed methodology is evaluated using the pipeline shown in Figure 4, including the following testing stages:

(i)   Foot Infrared Thermal Data Acquisition and Preprocessing.
(ii) Architecture Set-Up of tested Deep models for foot segmentation. Three DL architectures are contrasted using our CRFFg: U-Net, Fully Convolutional Network (FCN), and ResUNet.
(iii) Assessment of semantic segmentation accuracy. In this study, we examine how data augmentation affects the performance of tested deep learning algorithms.
(iv) Relevance-maps extraction from our Layer-Wise weighted CAMs to provide interpretability.

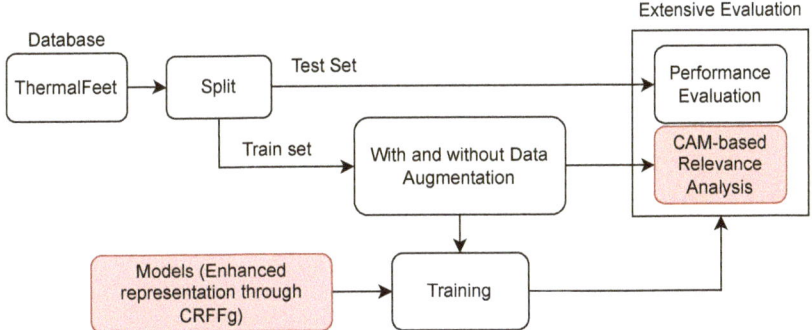

Figure 4. Foot segmentation from thermal images using our CRFFg-based deep learning enhancement holding layer-wise weighted CAM interpretability.

3. Experimental Set-Up

The proposed deep learning model for semantic segmentation enhances foot thermal images' interpretability, achieving competitive segmentation performance. To this end, we evaluate the impact of incorporating a convolutional representation of CRFFg and layer-wise weighted CAM into three well-known deep-learning architectures.

3.1. Protocol for Infrared Thermal Data Acquisition: ThermalFeet Dataset

The protocol for data acquisition was designed by the physician staff at "SES Hospital Universitario de Caldas" to standardize the data collection of infrared thermal images acquired from pregnant women who underwent epidural anesthesia during labor. This protocol is in accordance with the occupational risks associated with assisting local anesthetics via epidural neuraxial as specified by the hospital's administration, following previously implemented protocols [8,45–48].

Patient monitoring includes the necessary equipment for taking vital signs and a metal stretcher with foam cushion and plastic exterior covered only with a white sheet. The continuous monitoring device is placed 1.5 m from the stretcher in the same room, as shown in Figure 5. Before the epidural procedure, anesthesiologists assess each patient clinically and provide written and verbal information about the trial before obtaining her written consent. The patient's body temperature, heart rate, oxygen saturation, and non-invasive blood pressure are monitored every five minutes. Skin temperature values are recorded during the procedure. Sensitivity responses are evaluated using superficial touch and cold tests with cotton wool soaked in water applied to the previously determined dermatomes. The temperature test records the verbal response as Yes or No for superficial touch and Cold or No Cold.

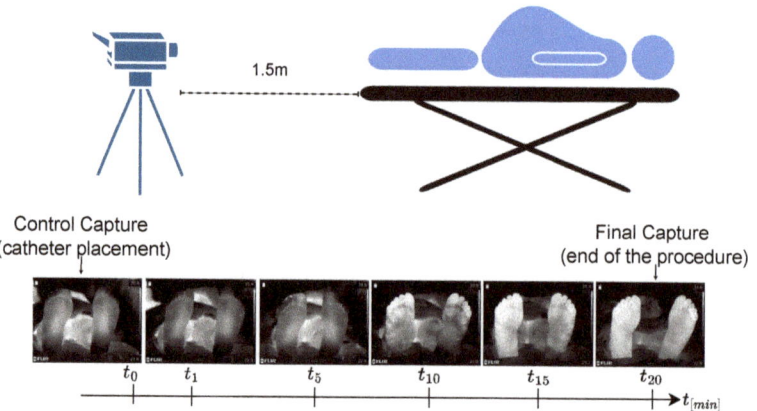

Figure 5. Regional analgesia monitoring protocol using local anesthetics via epidural neuraxial and thermal images.

The protocol timeline for acquiring infrared thermal images is as follows: Initially, the woman is asked to be in a supine position before the first thermal image (T0) is captured once the first dose of the analgesic mixture is administered. A single thermal picture is taken at the placement of the operated catheter (0.45 mm; Perifix, Braun®, Kronberg, Germany) positioned within the space selected for injecting epidural anesthesia in the cervical region (at L2 to L3 or L3 to L4), measuring a few millimeters.

Within the next 25 min, one thermographic recording of the lower extremity is taken every five minutes (T1–T5). The catheter remains in the epidural space taped to the skin so that one image is captured every five minutes until six pictures have been collected. Though the clinical protocol demands images of both feet taken in a fixed corporal position, this condition is barely achievable due to the difficulty of labor procedures and contractions.

The data was collected under two different hardware specifications: (i) A set of 196 images captured from 22 pregnant women during labour using a FLIR A320 infrared camera with a resolution of 640 × 480 and a spectral range within 7.5 to 13 µm. (ii) A set of 128 images with improved sensitivity and flexibility taken using a FLIR E95 thermal camera, having a resolution of 640 × 480 and spectral range within 7.5 to 14 µm. In this study, 166 thermal images are selected from both sets as fulfilling the quality criteria of validation, as detailed in [24]. An anesthesiologist manually segmented the region of interest. The dataset is publicly available at https://gcpds-image-segmentation.readthedocs.io/en/latest/notebooks/02-datasets.html (accessed on 5 April 2023).

3.2. Set-Up of Compared Deep Learning Architectures

The following deep learning architectures are contrasted and enhanced using our CRFFg approach:

- Fully Convolutional Network (FCN) [11]: This architecture is based on the VGG (Very Deep Convolutional Network) [49] model to recognize large-scale images. By using only convolutional layers, FCN models can deliver a segmentation map with pixel-level accuracy while reducing the computational burden.
- U-Net [13]: This architecture unfolds into two parts: The encoder consists of convolutional layers to reduce the spatial image dimensions. The decoder holds layers to upsample the encoded features back to the original image size.
- ResUNet [36]: This model extends the U-Net architecture by incorporating residual connections to improve performance. Deep learning training is improved by residual connections, which allow gradients to flow directly through the network.

Figure 6 presents the mentioned architectures, illustrating their unique layers, blocks, and the dimensions and filters associated. Different colors represent the different blocks or layers, and the spatial dimension of each level is also indicated. We estimate the effectiveness of incorporating the CRFFg layer for comparison purposes in FCN, U-Net, and ResUNet architectures. However, each evaluated CRFFg layer arrangement differs from another in the semantic segmentation features that feed the decoder, as detailed in [50–52]. Then, the CRFFg layer is placed at skip connections to enhance the feature fusion between encoders and decoders.

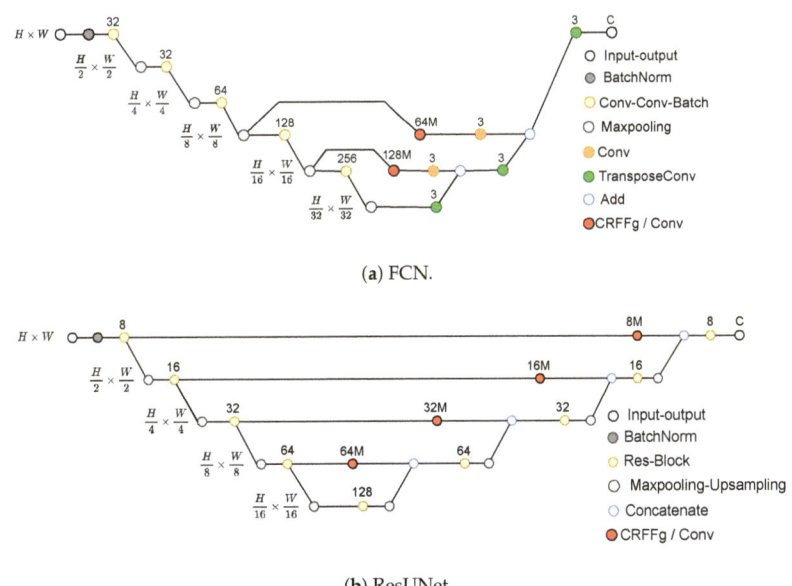

(a) FCN.

(b) ResUNet.

Figure 6. *Cont.*

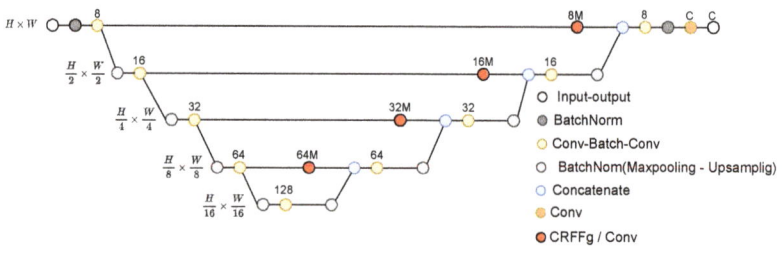

(c) U-Net.

Figure 6. Tested semantic segmentation architectures. Our CRFFg approach aims to enhance the data representation (see red dots).

To evaluate the performance difference with the proposed CRFFg-layer strategy, we utilize a standard convolutional layer featuring an equal number of filters and a ReLU activation function at the same position within the architecture. In particular, we analyze the influence of the CRFFg layer dimension on segmentation performance, testing two multiplication values (one and three). Besides, to study the impact of CRFFg, we set the hyperparameters of all models variation from FCN, U-Net, and ResUNet architectures the same. The number of epochs is 200, and the batch size is 16. Additionally, the scale value of the CRFFg, Δ, is set as described in the standard RFF's Tensorflow implementation for simplicity. Regarding the weights, they are trained using gradient descent with backpropagation. The selected optimizer is Adam due to its faster convergence, adaptive learning rate, reduced sensitivity to hyperparameters, and combining benefits of convex optimization [53]. The learning rate is initialized as $1e-3$, and a dice-based loss is employed in Equation (2), as follows:

$$\mathcal{L}_{Dice}(M_n, \hat{M}_n) = 2 \frac{\mathbf{1}^\top (M_n \odot \hat{M}_n)\mathbf{1} + \epsilon}{\mathbf{1}^\top M_n \mathbf{1} + \mathbf{1}^\top \hat{M}_n \mathbf{1} + \epsilon}, \quad (14)$$

where $\epsilon = 1$ avoids numerical instability. All experiments are carried out in Python 3.8, with the Tensorflow 2.4.1 API, on a Google Colaboratory environment (code repository: https://github.com/aguirrejuan/Foot-segmentation-CRFFg, accessed on 25 April 2023).

3.3. Training Details and Quantitative Assessment

With the aim to prevent overfitting and improve the generalization of trained models, the data augmentation procedure is performed on each image with horizontal flip enabled since feet are mostly symmetrical on the horizontal axis, specifically left-right and right-left on each foot. Hence, vertical overturn is disabled to prevent unrealistic upside-down foot representations. In the augmentation procedure, the images are rotated seven times within a range of -15 to 15 degrees, translated by 10% right to left, and zoomed in and out by 15%, as described in [20].

Moreover, the following metrics are used to measure segmentation performance [54]:

$$D = \frac{2|M \cap \hat{M}|}{|M| + |\hat{M}|} = \frac{2T_P}{2T_P + F_P + F_N} \quad (15a)$$

$$J = \frac{|M \cap \hat{M}|}{|M \cup \hat{M}|} = \frac{T_P}{F_N + F_P + T_P} \quad (15b)$$

$$S_e = \frac{T_P}{T_P + F_N} \quad (15c)$$

$$S_p = \frac{T_N}{T_N + F_P} \quad (15d)$$

where T_P, F_N, and F_P represent the true positive, false negative, and false positive predictions, respectively, for comparing the actual and estimated label masks M_n and $\hat{M}_n$ for a given input image I_n. In addition, the introduced layer-wise, weighted CAM-based interpretability measures are computed for CAM-Dice, CAM-based Cumulative Relevance, and Mask-based Cumulative Relevance (see Equations (12) and (13)).

As for the validation strategy, we selected the hold-out cross-validation strategy with the following partitions: 80% of the samples for training, 10% for validation, and 10% for testing.

4. Results and Discussion

4.1. Visual Inspection Results

Figure 7 shows results obtained from thermalFeet database without data augmentation, where each row represents a different architecture: FCN in the first row, U-Net in the second row, and ResUNet in the third row. As expected, the performance of the models under a small-size dataset is poor. The regions of faster change in temperature, which characterize the dataset, are where the models struggle more. At first glance, we observe that the FCN architecture is the one that struggles the most, having high false positives regions in regions that exhibit low-high temperatures.

Figure 8 shows results obtained incorporating data augmentation. The positive impact of the data augmentation on the resulting segmentation of all the models is visible. Moreover, FCN architectures produce smoother borders and fewer false positives than other architectures. This can be explained due to the high receptive field that possesses the FCN architecture, allowing it to capture complex and heterogeneous regions (the variability of the temperatures) that compose the feet.

Notably, when comparing FCN models with a multiplication factor of 1 (M1), the model with our CRFFg (blue) generally outperforms in terms of pixel membership prediction (sensitivity). However, this trend only holds when the multiplication factor is increased to 3 (M3), probably because the large model is a propensity to overfit, making the prediction less confident in new data points. On the other hand, U-Net models blunder with regions that exhibit fast temperature changes. The same characteristic the FCN possesses can explain this, but the U-Net does not have a high receptive field that allows it to characterize high heterogeneous feet. As a result, among the U-Net approaches, U-Net CRFFg S-M1 performs satisfactorily with low false positives and high false negatives. At the same time, its direct competitor, U-Net S-M1, shows the opposite trend. Similarly, using CRFFg in the other U-Net alternatives reduces the number of false positives. Finally, the ResUNet architecture has the same behavior as the U-Net but with smoother borders, which can be explained due to the multiple stack layers at the ResBlock, which increase multiple steps of representation, allowing to capture of helpful representation. The ResUNet S-M1 works better on average; adding layers at the skip connections appears to reduce performance, creating false positives and false negatives. The latter can be explained due to the small size of the dataset. Specifically, using CRFFg with ResUNet does not result in noteworthy improvements.

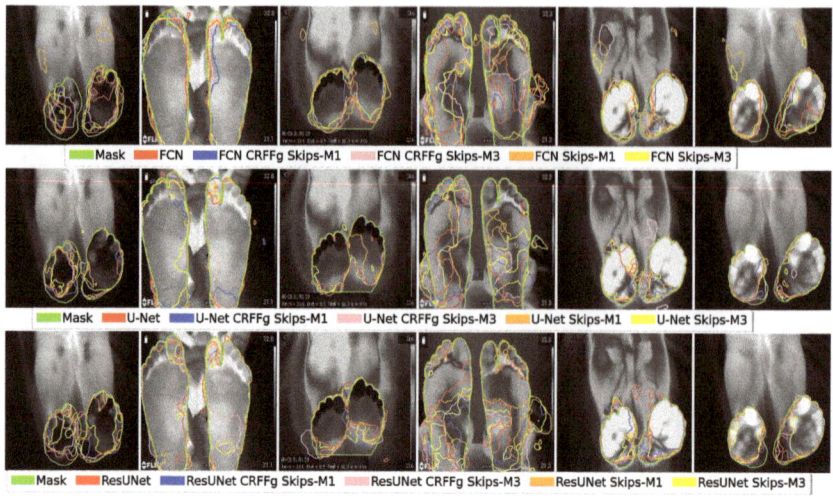

Figure 7. Visual inspection of the results on thermalFeet database without data augmentation. Our CRFFg-based enhancements are also presented. The first row shows the results for the FCN architecture, the second row for U-Net, and the third row for ResUNet. A unique color differentiates each model within an architecture. M1 and M3 represent CRFFg's dimension as a multiplication factor of the enhanced layer's size.

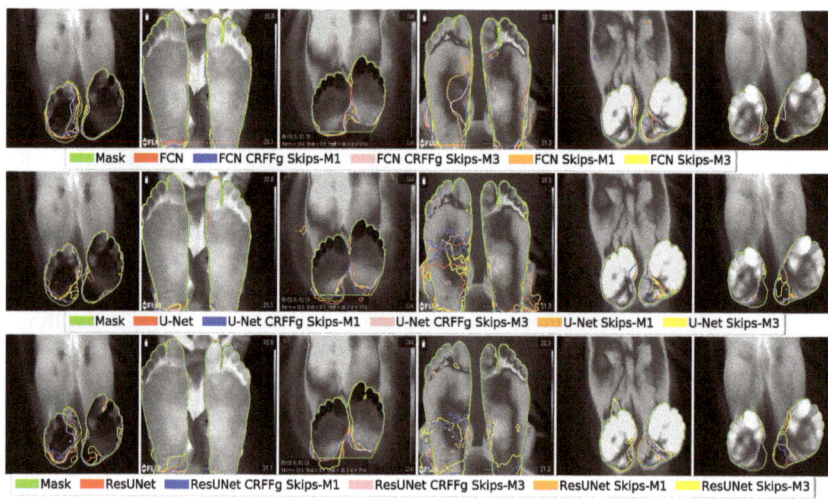

Figure 8. Visual inspection of the results on thermalFeet database with data augmentation. Our CRFFg-based enhancements are also presented. The first row shows the results for the FCN architecture, the second row for U-Net, and the third row for ResUNet. A unique color differentiates each model within an architecture. M1 and M3 represent CRFFg's dimension as a multiplication factor of the enhanced layer's size.

4.2. Method Comparison Results of Semantic Segmentation Performance

Figure 9 illustrates the learning curves, e.g., training loss vs. epochs, of the compared models. Upon visual inspection, notable differences between the curves with and without data augmentation can be observed. When data augmentation is not applied, the algorithms exhibit higher validation loss in the initial 40 epochs. Regardless, they subsequently demonstrate a downward trend in validation loss. It is essential to mention that the learning curves exhibit increased noise, likely due to the limited size of the dataset. The limited dataset challenges the models to capture generalized features early in training. Moreover, in the validation partition without data augmentation, some models display a phenomenon known as double descent [55], where layers at different locations in the networks may learn at different rates [56]. In contrast, the training and validation losses consistently decrease in the data augmentation scenario, albeit with minor noise in the validation partition.

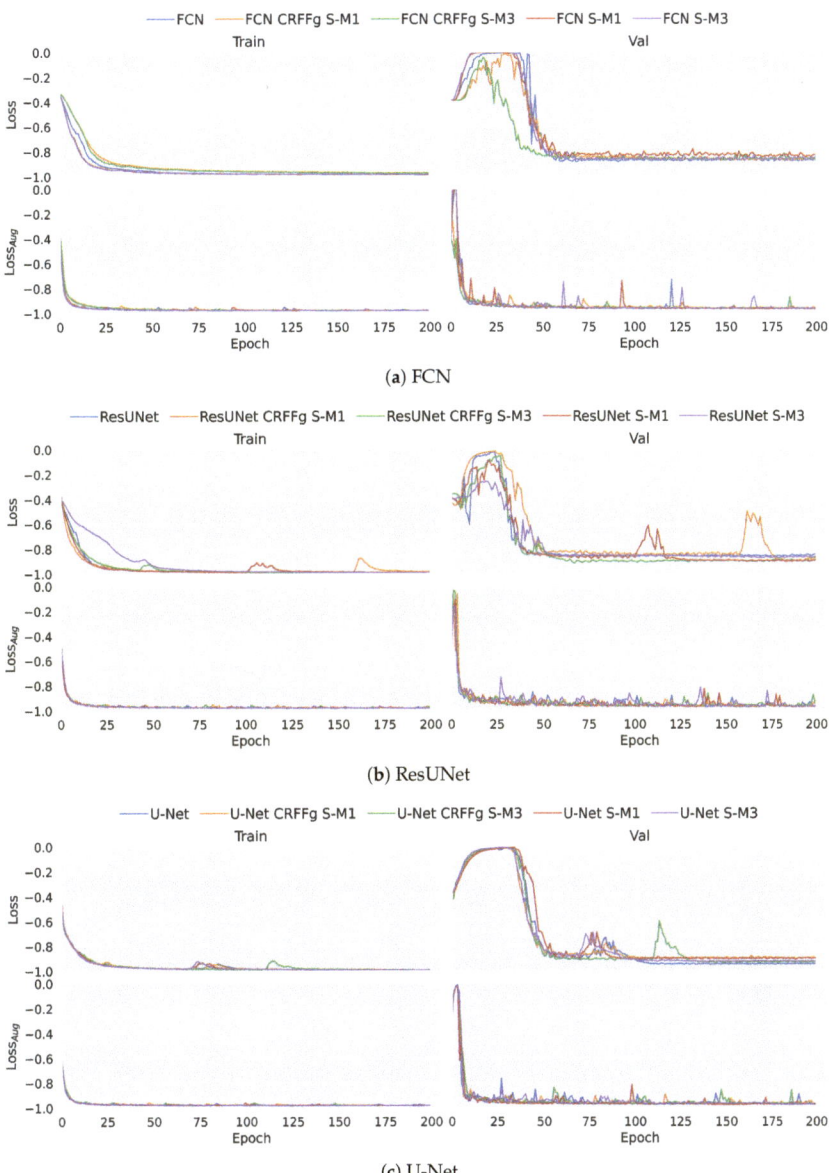

Figure 9. Training neural network loss vs. epochs corresponding to the various models examined are presented. M1 and M3 signify the dimensions of the CRFFg layer, expressed as multiplication factors of the enhanced layer's size.

It is worth noting that both the FCN CRFFg S-M3 and, to a lesser extent, the FCN CRFFg S-M1 tend to exhibit faster decreases in validation loss during early iterations. This conduct can be attributed to the generalization capabilities of the RFF from kernel methods. On the other hand, in the ResUNet architectures, although it needs to be clarified, the ResUNet S-M3 tends to experience an early decline, even though it also reaches its minimum early, which is not the minimum among the approaches. Conversely, no apparent differences are observed within the U-Net architectures. Notably, the models in the data augmentation scenario are similar.

In turn, Figure 10a displays the values of semantic segmentation performance for thermalFeet dataset achieved by each compared deep learning architecture: FCN (colored in blue), ResNet (red), U-Net (green). For interpretation purposes, the results are presented for the evaluation measures separately. As seen, the specificity estimates are very close to the maximal value and show the lowest variability. This result can be explained by the relatively small feet sizes compared with the background, making their correct detection and segmentation more difficult. On the contrary, sensitivity assessments are of less value and have much more variability, accounting for the diversity in the regions of interest (i.e., size, shape, and location). Due to the changing behavior of thermal patterns and the limited datasets available, learners have difficulty obtaining an accurate model.

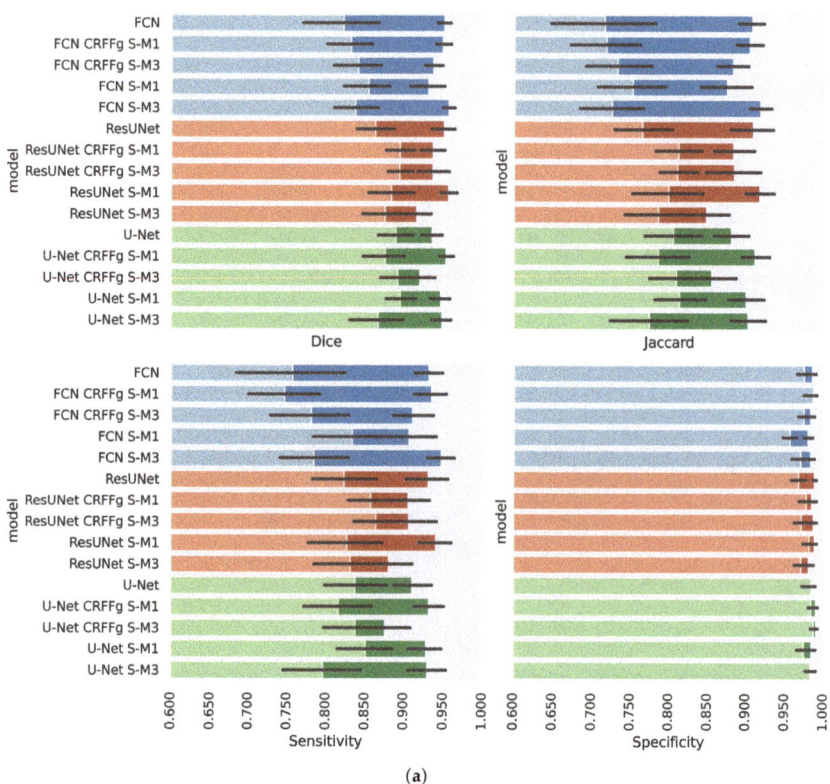

(a)

Figure 10. *Cont.*

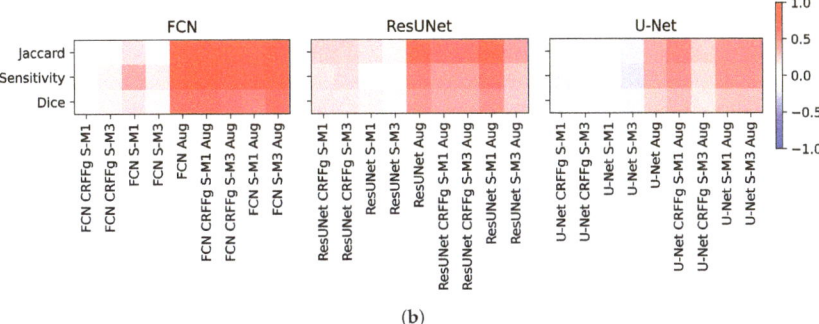

(b)

Figure 10. Results of the comparison between methods. The segmentation performance of ThermalFeet is evaluated using baseline models FCN, UNet, and ResUNet, and compared to our proposal that incorporates CRFFg-based enhancements. M1 and M3 represent CRFFg's dimension as a multiplication factor of the enhanced layer's size. Aug stands for data augmentation. (**a**) Segmentation performance results on ThermalFeet database. The three types of architecture used in this study (FCN, U-Net, ResUNet) are differentiated by color. The type of variation in the architecture is indicated by the marker used; (**b**) The improvement of each strategy, normalized with respect to the baseline performance of each architecture.

Regarding overlapping between estimated thermal masks, the Dice value is acceptable but with higher variance values for FCN, implying that other tested models segment complex shapes more accurately. As expected, the Jaccard index mean values resemble the Dice assessments, although with increased variance, which highlights the mismatch between the ground truth and the predicted mask even more.

A comparison between the segmentation metric value achieved by the baseline architecture (without any modifications) and the value estimated for every evaluated semantic segmentation strategy is presented in Figure 10b. Note that specificity is removed because its estimates are obtained with minimal variations.

As seen, the performance improvement depends on the learner model size (also called algorithm complexity). Namely, the baseline architecture of FCN holds 1,197,375 parameters, baseline ResUnet— 643,549, and baseline Unet—494,093. Thus, the FCN model contains the largest tuning parameter set and achieves the poorest performance, but it benefits the most from the evaluated architectures. As data augmentation is also applied, this finding becomes more evident. It may be pointed out that adding new data decreases model overfitting inherent to massive model sizes. Likewise, the following ResUnet model takes advantage of the enhanced architecture strategy using our CRFFg and improves performance. It increases more by generating new data points, however, to a lesser extent. Lastly, the learner with the lowest parameter set gets almost no benefits or is negatively affected by the strategies considered for architecture enhancement. Still, the strategies taken into account combined with expanded training data sizes can be improved, though very modestly. See Table A1, Appendix A, for the detailed segmentation performance results concerning the studied approaches.

4.3. Results of Assessing the Proposed CAM-Based Relevance Analysis Measures

We aim to evaluate the tested deep learning models for assessing the contribution of CAM-based representations to interpretability. To this end, we plot the pairwise relationship between the essential explanation elements (background and foreground) and the above-proposed measure for assessing the CAM-based relevance of performed image segmentation masks. Figure 11 displays the scatter plots obtained by each segmentation learner. CAMs extracted by the learner contribute more to the interpretability of regions of interest if the measure value tends toward the top-right corner. Moreover, we focus on the contribution of CAM representations to segmenting between background and foreground, utilizing the patient's feet as critical identification features.

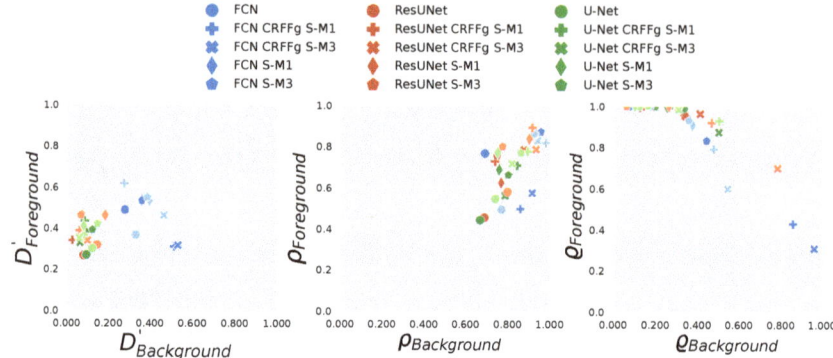

Figure 11. Results of Interpretability Measures on ThermalFeet. The three types of architecture used in this study (FCN, U-Net, ResUNet) are differentiated by color. The type of variation in the architecture is indicated by the marker used. M1 and M3 represent CRFFg's dimension as a multiplication factor of the enhanced layer's size.

The findings from the modified CAM-Dice results can be split into two groups (refer to the left plot in Figure 11). One group involves ResUnet and UNet architectures, and the other showcases the better performance, featuring FCN architectures. It is also important to mention that the data augmentation strategy does not significantly boost interpretability as much as it enhances segmentation performance measures. Looking at the CAM-based Cumulative Relevance (refer to the middle plot in Figure 11), it is apparent that models with refined representations at skip connections surpass the baseline models. Even though there is no substantial difference between models with these enhancements, most models are situated in the top-right corner. This position suggests that the primary relevance is focused on the area of interest. Significantly, relevance seems to accumulate more in the background than in the foreground, which is logical, considering the relative sizes of both areas. In Figure 11, the Mask-based Cumulative Relevance plot on the right side demonstrates that most models tend to exhibit high-foreground-low-background relevance. This pattern leads to a bias favoring the foreground class, as reflected in the more robust activation of CAMs for the foreground class. However, it is interesting that models employing CRFFg perform better in separating classes situated towards the top-right corner, suggesting superior capabilities in differentiating foreground and background classes.

Figure 12 displays examples of CAMs extracted by the best models per architecture under the Mask-based Cumulative Relevance for feet (colored in green) and background (red color), respectively. As seen, the higher weight is located at the last part of the decoder, where the higher values of semantic information are found. Besides, the weights for the background class are also less than for the foreground class, showing that the models emphasize the latter while preserving the relevance weights for the former.

In particular, FCN CRFFg S-M3 is the best FCN model, as shown in Figure 12a, and extracts most of the weights in three layers (i.e., l3, l4, and l5), meaning that other layers do not contribute to the class foreground. On the other hand, this architecture leads to CAMs with lower values for background class (see examples on the right). This behavior can be explained because the FCN architecture holds an extensive receptive field. Hence, the FCN CRFFg S-M3 model enables capturing more global information crucial for segmentation and concentrating weights in a few layers.

In the case of ResUNet, ResUNet CRFFg S-M3 performs the most efficiently, as shown in Figure 12b. Since the receptive field decreases, the ResUNet architecture distributes the contribution more evenly among the extracted CAM representations. However, the more significant values remain in the l3, l4, and l5 layers. There is also activation of weights for the background class that can be explained, firstly, since the CRFFg configuration helps capture complex non-linear dependencies. Secondly, the local receptive field allows class separation.

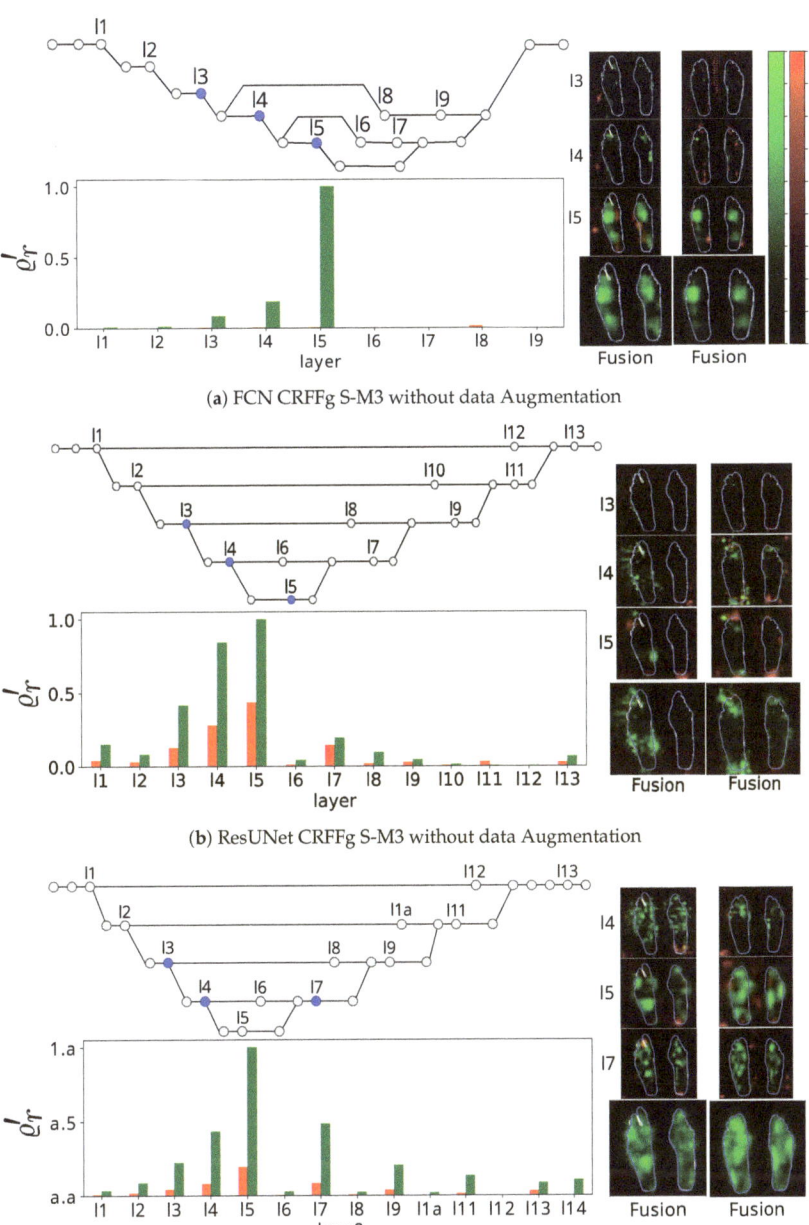

(a) FCN CRFFg S-M3 without data Augmentation

(b) ResUNet CRFFg S-M3 without data Augmentation

(c) U-Net CRFFg S-M3 with data Augmentation

Figure 12. Salient relevance analysis results. Best models concerning the Mask-based Cumulative Relevance, ϱ_r measure, are presented for FCN, UNet, and ResUNet with our CRFFg-based enhancement.

Lastly, the CRFFg S-M3 model is the most effective for the U-Net architecture, with a performance similar to the outperforming ResUNet architecture, as shown in Figure 12c. However, several differences in the Fusion CAMs extracted by U-Net CRFFg S-M3 show high activation within the feet, suggesting that this model is not only sensitive to the foreground class. In addition, it captures more global features from feet.

5. Concluding Remarks

We introduce an innovative semantic segmentation approach that enhances interpretability by incorporating Convolutional Random Fourier Features and layer-wise weighted class activation maps. Our approach has been tested on a unique dataset of thermal foot images from pregnant women who have received epidural anesthesia, which is small but exhibits considerable variability. Besides, our strategy is two-pronged. Firstly, we introduce a novel Random Fourier Features layer, CRFFg, for handling image data, aiming to enhance three renowned architectures - FCN, UNet, and ResUNet. Secondly, we introduce three new quantitative measures to assess the interpretability of any deep learning model used for segmentation tasks. Our validation results indicate that the proposed approach boosts explainability and maintains competitive foot segmentation performance. In addition, the dataset used is tailored explicitly for epidural insertion during childbirth, reinforcing the practical relevance of our methodology.

There are, however, several observations worth mentioning:

Data acquisition tailored for Epidural. Epidural anesthesia involves the delivery of medicines that numb body parts to relieve pain, and the acquisition of data is usually performed under uncontrolled conditions with strong maternal artifacts. Moreover, it is impossible to fix a timeline for data collection. In addition, a timeline for gathering data cannot be set correctly. To the extent of our knowledge, this is the first time a protocol has been presented to regulate the data collection of infrared thermal images acquired from pregnant women who underwent epidural anesthesia during labor. As a result, data were assembled under real-world conditions that contained 196 thermal images fulfilling validation quality criteria.

Deep learning models for image semantic segmentation. Combined with machine learning, thermal imaging has proven helpful for performing semantic segmentation as a powerful method of dense prediction to adverse lighting conditions, providing better performance compared to their traditional counterparts. State-of-the-art medical image segmentation models include variants of U-Net models. A major reason for their success is that they employ skip connections, combining deep, semantic, and coarse-grained feature maps from the decoder subnetwork with shallow, low-level, fine-grained feature maps from the encoder subnetwork. They recover fine-grained details of target objects despite complex backgrounds [57]. Nevertheless, the collected image data from epidural anesthesia is insufficient for training the most commonly-known deep learners, which may result in overfitness to the training set. We address this issue by employing data augmentation addresses that artificially increase training data inputs to feed three tested architectures of deep learning models (FCN, U-Net, ResUNet), thus improving segmentation accuracy results. As seen in Figure 10b, the segmentation accuracy gain depends on the learner model complexity used: The fewer parameters the learner holds, the more the effectivity of data augmentation. Thus, the UNet learner with the lowest parameter set gets almost no benefit.

Strategies for enhancing the performance of deep learning-based segmentation. Three deep-learning architectures are explored to increase the interpretability of semantic segmentation results at competitive accuracy, ranked in decreased order of computational complexity as follows: FCN, ResUNet, and U-Net. Regarding the accuracy of semantic models, the data augmentation yields a sensibility metric value dependent on the model complexity: the more parameters the architecture holds, the higher the segmentation accuracy improvement. Thus, FCN benefits more from artificial data than ResUNet and U-Net. In the same way, both overlapping metrics (Jaccard and Dice) depend on the complexity of models. By contrast, the specificity reaches very high values regardless of trained deep learning because the background texture's homogeneity saturates most captured thermal images. Nonetheless, the proposed modifications to architectures are not a solid argument for influencing their performed accuracy of semantic segmentation. In terms of enhancing explainability, the weak influence of data augmentation is the first finding to be drawn, as seen in the scatterplots of Figure 11. All tested models produce more significant CAM activations from layers with a wider receptive field. Moreover, the CRFFg layer also improves the representation of the foreground and background. It is also important to note the metrics developed for assessing the explainability of CAM representations, allowing scalability to larger image sets without visual inspection.

In terms of future research, the authors intend to integrate Vision Transformers and attention mechanisms for semantic segmentation into the CRFFg-based representation [58]. Besides, we propose to include variational autoencoders and transfer learning strategies within our framework to prevent overfitting and enhance data interpretability [33,59].

Author Contributions: Conceptualization, J.C.A.-A., A.M.Á.-M. and G.C.-D.; methodology, J.C.A.-A., A.M.Á.-M. and G.C.-D.; software, J.C.A.-A.; validation, J.C.A.-A., A.M.Á.-M. and G.C.-D.; formal analysis, J.C.A.-A. and G.C.-D.; investigation, J.C.A.-A., A.M.Á.-M. and G.C.-D.; resources, A.M.Á.-M. and G.C.-D.; data curation, J.C.A.-A.; writing–original draft preparation, J.C.A.-A., A.M.Á.-M. and G.C.-D.; writing–review and editing, A.M.Á.-M. and G.C.-D.; visualization, J.C.A.-A.; supervision, A.M.Á.-M. and G.C.-D.; project administration, A.M.Á.-M.; funding acquisition, A.M.Á.-M. and G.C.-D. All authors have read and agreed to the published version of the manuscript.

Funding: Under grants provided by the projects: "Prototipo de visión por computador para la identificación de problemas fitosanitarios en cultivos de plátano en el departamento de Caldas" (Hermes 51175) funded by Universidad Nacional de Colombia, and "Desarrollo de una herramienta de visión por computador para el análisis de plantas orientado al fortalecimiento de la seguridad alimentaria" (Hermes 54339) funded by Universidad Nacional de Colombia and Universidad de Caldas.

Institutional Review Board Statement: This study uses anonymized public datasets with institutional review board statement as presented in https://gcpds-image-segmentation.readthedocs.io/en/latest/notebooks/02-datasets.html (accessed on 5 April 2023).

Informed Consent Statement: This study uses anonymized public datasets as presented in https://gcpds-image-segmentation.readthedocs.io/en/latest/notebooks/02-datasets.html (accessed on 5 April 2023).

Data Availability Statement: Dataset is publicly available at: https://gcpds-image-segmentation.readthedocs.io/en/latest/notebooks/02-datasets.html (accessed on 5 April 2023).

Conflicts of Interest: The authors declare no conflict of interest.

Appendix A. Method Comparison from Absolute Semantic Segmentation Performances

Table A1 presents the absolute semantic segmentation results acquired from thermalFeet database. For clarity, the rank position of each method is also included. As can be seen, our enhancement based on CRFFg boosts the segmentation performance. Notably, ResUNet CRFFg S-M1 outperforms the tested approaches concerning the measured quantitative assessments.

Table A1. Absolute Semantic Segmentation Results on thermalFeet database. WODA: Without Data Augmentation, WDA: With Data Augmentation. The average ± standard deviation performance is displayed regarding the test partitions. M1 and M3 stand for CRFFg's dimension as a multiplication factor of the enhanced layer's size.

Approach	Measure	WODA	Rank	WDA	Rank
FCN	Dice	0.9527 ± 0.0238	3.0	0.8646 ± 0.0624	10.0
	Jaccard	0.9106 ± 0.0424	3.0	0.7668 ± 0.0969	10.0
	Sensitivity	0.9352 ± 0.0482	4.0	0.8260 ± 0.1098	6.0
	Specificity	0.9857 ± 0.0105	7.0	0.9697 ± 0.0186	13.0
FCN CRFFg S-M1	Dice	0.9530 ± 0.0257	2.0	0.8510 ± 0.0623	12.0
	Jaccard	0.9113 ± 0.0456	2.0	0.7456 ± 0.0913	12.0
	Sensitivity	0.9424 ± 0.0526	3.0	0.8016 ± 0.0999	13.0
	Specificity	0.9810 ± 0.0158	12.0	0.9697 ± 0.0233	14.0
FCN CRFFg S-M3	Dice	0.9480 ± 0.0224	5.0	0.8346 ± 0.0916	15.0
	Jaccard	0.9021 ± 0.0403	5.0	0.7262 ± 0.1284	15.0
	Sensitivity	0.9340 ± 0.0423	6.0	0.7771 ± 0.1325	15.0
	Specificity	0.9804 ± 0.0168	13.0	0.9714 ± 0.0246	10.0
FCN S-M1	Dice	0.9469 ± 0.0273	6.0	0.8421 ± 0.0870	14.0
	Jaccard	0.9003 ± 0.0486	6.0	0.7367 ± 0.1254	14.0
	Sensitivity	0.9286 ± 0.0518	7.0	0.7867 ± 0.1422	14.0
	Specificity	0.9843 ± 0.0109	9.0	0.9714 ± 0.0207	9.0
FCN S-M3	Dice	0.9519 ± 0.0281	4.0	0.8470 ± 0.0737	13.0
	Jaccard	0.9096 ± 0.0499	4.0	0.7414 ± 0.1070	13.0
	Sensitivity	0.9341 ± 0.0543	5.0	0.8160 ± 0.1152	9.0
	Specificity	0.9865 ± 0.0107	6.0	0.9604 ± 0.0300	15.0
ResUNet	Dice	0.9348 ± 0.0502	11.0	0.8569 ± 0.0779	11.0
	Jaccard	0.8816 ± 0.0868	11.0	0.7575 ± 0.1152	11.0
	Sensitivity	0.9029 ± 0.0825	12.0	0.8152 ± 0.1316	11.0
	Specificity	0.9896 ± 0.0067	2.0	0.9712 ± 0.0180	12.0
ResUNet CRFFg S-M1	Dice	0.9456 ± 0.0317	7.0	0.8851 ± 0.0449	4.0
	Jaccard	0.8984 ± 0.0560	7.0	0.7968 ± 0.0709	4.0
	Sensitivity	0.9472 ± 0.0540	1.0	0.8283 ± 0.0853	5.0
	Specificity	0.9725 ± 0.0230	14.0	0.9841 ± 0.0123	3.0
ResUNet CRFFg S-M3	Dice	0.9111 ± 0.0602	15.0	0.8969 ± 0.0444	**1.0**
	Jaccard	0.8420 ± 0.0951	15.0	0.8160 ± 0.0737	1.0
	Sensitivity	0.9075 ± 0.0607	11.0	0.8675 ± 0.0803	1.0
	Specificity	0.9663 ± 0.0346	15.0	0.9712 ± 0.0244	11.0
ResUNet S-M1	Dice	0.9558 ± 0.0279	1.0	0.8865 ± 0.0676	3.0
	Jaccard	0.9167 ± 0.0498	1.0	0.8026 ± 0.1061	3.0
	Sensitivity	0.9459 ± 0.0482	2.0	0.8403 ± 0.1123	2.0
	Specificity	0.9831 ± 0.0152	10.0	0.9750 ± 0.0287	8.0
ResUNet S-M3	Dice	0.9237 ± 0.0411	14.0	0.8677 ± 0.0894	9.0
	Jaccard	0.8610 ± 0.0713	14.0	0.7763 ± 0.1281	9.0
	Sensitivity	0.8875 ± 0.0756	14.0	0.8179 ± 0.1333	8.0
	Specificity	0.9846 ± 0.0128	8.0	0.9755 ± 0.0217	7.0
U-Net	Dice	0.9371 ± 0.0312	10.0	0.8713 ± 0.0756	8.0
	Jaccard	0.8832 ± 0.0551	10.0	0.7796 ± 0.1145	8.0
	Sensitivity	0.9120 ± 0.0571	10.0	0.8107 ± 0.1248	12.0
	Specificity	0.9811 ± 0.0199	11.0	0.9847 ± 0.0130	2.0
U-Net CRFFg S-M1	Dice	0.9448 ± 0.0297	8.0	0.8827 ± 0.0617	5.0
	Jaccard	0.8969 ± 0.0528	8.0	0.7954 ± 0.0965	5.0
	Sensitivity	0.9160 ± 0.0561	9.0	0.8383 ± 0.1062	4.0
	Specificity	0.9902 ± 0.0057	1.0	0.9780 ± 0.0124	5.0

Table A1. Cont.

Approach	Measure	WODA	Rank	WDA	Rank
U-Net CRFFg S-M3	Dice	0.9252 ± 0.0404	13.0	0.8821 ± 0.0645	6.0
	Jaccard	0.8634 ± 0.0694	13.0	0.7948 ± 0.1004	6.0
	Sensitivity	0.8831 ± 0.0730	15.0	0.8231 ± 0.1110	7.0
	Specificity	0.9893 ± 0.0066	3.0	0.9873 ± 0.0088	1.0
U-Net S-M1	Dice	0.9400 ± 0.0364	9.0	0.8898 ± 0.0536	2.0
	Jaccard	0.8890 ± 0.0635	9.0	0.8056 ± 0.0861	2.0
	Sensitivity	0.9162 ± 0.0619	8.0	0.8384 ± 0.0904	3.0
	Specificity	0.9866 ± 0.0086	5.0	0.9777 ± 0.0208	6.0
U-Net S-M3	Dice	0.9293 ± 0.0419	12.0	0.8767 ± 0.0772	7.0
	Jaccard	0.8707 ± 0.0728	12.0	0.7883 ± 0.1152	7.0
	Sensitivity	0.8934 ± 0.0792	13.0	0.8152 ± 0.1181	10.0
	Specificity	0.9878 ± 0.0098	4.0	0.9805 ± 0.0189	4.0

References

1. Brown, D.T.; Wildsmith, J.A.W.; Covino, B.G.; Scott, D.B. Effect of Baricity on Spinal Anaesthesia with Amethocaine. *BJA Br. J. Anaesth.* **1980**, *52*, 589–596. [CrossRef] [PubMed]
2. McCombe, K.; Bogod, D. Regional anaesthesia: Risk, consent and complications. *Anaesthesia* **2021**, *76*, 18–26. [CrossRef] [PubMed]
3. Chae, Y.; Park, H.J.; Lee, I.S. Pain modalities in the body and brain: Current knowledge and future perspectives. *Neurosci. Biobehav. Rev.* **2022**, *139*, 104744. . [CrossRef] [PubMed]
4. Curatolo, M.; Petersen-Felix, S.; Arendt-Nielsen, L. Assessment of regional analgesia in clinical practice and research. *Br. Med Bull.* **2005**, *71*, 61–76. [CrossRef]
5. Bruins, A.; Kistemaker, K.; Boom, A.; Klaessens, J.; Verdaasdonk, R.; Boer, C. Thermographic skin temperature measurement compared with cold sensation in predicting the efficacy and distribution of epidural anesthesia. *J. Clin. Monit. Comput.* **2018**, *32*, 335–341. [CrossRef] [PubMed]
6. Haren, F.; Kadic, L.; Driessen, J. Skin temperature measured by infrared thermography after ultrasound-guided blockade of the sciatic nerve. *Acta Anaesthesiol. Scand.* **2013**, *57*. [CrossRef]
7. Stevens, M.F.; Werdehausen, R.; Hermanns, H.; Lipfert, P. Skin temperature during regional anesthesia of the lower extremity. *Anesth. Analg.* **2006**, *102*, 1247–1251. [CrossRef]
8. Werdehausen, R.; Braun, S.; Hermanns, H.; Freynhagen, R.; Lipfert, P.; Stevens, M.F. Uniform Distribution of Skin-Temperature Increase After Different Regional-Anesthesia Techniques of the Lower Extremity. *Reg. Anesth. Pain Med.* **2007**, *32*, 73–78. [CrossRef]
9. Zhang, L.; Nan, Q.; Bian, S.; Liu, T.; Xu, Z. Real-time segmentation method of billet infrared image based on multi-scale feature fusion. *Sci. Rep.* **2022**, *12*, 6879. [CrossRef]
10. Kütük, Z.; Algan, G. Semantic Segmentation for Thermal Images: A Comparative Survey. In Proceedings of the IEEE/CVF Conference on Computer Vision and Pattern Recognition (CVPR) Workshops, New Orleans, LA, USA, 19–20 June 2022. [CrossRef]
11. Long, J.; Shelhamer, E.; Darrell, T. Fully Convolutional Networks for Semantic Segmentation. *arXiv* **2014**, arXiv:1411.4038. [CrossRef]
12. Bi, L.; Kim, J.; Kumar, A.; Fulham, M.J.; Feng, D. Stacked fully convolutional networks with multi-channel learning: Application to medical image segmentation. *Vis. Comput.* **2017**, *33*, 1061–1071. [CrossRef]
13. Ronneberger, O.; Fischer, P.; Brox, T. U-Net: Convolutional Networks for Biomedical Image Segmentation. *arXiv* **2015**, arXiv:1505.04597. [CrossRef]
14. Kumar, V.; Webb, J.M.; Gregory, A.; Denis, M.; Meixner, D.D.; Bayat, M.; Whaley, D.H.; Fatemi, M.; Alizad, A. Automated and real-time segmentation of suspicious breast masses using convolutional neural network. *PLoS ONE* **2018**, *13*, e0195816. [CrossRef] [PubMed]
15. Zhou, Z.; Siddiquee, M.M.R.; Tajbakhsh, N.; Liang, J. UNet++: A Nested U-Net Architecture for Medical Image Segmentation. *arXiv* **2018**, arXiv:1807.10165. [CrossRef]
16. Badrinarayanan, V.; Kendall, A.; Cipolla, R. SegNet: A Deep Convolutional Encoder–decoder Architecture for Image Segmentation. *arXiv* **2016**, arXiv:1511.00561. [CrossRef]
17. He, K.; Gkioxari, G.; Dollár, P.; Girshick, R. Mask R-CNN. *arXiv* **2018**, arXiv:1703.06870. [CrossRef]
18. Ren, S.; He, K.; Girshick, R.; Sun, J. Faster R-CNN: Towards Real-Time Object Detection with Region Proposal Networks. *arXiv* **2016**, arXiv:1506.01497. [CrossRef]
19. Zhao, H.; Shi, J.; Qi, X.; Wang, X.; Jia, J. Pyramid Scene Parsing Network. *arXiv* **2017**, arXiv:1612.01105. [CrossRef]
20. Arteaga-Marrero, N.; Hernández, A.; Villa, E.; González-Pérez, S.; Luque, C.; Ruiz-Alzola, J. Segmentation Approaches for Diabetic Foot Disorders. *Sensors* **2021**, *21*, 934. [CrossRef]

21. Fischler, M.A.; Bolles, R.C. Random Sample Consensus: A Paradigm for Model Fitting with Applications to Image Analysis and Automated Cartography. *Commun. ACM* **1981**, *24*, 381–395. [CrossRef]
22. Bouallal, D.; Bougrine, A.; Douzi, H.; Harba, R.; Canals, R.; Vilcahuaman, L.; Arbanil, H. Segmentation of plantar foot thermal images: Application to diabetic foot diagnosis. In Proceedings of the 2020 International Conference on Systems, Signals and Image Processing (IWSSIP), Niteroi, Brazil, 1–3 July 2020; pp. 116–121. [CrossRef]
23. Bougrine, A.; Harba, R.; Canals, R.; Ledee, R.; Jabloun, M. On the segmentation of plantar foot thermal images with deep learning. In Proceedings of the 2019 27th European Signal Processing Conference (EUSIPCO), A Coruna, Spain, 2–6 September 2019. [CrossRef]
24. Mejia-Zuluaga, R.; Aguirre-Arango, J.C.; Collazos-Huertas, D.; Daza-Castillo, J.; Valencia-Marulanda, N.; Calderón-Marulanda, M.; Aguirre-Ospina, Ó.; Alvarez-Meza, A.; Castellanos-Dominguez, G. Deep Learning Semantic Segmentation of Feet Using Infrared Thermal Images. In Proceedings of the Advances in Artificial Intelligence—IBERAMIA, Cartagena de Indias, Colombia, 23–25 November 2022; Bicharra Garcia, A.C., Ferro, M., Rodríguez Ribón, J.C., Eds.; Springer International Publishing: Cham, Switzerland, 2022; pp. 342–352.
25. Dosovitskiy, A.; Beyer, L.; Kolesnikov, A.; Weissenborn, D.; Zhai, X.; Unterthiner, T.; Dehghani, M.; Minderer, M.; Heigold, G.; Gelly, S.; et al. An Image is Worth 16x16 Words: Transformers for Image Recognition at Scale. *arXiv* **2020**, arXiv:2010.11929. [CrossRef].
26. Chen, J.; Lu, Y.; Yu, Q.; Luo, X.; Adeli, E.; Wang, Y.; Lu, L.; Yuille, A.L.; Zhou, Y. TransUNet: Transformers Make Strong Encoders for Medical Image Segmentation. *arXiv* **2021**, arXiv:2102.04306. [CrossRef].
27. Cao, H.; Wang, Y.; Chen, J.; Jiang, D.; Zhang, X.; Tian, Q.; Wang, M. Swin-Unet: Unet-like Pure Transformer for Medical Image Segmentation. In *Computer Vision—ECCV 2022 Workshops*; Springer: Cham, Switzerland, 2021; pp. 205–218. [CrossRef]
28. Zhang, Y.; Liu, H.; Hu, Q. TransFuse: Fusing Transformers and CNNs for Medical Image Segmentation. In *Medical Image Computing and Computer Assisted Intervention—MICCAI 2021*; Lecture Notes in Computer Science; Springer: Cham, Switzerland, 2021; pp. 14–24. [CrossRef]
29. Li, S.; Sui, X.; Luo, X.; Xu, X.; Liu, Y.; Goh, R. Medical Image Segmentation Using Squeeze-and-Expansion Transformers. *Ijcai Int. Jt. Conf. Artif. Intell.* **2021**, 807–815. [CrossRef]
30. Luo, X.; Hu, M.; Song, T.; Wang, G.; Zhang, S. Semi-Supervised Medical Image Segmentation via Cross Teaching between CNN and Transformer. In Proceedings of the International Conference on Medical Imaging with Deep Learning, Zurich, Germany, 6–8 July 2022; pp. 820–833.
31. Kirillov, A.; Mintun, E.; Ravi, N.; Mao, H.; Rolland, C.; Gustafson, L.; Xiao, T.; Whitehead, S.; Berg, A.C.; Lo, W.Y.; et al. Segment Anything. *arXiv* **2023**, arXiv:2304.02643. [CrossRef].
32. Ruan, B.K.; Shuai, H.H.; Cheng, W.H. Vision Transformers: State of the Art and Research Challenges. *arXiv* **2022**, arXiv:2207.03041. [CrossRef].
33. Karimi, D.; Warfield, S.K.; Gholipour, A. Critical Assessment of Transfer Learning for Medical Image Segmentation with Fully Convolutional Neural Networks. *arXiv* **2022**, arXiv:2006.00356. [CrossRef].
34. Alzubaidi, L.; Fadhel, M.A.; Al-Shamma, O.; Zhang, J.; Santamaría, J.; Duan, Y.; R. Oleiwi, S. Towards a better understanding of transfer learning for medical imaging: A case study. *Appl. Sci.* **2020**, *10*, 4523. [CrossRef]
35. Guan, H.; Liu, M. Domain Adaptation for Medical Image Analysis: A Survey. *IEEE Trans. Biomed. Eng.* **2022**, *69*, 1173–1185. [CrossRef]
36. Anas, E.M.A.; Nouranian, S.; Mahdavi, S.S.; Spadinger, I.; Morris, W.J.; Salcudean, S.E.; Mousavi, P.; Abolmaesumi, P. Clinical Target-Volume Delineation in Prostate Brachytherapy Using Residual Neural Networks. In Proceedings of the Medical Image Computing and Computer Assisted Intervention—MICCAI 2017, Quebec City, QC, Canada, 11–13 September 2017; Descoteaux, M., Maier-Hein, L., Franz, A., Jannin, P.; Collins, D.L.; Duchesne, S., Eds.; Springer International Publishing: Cham, Switzerland, 2017; pp. 365–373.
37. Zhou, B.; Khosla, A.; Lapedriza, A.; Oliva, A.; Torralba, A. Learning Deep Features for Discriminative Localization. *arXiv* **2015**, arXiv:1512.04150. [CrossRef].
38. Zhang, A.; Lipton, Z.C.; Li, M.; Smola, A.J. Dive into deep learning. *arXiv Prepr.* **2021**, arXiv:2106.11342.
39. Rahimi, A.; Recht, B. Random features for large-scale kernel machines. *Adv. Neural Inf. Process. Syst.* **2007**, *20*, 1–8.
40. Rudin, W., Bers, L., Courant, R., Stoker, J. J.; Henney, D. R. *Fourier Analysis on Groups*; Interscience Tracts in Pure and Applied Mathematics; Interscience: New York, NY, USA, 1976; p. 19.
41. Álvarez-Meza, A.M.; Cárdenas-Pe na, D.; Castellanos-Dominguez, G. Unsupervised Kernel Function Building Using Maximization of Information Potential Variability. In Proceedings of the Progress in Pattern Recognition, Image Analysis, Computer Vision, and Applications, Puerto Vallarta, Mexico, 2–5 November 2014 ; Bayro-Corrochano, E., Hancock, E., Eds.; Springer International Publishing: Cham, Switzerland, 2014; pp. 335–342.
42. Bronstein, M.M.; Bruna, J.; Cohen, T.; Velickovic, P. Geometric Deep Learning: Grids, Groups, Graphs, Geodesics, and Gauges. *arXiv* **2021**, arXiv:2104.13478, [CrossRef].
43. Jiang, P.T.; Zhang, C.B.; Hou, Q.; Cheng, M.M.; Wei, Y. LayerCAM: Exploring hierarchical class activation maps for localization. *IEEE Trans. Image Process.* **2021**, *30*, 5875–5888. [CrossRef] [PubMed]

44. Jimenez-Casta no, C.A.; Álvarez-Meza, A.M.; Aguirre-Ospina, O.D.; Cárdenas-Pe na, D.A.; Orozco-Gutiérrez, Á.A. Random fourier features-based deep learning improvement with class activation interpretability for nerve structure segmentation. *Sensors* **2021**, *21*, 7741. [CrossRef]
45. Galvin, E.M.; Niehof, S.; Medina, H.J.; Zijlstra, F.J.; van Bommel, J.; Klein, J.; Verbrugge, S.J.C. Thermographic temperature measurement compared with pinprick and cold sensation in predicting the effectiveness of regional blocks. *Anesth. Analg.* **2006**, *102*, 598–604. [CrossRef] [PubMed]
46. Chestnut, D.H.; Wong, C.A.; Tsen, L.C.; Kee, W.M.D.N.; Beilin, Y.; Mhyre, J. *Chestnut's Obstetric Anesthesia: Principles and Practice E-Book: Expert Consult-Online and Print*; Saunders: Philadelphia, PA, USA, 2014.
47. Asghar, S.; Lundstrøm, L.H.; Bjerregaard, L.S.; Lange, K.H.W. Ultrasound-guided lateral infraclavicular block evaluated by infrared thermography and distal skin temperature. *Acta Anaesthesiol. Scand.* **2014**, *58*, 867–874. [CrossRef]
48. Lange, K.H.; Jansen, T.; Asghar, S.; Kristensen, P.; Skjønnemand, M.; Nørgaard, P. Skin temperature measured by infrared thermography after specific ultrasound-guided blocking of the musculocutaneous, radial, ulnar, and median nerves in the upper extremity. *Br. J. Anaesth.* **2011**, *106*, 887–895. [CrossRef]
49. Simonyan, K.; Zisserman, A. Very Deep Convolutional Networks for Large-Scale Image Recognition. *arXiv* **2014**, arXiv:1409.1556. [CrossRef].
50. Ibtehaz, N.; Rahman, M.S. MultiResUNet: Rethinking the U-Net architecture for multimodal biomedical image segmentation. *Neural Netw.* **2020**, *121*, 74–87. [CrossRef]
51. Zhou, Z.; Siddiquee, M.M.R.; Tajbakhsh, N.; Liang, J. UNet++: A Nested U-Net Architecture for Medical Image Segmentation. In *DLMIA 2018, ML-CDS 2018: Deep Learning in Medical Image Analysis and Multimodal Learning for Clinical Decision Support*; Lecture Notes in Computer Science; Springer International Publishing: Cham, Switzerland, 2018; pp. 3–11. [CrossRef]
52. Wang, X.; Wang, L.; Zhong, X.; Bai, C.; Huang, X.; Zhao, R.; Xia, M. PaI-Net: A modified U-Net of reducing semantic gap for surgical instrument segmentation. *IET Image Process.* **2021**, *15*, 2959–2969. [CrossRef]
53. Haji, S.H.; Abdulazeez, A.M. Comparison of optimization techniques based on gradient descent algorithm: A review. *PalArch J. Archaeol. Egypt/Egyptol.* **2021**, *18*, 2715–2743.
54. Van der Walt, S.; Schönberger, J.L.; Nunez-Iglesias, J.; Boulogne, F.; Warner, J.D.; Yager, N.; Gouillart, E.; Yu, T. scikit-image: Image processing in Python. *PeerJ* **2014**, *2*, e453. [CrossRef]
55. Nakkiran, P.; Kaplun, G.; Bansal, Y.; Yang, T.; Barak, B.; Sutskever, I. Deep Double Descent: Where Bigger Models and More Data Hurt. *arXiv* **2019**, arXiv:1912.02292. [CrossRef].
56. Heckel, R.; Yilmaz, F.F. Early Stopping in Deep Networks: Double Descent and How to Eliminate it. *arXiv*, **2020**, arXiv:2007.10099. [CrossRef].
57. Zhou, Z.; Siddiquee, M.M.R.; Tajbakhsh, N.; Liang, J. UNet++: Redesigning Skip Connections to Exploit Multiscale Features in Image Segmentation. *IEEE Trans. Med. Imaging* **2020**, *39*, 1856–1867. [CrossRef] [PubMed]
58. Peng, H.; Pappas, N.; Yogatama, D.; Schwartz, R.; Smith, N.A.; Kong, L. Random feature attention. *arXiv* **2021**, arXiv:2103.02143.
59. Nguyen, T.P.; Pham, T.T.; Nguyen, T.; Le, H.; Nguyen, D.; Lam, H.; Nguyen, P.; Fowler, J.; Tran, M.T.; Le, N. EmbryosFormer: Deformable Transformer and Collaborative Encoding-Decoding for Embryos Stage Development Classification. In Proceedings of the Proceedings of the IEEE/CVF Winter Conference on Applications of Computer Vision, Waikola, HI, USA, 2–7 January 2023; pp. 1981–1990.

Disclaimer/Publisher's Note: The statements, opinions and data contained in all publications are solely those of the individual author(s) and contributor(s) and not of MDPI and/or the editor(s). MDPI and/or the editor(s) disclaim responsibility for any injury to people or property resulting from any ideas, methods, instructions or products referred to in the content.

Article

Regional Contribution in Electrophysiological-Based Classifications of Attention Deficit Hyperactive Disorder (ADHD) Using Machine Learning

Nishant Chauhan and Byung-Jae Choi *

Department of Electronic Engineering, Daegu University, Gyeongsan 38453, Republic of Korea
* Correspondence: bjchoi@daegu.ac.kr

Abstract: Attention deficit hyperactivity disorder (ADHD) is a common neurodevelopmental condition in children and is characterized by challenges in maintaining attention, hyperactivity, and impulsive behaviors. Despite ongoing research, we still do not fully understand what causes ADHD. Electroencephalography (EEG) has emerged as a valuable tool for investigating ADHD-related neural patterns due to its high temporal resolution and non-invasiveness. This study aims to contribute to diagnostic accuracy by leveraging EEG data to classify children with ADHD and healthy controls. We used a dataset containing EEG recordings from 60 children with ADHD and 60 healthy controls. The EEG data were captured during cognitive tasks and comprised signals from 19 channels across the scalp. Our primary objective was to develop a machine learning model capable of distinguishing ADHD subjects from controls using EEG data as discriminatory features. We employed several well-known classifiers, including a support vector machine, random forest, decision tree, AdaBoost, Naive Bayes, and linear discriminant analysis, to discern distinctive EEG patterns. To further enhance classification accuracy, we explored the impact of regional data on the classification outcomes. We arranged the EEG data according to the brain regions from which they were derived (namely frontal, temporal, central, parietal, and occipital) and examined their collective effects on the accuracy of our classifications. Notably, we considered combinations of three regions at a time and found that certain combinations led to enhanced accuracy. Our findings underscore the potential of EEG-based classification in distinguishing children with ADHD from healthy controls. The Naive Bayes classifier yielded the highest accuracy (84%) when applied to specific region combinations. Moreover, we evaluated the classification performance based on hemisphere-specific EEG data and found promising results, particularly when using the right hemisphere region channels.

Keywords: electrophysiological (EEG) signals; support vector machine; random forest; decision tree; AdaBoost; Naive Bayes; linear discriminant analysis (LDA); machine learning; attention deficit hyperactivity disorder (ADHD)

1. Introduction

Attention deficit hyperactivity disorder (ADHD) is a neurodevelopmental condition prevalent in children that often persists into adulthood [1,2]. Global studies indicate a prevalence of approximately 5–12% among school-going children, with a higher prevalence observed in males [3]. This disorder encompasses three subtypes (predominantly inattentive, predominantly hyperactive–impulsive, and combined) and is characterized by primary symptoms such as inattention, impulsivity, and hyperactivity [4]. Early detection and intervention can greatly benefit children with ADHD, their parents, and their communities. Presently, clinical interviews, observations, and ratings from various sources, including parents and teachers, are employed for diagnosis [5]. Traditional clinical procedures are time-consuming and prone to ambiguity, underscoring the need for objective diagnostic methods based on biological signals that reflect ADHD behaviors.

Electroencephalography (EEG) has emerged as a valuable tool for investigating ADHD-related neural patterns due to its non-invasiveness and high temporal resolution [6]. EEG signals have been utilized in the diagnosis of various neurological disorders by extracting distinctive features and employing different classifiers in automated detection systems. EEG signals have been employed for the automatic detection of neurophysiological conditions such as alcoholism [7], dementia [8], epilepsy [9], schizophrenia [10], Parkinson's disease [11], and depressive disorders [12]. The EEG patterns of children with ADHD display differences in complexity, randomness, amplitude, and frequency compared with those of typically developing children. Researchers have applied diverse feature extraction techniques and classifiers to utilize EEG signals for ADHD identification [13–15].

EEG is widely recognized as a functional imaging technique that measures the electrical activity of the brain. It offers valuable insights into neural processes, cognitive functions, and neurological disorders through the recording of voltage fluctuations produced by neuronal activity. This functional imaging modality provides real-time information on brain dynamics, making it an essential tool in neuroscience research [16]. EEG is increasingly acknowledged as a functional imaging modality, and researchers have capitalized on its exceptional temporal resolution to reveal swift alterations in neural dynamics [17]. Its versatility is evident in its wide-ranging applications, which include obtaining real-time neurofeedback for anesthesia optimization [18] and investigating spatiotemporal dynamics in functional connectivity [19]. Notably, ongoing discussions surround the classification of EEG as a functional imaging technique because of its unique temporal capabilities that enhance our understanding of neural processes and complement other imaging methods. This article aims to contribute to this ongoing discourse by offering a comprehensive investigation that employs EEG in the domain of ADHD classification.

The accessibility of machine learning models has spurred interest in their application to psychiatric disorder research. Machine learning models, which are mathematical constructs capable of learning intricate patterns within existing datasets, have the potential to predict outcomes in new datasets and emphasize key variables in such predictions [16]. For instance, in [20], SVM was utilized to classify ADHD based on EEG signals, and this yielded promising results in distinguishing between ADHD and healthy controls. Our proposed method builds upon this foundation by not only employing SVM but also integrating a comprehensive feature-selection strategy based on distinct brain regions. This enables our model to capture intricate neural patterns that could be missed using a single algorithm. Similarly, [21] employed RF and DT algorithms to identify ADHD in children using event-related potentials (ERPs), highlighting the potential of these classifiers to capture distinctive neural patterns linked to the disorder. We extend this approach by incorporating both RF and DT classifiers within a unified framework and thereby enhancing the robustness of our classification model. Neural networks were used to diagnose ADHD from EEG data in [22]. In contrast, our method offers interpretable insights by considering specific brain regions, thus making it more transparent and clinically applicable. A systematic review in [23] emphasized the application of SVM, RF, and DT in the EEG-signal-based analysis of mental tasks, further reinforcing their relevance in ADHD classification studies. Notably, the integration of electrodermal activity (EDA) features in classification models has shown potential [24]; SVMs have been employed in a multiclass brain–computer interface classification study, underscoring their ability to enhance ADHD classification accuracy. Our approach, however, surpasses this by focusing solely on EEG data, which are more directly relevant to the neural patterns associated with ADHD.

In this paper, we focus on exploring the significance and contribution of different brain regions in classifying children with ADHD and healthy controls. Various brain regions exhibit distinct electrical activity levels, influenced by EEG acquisition conditions and inter-regional connectivity. The frontal region is essential for attention and concentration, reasoning, and judgment. The parietal and central regions process senses and motor movements, the temporal region is responsible for memory and language understanding, and the occipital lobe governs vision and object recognition. Additionally, the brain's two hemispheres, left and right, control different functions. Individuals with ADHD often exhibit information-processing deficits in the right hemisphere, leading to self-reported inattention symptoms. Our research intends to investigate the impact of particular brain regions, interhemispheric regions, or combinations of these regions on obtaining significant accuracy rates in distinguishing ADHD patients from healthy controls.

Using EEG data from both ADHD patients and a healthy control group, we employ various machine learning models, including support vector machine (SVM), random forest (RF), decision tree (DT), AdaBoost, Naive Bayes, and linear discriminant analysis (LDA). Our investigation explores distinctive EEG patterns linked to ADHD, investigating the impact of individual brain regions and their combinations on achieving high classification accuracy. By investigating regional differences in EEG activity between patients with ADHD and healthy controls, this study aims to enhance our understanding of this disorder.

2. Materials and Methods

Figure 1 illustrates the framework for the machine learning-based classification of ADHD. To conduct a comparative analysis, various machine learning algorithms were applied to state-of-the-art ADHD datasets. The experimentation was performed using the Python language, utilizing the Scikit-learn library for implementing machine learning models. Specifically, six ML algorithms, namely SVM [25], RF [26], DT [27], AdaBoost [28], Naive Bayes [29], and LDA [30] were employed to the ADHD dataset. Extensive experimentation was carried out to achieve two main objectives:

1. Our goal is to highlight brain regions or combinations that are crucial for accurately classifying ADHD and healthy controls;
2. We aim to identify the optimal machine learning algorithm that effectively utilizes regional combinations of EEG data for enhanced ADHD classification accuracy.

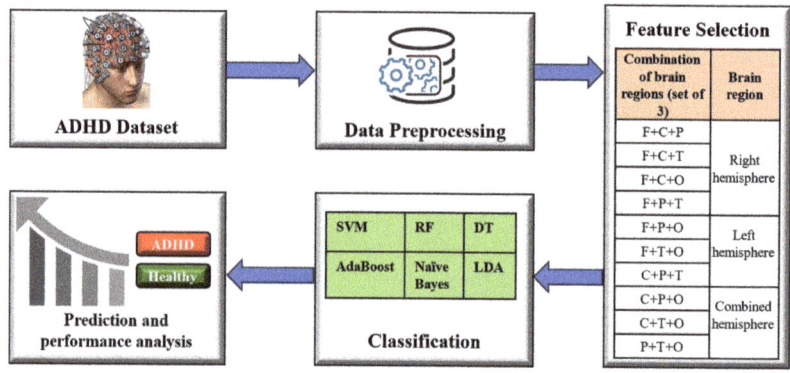

Figure 1. Classification Framework for ADHD and healthy controls.

2.1. Dataset

In this study, we utilized a publicly available EEG dataset accessed on 8 August 2023, from the following source: https://ieee-dataport.org/open-access/eegdata-adhd-control-children) [31].

This dataset comprises EEG recordings from two groups: 61 children diagnosed with ADHD (48 boys and 13 girls, average age of 9.61 ± 1.75 years) and 60 healthy children (50 boys and 10 girls, average age of 9.85 ± 1.77 years). ADHD diagnoses were made by an experienced child and adolescent psychiatrist based on DSM-IV criteria [32]. Healthy children from elementary schools were recruited, excluding those with a history of significant neurological conditions, brain injuries, major medical ailments, learning or speech difficulties, other psychiatric disorders, or use of benzodiazepine or barbiturate medications.

During data collection, EEG recordings were taken at a sampling frequency of 128 Hz. Given the challenge of visual attention deficits in ADHD children, EEG data were captured during visual attention tasks. The children were presented with a series of cartoon character images and asked to count the characters. Each image contained a random assortment of characters ranging from 0 to 16, and the images were designed to be sufficiently large for easy viewing and counting. Images were rapidly displayed following the child's response, ensuring a consistent stimulus presentation. Consequently, the duration of EEG recordings varied based on each child's performance in this cognitive visual task, specifically their response speed.

The EEG data were acquired using an SD-C24 machine with 19 channels, following the 10–20 electrode placement system. Figure 2 illustrates the 10–20 standard electrode positions for EEG recording [33]. The 10–20 standard system is a widely employed methodology for the consistent placement of electrodes during electroencephalography (EEG) recordings. This system involves dividing the scalp into specific regions and designating unique labels to key points based on a standardized percentage of distances between prominent anatomical landmarks, such as the nasion and inion. Electrodes are positioned precisely at these predetermined locations, ensuring uniform and reproducible EEG signal collection from diverse brain regions. The nomenclature "10–20" denotes the standardized distances of either 10% or 20% between electrode placements, creating a systematic grid that fosters consistent electrode positioning across different subjects and research settings. This approach not only guarantees methodological rigor in data acquisition but also facilitates meaningful cross-subject and cross-study comparisons, making it an integral tool in the field of EEG-based research. The topoplot (topographic map of a scalp) was generated using the MNE-python software package [34]. Different color shades depict the five brain regions: frontal, central, occipital, temporal, and parietal, each corresponding to specific channels.

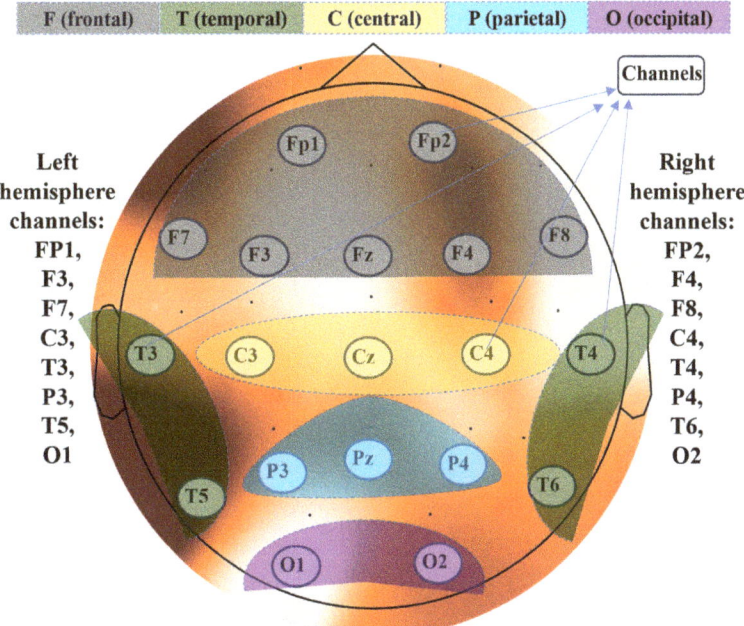

Figure 2. EEG electrode positions according to the 10–20 standard system. Different shades indicate distinct regions and their corresponding channels. The figure displays the left and right hemispheres along with the channels included within each hemisphere.

2.2. Data Preprocessing

EEG signals are prone to the presence of numerous artifacts and sources of noise, necessitating their thorough preprocessing before they can be reliably utilized in analyses. The datasets employed in this study were preprocessed in advance to mitigate such concerns. A comprehensive breakdown of the preprocessing procedures is provided below [35].

The dataset owners used a modified iteration of Makoto's preprocessing pipeline, adjusted for use with EEGLab functions (version 14.1.1; Delorme & Makeig, 2004, San Diego, CA, USA) and executed using MATLAB 2018a. Initially, artifacts stemming from eye movements and muscle activity were carefully removed through manual visual inspection. Channels containing erroneous or inaccurately captured data were excluded, and interpolation using neighboring channel signals was employed to restore the missing information. To eliminate artifacts, a band-pass finite impulse response (FIR) filter spanning from 0.5 Hz to 48 Hz was administered to the continuous EEG data. Subsequently, the CleanLine plugin was applied to further suppress line noise. To enhance artifact rejection, the EEG data underwent decomposition using independent component analysis (ICA). Independent components (ICs) associated with eye blinks and muscle artifacts were identified based on their spectral properties, scalp maps, and temporal characteristics. These components were manually excluded. Following a thorough cleaning, the time series were filtered across classical EEG frequency bands [delta (0.5–4 Hz), theta (4–8 Hz), alpha (8–13 Hz), beta (13–30 Hz), and lower gamma (30–45 Hz)] using a zero-phase shift FIR filter that maintains phase coherence. Throughout the filtering stages, a zero-phase Hamming-windowed filter was incorporated through the first plugin from EEGLab, featuring a -6 dB cutoff frequency.

For each subject, the time series were segmented into 1024 sample (8 s) segments, albeit with variable counts due to task-specific timings. The minimum task duration was 50 s for a participant from the control group, while the maximum duration reached 285 s for a subject with ADHD. The mean segment count stood at 13.18 (standard deviation = 3.15) for the control group and 16.14 segments (standard deviation = 6.42) for the ADHD group.

2.3. Feature Selection

The EEG data collected from 19 channels (listed in Table 1) were utilized as input features for the classifiers. These features were used in various combinations to train the classifiers. The entire feature set was partitioned into different combinations of frontal (F), central (C), parietal (P), temporal (T), and occipital (O) regions. Detailed information about these regions and their corresponding channels is provided in Table 1.

Table 1. Brain regions and their corresponding channels.

Regions	Combined Channels
F (frontal)	Fz, FP1, FP2, F3, F4, F7, F8
C (central)	Cz, C3, C4
P (parietal)	Pz, P3, P4
T (temporal)	T3, T4, T5, T6
O (occipital)	O1, O2
Right hemisphere	FP2, F4, F8, C4, T4, P4, T6, O2
Left hemisphere	C3, T3, FP1, F3, F7, P3, T5, O1

2.4. Classification

After obtaining the various combinations of feature channels, the features were set as inputs to six different machine learning algorithms: SVM, RF, DT, AdaBoost, Naive Bayes, and LDA. Here is a detailed explanation of how each of these algorithms works:

2.4.1. Support Vector Machine (SVM)

SVM is a robust supervised learning algorithm employed for both classification and regression tasks. Its fundamental principle involves identifying an optimal hyperplane that effectively separates data points belonging to different classes. In the context of binary classification, SVM endeavors to determine the hyperplane that maximizes the margin or distance between classes. The data points nearest to the hyperplane are known as support vectors and significantly influence the delineation of the decision boundary. SVM efficiently manages high-dimensional feature spaces and excels in handling non-linearly separable data through the use of kernel functions.

2.4.2. Decision Tree (DT)

A decision Tree is a versatile machine learning algorithm used primarily for classification and regression tasks. It operates by recursively partitioning the data based on feature values to create a hierarchical structure resembling a tree. At every node of the tree, a decision is made to determine which feature it will use for splitting, with the goal of minimizing uncertainty within each branch. The leaves of the tree correspond to the final class predictions. DTs can efficiently handle both numerical and categorical features, and their intuitive structure makes them interpretable and useful for feature selection.

2.4.3. Random Forest (RF)

RF is an ensemble learning technique that combines multiple decision trees to pro-duce robust predictions. Each tree is individually constructed using a subset of the data and a random assortment of features. During the training process, each tree recursively partitions the data into subsets by considering the selected features. The final prediction is derived by combining the predictions from each individual tree. RF effectively counteracts overfitting concerns and adeptly manages high-dimensional data. Its proficiency lies in its accuracy and capacity to capture intricate relationships within the data.

2.4.4. AdaBoost (Adaptive Boosting)

AdaBoost is a type of ensemble learning that uses multiple weak learners to build a strong classifier. It assigns higher weights to misclassified data points in each iteration, with a focus on challenging-to-classify samples. In subsequent iterations, it allocates more attention to misclassified instances, thereby refining the model's predictive performance. Through an iterative process of adjusting sample weights, AdaBoost develops a potent classifier capable of accurate classification. Its adaptability to varying complexities of data and potential for boosting model accuracy are notable attributes.

2.4.5. Naive Bayes

Naive Bayes is a probabilistic classification algorithm based on Bayes' theorem. It assumes that features are conditionally independent when you know the class label, which is why it is called "naive". Naive Bayes computes the probability of a data point belonging to a certain class based on its feature values and prior class probabilities. Despite its simplistic assumptions, Naive Bayes often performs remarkably well, particularly when the independence assumption is not drastically violated. It is computationally efficient, requires relatively small amounts of training data, and is particularly useful for text classification tasks.

2.4.6. Linear Discriminant Analysis (LDA)

LDA is a dimensionality reduction technique often used in the context of classification. It seeks to find the linear combinations of features that best separate different classes while minimizing the variance within each class. LDA essentially projects data onto a lower-dimensional space, maximizing class separability. It is particularly useful when classes have distinct distributions, and it is known for its effectiveness in reducing overfitting in high-dimensional data.

2.5. Performance Analysis

The comparative analysis involved assessing performance metrics across different combinations of features, regions, and classifiers. The primary objective was to identify the optimal combinations of channels/features and classifiers that demonstrated superior results in terms of two key performance metrics: accuracy and the area under the curve (AUC).

Accuracy performance, commonly utilized in classification tasks, is determined by the ratio of correctly predicted instances (both true positives and true negatives) to the total number of instances in the dataset. This can be mathematically formulated using the following Equation (1) [36]:

$$\text{Accuracy} = \frac{TP + TN}{TP + TN + FP + FN} \quad (1)$$

Here, *TP*: True positive, *TN*: True negative, *FP*: False positive, and *FN*: False negative.

The AUC score is a widely used metric for evaluating the performance of binary classification models. It measures the classifier's capacity to differentiate between positive and negative instances by plotting the True Positive Rate (TPR) against the False Positive Rate (FPR) across various discrimination thresholds. A higher AUC value signifies superior classification performance. The AUC score can be computed using Equation (2) as provided below [36]:

$$\text{AUC} = \int_0^1 TPR\left(FPR^{-1}(x)\right) dx \quad (2)$$

Here, *TPR and FPR* can be measured using Equations (3) and (4):

$$TPR = \frac{TP}{TP + FN} \quad (3)$$

$$FPR = \frac{FP}{FP + TN} \quad (4)$$

In the next section, we will present a detailed overview of the experimental results and provide a thorough analysis.

3. Results and Discussion

In this study, we assessed how effectively different classifiers performed with various combinations of brain regions, as shown in Table 2. Among all of the classifiers examined, SVM displayed interesting accuracy in several combinations, particularly the highest accuracy of 76% attained for the brain region combination F+C+P. RF displayed competitive accuracy, reaching 72% accuracy in the F+C+O and F+P+O combinations. DT exhibited varying accuracy, with the highest value of 64% in the F+C+P configuration. AdaBoost demonstrated its effectiveness with the F+C+T combination, yielding 72% accuracy. Surprisingly, Naïve Bayes displayed relatively lower accuracy across the board, indicating limitations in capturing the non-linear relationships present in the data. Furthermore, it is noteworthy that the Naïve Bayes classifier, despite demonstrating lower accuracy in some configurations, showcased a remarkable accuracy of 84% in the F+T+O combination. LDA consistently exhibited moderate accuracy, with its highest value of 56% obtained in the P+T+O combination. Overall, the classifiers' performances were influenced by the specific combination of brain regions, highlighting the importance of tailored feature selection in optimizing classification accuracy.

Figure 3 shows a graph that summarizes how accurate different classifiers were with different combinations of three brain regions. The values depicted in the graph correspond to the accuracy scores presented in Table 1. Notably, the Naïve Bayes classifier demonstrates a sufficient accuracy of 84% in both the F+T+O and F+P+T combinations due to its ability to effectively capture and model complex relationships within these specific combinations of brain regions. The amalgamation of these brain region combinations likely encompasses unique neural patterns that are indicative of ADHD-related activity. Additionally, the Naïve Bayes classifier's proficiency in modeling probabilistic relationships between features and classes makes it well-suited for capturing the nuanced distinctions present in these particular combinations of brain regions. This graph provides an intuitive representation of the performance trends of different classifiers across distinct regional sets, offering insights into their varying strengths in capturing relevant features and interactions within the EEG data.

Table 2. Classification accuracy using a set of 3 regional combinations.

Combination of Brain Regions (Set of 3)	Classifiers (Accuracy)					
	SVM	RF	DT	AdaBoost	Naïve Bayes	LDA
F+C+P	76	56	64	64	32	44
F+C+T	68	64	60	72	60	36
F+C+O	68	72	68	64	28	40
F+P+T	64	60	52	56	84	32
F+P+O	72	72	60	56	56	40
F+T+O	68	64	48	56	84	52
C+P+T	64	60	40	48	44	40
C+P+O	56	52	44	44	24	16
C+T+O	60	60	40	60	60	52
P+T+O	56	72	64	56	72	56

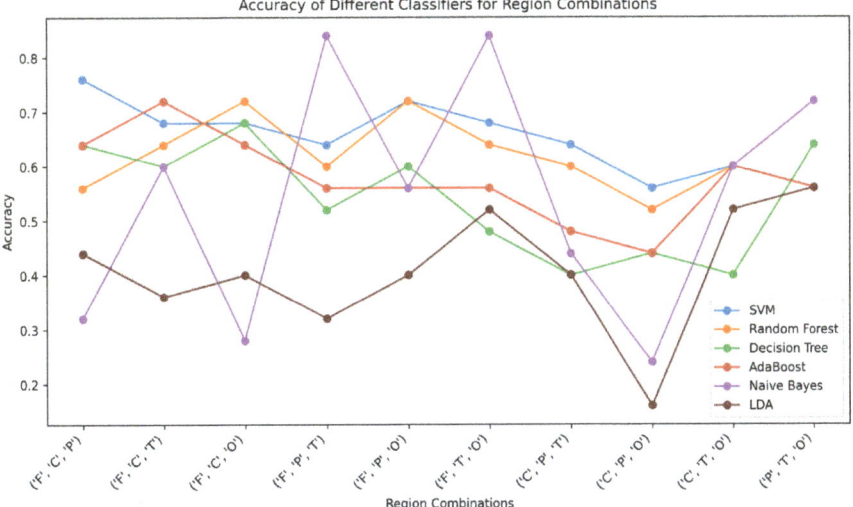

Figure 3. Accuracy of different classifiers across three brain region combinations.

As shown in Table 3, among the various combinations of brain regions and classifiers evaluated in the study, the combination "F+C+T" stands out as having achieved the highest AUC performance. This combination demonstrates the ability to effectively distinguish between ADHD and healthy control subjects with remarkable AUC scores, which is shown in Figure 4. Specifically, the classifiers RF and AdaBoost exhibit notable AUC values of 75.9% and 67.5%, respectively, when employed with this brain region combination. These elevated AUC scores indicate the classifiers' proficiency in capturing the distinctive patterns and features associated with ADHD and healthy control subjects within the selected brain regions. The significant performance of "F+C+T" in terms of AUC underscores its potential as a discriminative feature set, showcasing the effectiveness of RF and AdaBoost classifiers in this context. This finding serves as valuable insight into the optimal feature combination and classifier choice for accurate classification between the two subject groups.

Table 3. AUC performance of different brain region combinations and classifiers.

Combination of Brain Regions (Set of 3)	Classifiers (AUC)					
	SVM	RF	DT	AdaBoost	Naïve Bayes	LDA
F+C+P	31.1	62.3	63.9	63.6	38.9	35.7
F+C+T	77.2	75.9	61.3	67.5	59.7	36.6
F+C+O	29.2	71.4	70.4	71.4	38.9	31.1
F+P+T	72	73.3	54.2	66.8	80.5	34.4
F+P+O	37.6	69.4	61.3	64.9	69.4	32.4
F+T+O	77.9	79.2	49.6	66.2	80.5	56.4
C+P+T	66.2	62.2	38.6	48.7	38.9	37.1
C+P+O	44.8	58.1	46.1	38.9	2.5	12.9
C+T+O	66.8	67.2	42.5	62.3	45.4	58.4
P+T+O	63.6	74	63.9	59.7	68.1	55.19

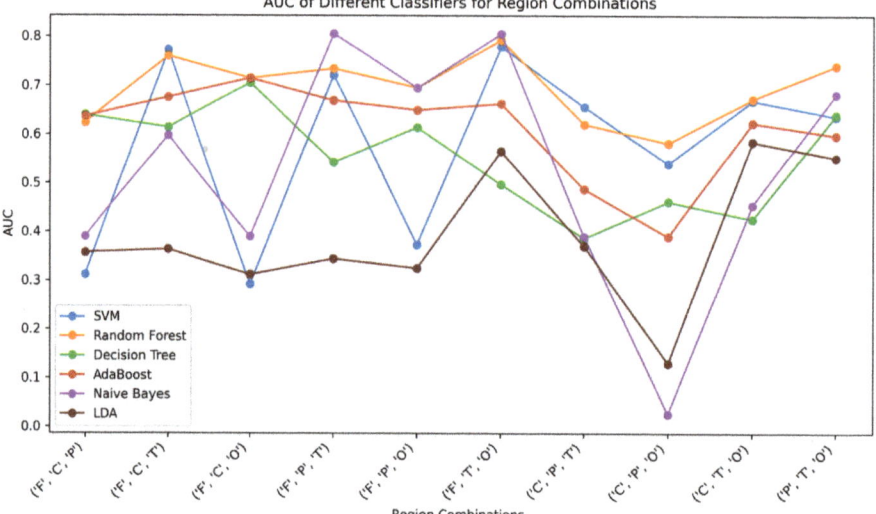

Figure 4. AUC performance of classifiers based on different brain region combinations.

Table 4 presents the accuracy performance of various classifiers across different brain region combinations, focusing on the right hemisphere, left hemisphere, and the fusion of both hemispheres. Notably, the analysis highlights the highest accuracy achieved by any classifier within each dataset, providing valuable insights into their efficacy in detecting ADHD-related patterns in region-specific EEG data. In the right hemisphere dataset, the Naïve Bayes classifier exhibited the highest accuracy of 84%, underscoring its ability to discern distinctive neural patterns associated with ADHD. For the left hemisphere dataset, the RF classifier achieved the highest accuracy of 64%, whereas the Naïve Bayes classifier demonstrated an accuracy of 28%. Remarkably, in the combined hemisphere dataset, the RF classifier attained the highest accuracy of 68% along with SVM, reaffirming the proficiency of these classifiers in capturing intricate relationships across neural regions. These outcomes underscore the variability in accuracy across classifiers and brain regions, providing valuable insights into their performance in classifying ADHD patterns based on hemisphere-specific neural activity. The accuracy plot depicting the performance of all classifiers based on hemisphere-specific regions is illustrated in Figure 5.

Table 4. Classification accuracy is based on specific regions.

Brain Region	Classifiers (Accuracy)					
	SVM	RF	DT	AdaBoost	Naïve Bayes	LDA
Right hemisphere	72	60	44	48	84	40
Left hemisphere	44	64	48	44	28	36
Combined hemisphere	68	68	44	64	32	40

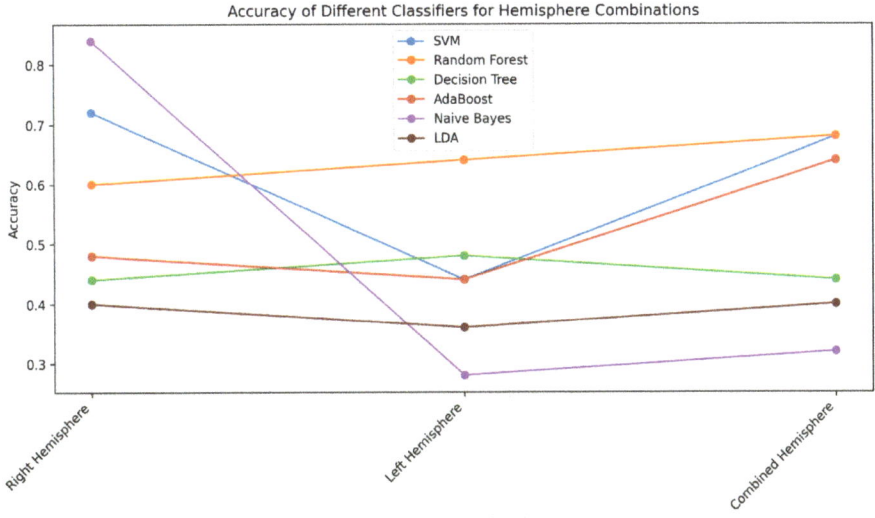

Figure 5. Accuracy plot for classifier performance based on hemisphere-specific regions.

Furthermore, the AUC values for the different hemisphere regions were meticulously evaluated, as illustrated in Table 5. In the "Right hemisphere", the highest AUC score of 86.3% was achieved with the SVM, closely trailed by Naïve Bayes at 85.7%. Intermediate AUC results were demonstrated with RF (59.4%) and AdaBoost (50.3%), while DT (42.2%) and LDA (34.4%) exhibited relatively lower AUC values. Shifting to the "Left hemisphere", RF excelled with an AUC of 63.3%, outperforming the other classifiers that registered comparatively lower AUC scores. Notably, when examining the "Combined hemisphere" scenario, RF continued its prominent performance by achieving the highest AUC of 72.7%, closely pursued by AdaBoost at 63.6%. In this consolidated context, SVM, Naïve Bayes, and DT showcased relatively lower AUC values. Collectively, these AUC findings offer insightful comparisons of the classifiers' capabilities in discerning EEG patterns across distinct brain hemisphere regions, with RF and AdaBoost emerging as top performers, particularly in the combined setting. Figure 6 displays a plot of AUC scores, which illustrates how all the classifiers performed when considering hemisphere-specific regions.

Table 5. AUC score based on hemispheres-specific regions.

Brain Region	Classifiers (AUC)					
	SVM	RF	DT	AdaBoost	Naïve Bayes	LDA
Right hemisphere	86.3	59.4	42.2	50.3	85.7	34.4
Left hemisphere	49.3	63.3	46.7	39.6	27.9	22.72
Combined hemisphere	28.5	72.7	44.1	63.6	48	28.5

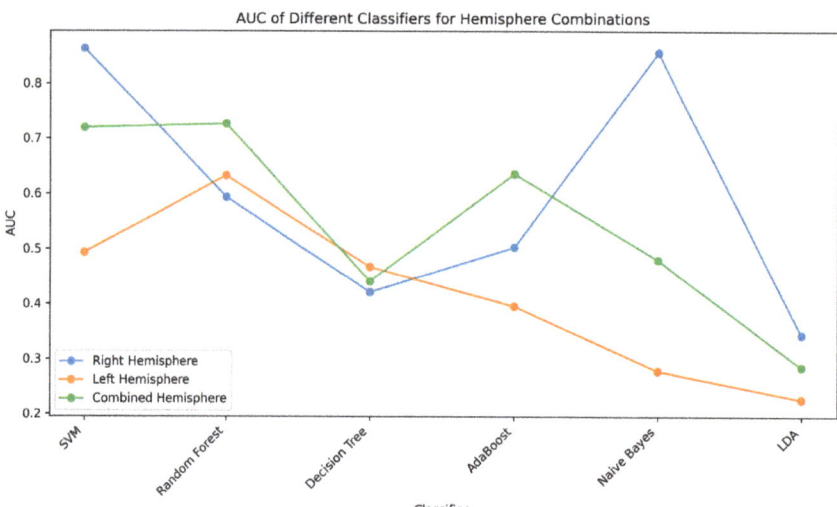

Figure 6. AUC score plot for classifier performance across hemisphere-specific regions.

The comprehensive analysis of classification performance across different brain region combinations, focusing on the right hemisphere, left hemisphere, and the combined hemisphere, has provided valuable insights into the discriminatory capabilities of various classifiers for ADHD-related patterns. In terms of accuracy, we observed an interesting variation based on distinct features obtained from different hemisphere regions. Notably, the right hemisphere dataset exhibited remarkable accuracy performance by the Naïve Bayes classifier, achieving the highest accuracy of 84%. This can be attributed to Naïve Bayes' probabilistic nature, which assumes feature independence given the class label. In cases where the underlying patterns are well-approximated by the independence assumption, Naïve Bayes excels, as seen in the distinctive neural patterns associated with ADHD discerned in the right hemisphere. On the other hand, in the left hemisphere, the RF classifier achieved the highest accuracy of 64%. RF's capacity to capture complex relationships among features played a significant role. Interestingly, in the combined hemisphere dataset, both the RF classifier and SVM achieved the highest accuracy of 68%. This underlines the proficiency of RF in capturing intricate relationships across neural regions, complemented by SVM's ability to define an optimal hyperplane for class separation. The AUC results further substantiate these findings. Notably, in the right hemisphere, Naïve Bayes and SVM achieved the highest AUC scores, capitalizing on the probabilistic and hyperplane-based approaches, respectively. RF's excellence in capturing complex patterns enabled it to attain the highest AUC in the left hemisphere. Remarkably, in the combined hemisphere, the RF classifier continued its prominent performance, emphasizing its adaptability to amalgamate neural region information. These findings collectively underscore the strengths of specific classifiers in discerning ADHD-related patterns, driven by their inherent characteristics and adaptability to distinct data distributions.

The varying performance of the Naive Bayes classifier across different brain region combinations can be attributed to its underlying assumption of feature independence. In cases where the features within a specific combination of brain regions are indeed conditionally independent given the class label, the Naive Bayes classifier excels by effectively capturing the underlying patterns; however, its performance diminishes when faced with brain region combinations characterized by intricate inter-feature relationships and nonlinear interactions. In such scenarios, the assumption of feature independence becomes less valid, leading to suboptimal classification accuracy. This disparity underscores the significance of tailored feature selection and highlights that the Naive Bayes classifier's effectiveness is contingent upon the specific nature of the data it encounters.

The computational complexity of the study can be categorized as moderate to high because we implemented and evaluated several machine learning models, such as SVM, RF, DT, AdaBoost, Naive Bayes, and LDA. The investigation into various combinations of brain regions and hemispheres, coupled with feature selection, further contributes to the complexity. Additionally, hyperparameter tuning, potential cross-validation, ensemble methods, feature scaling, and multiple evaluation metrics amplify the computational workload. For experimentation, Python 3.8.5 was employed on a Samsung PC with an Intel Core i5-8400 CPU @ 2.80GHz, 8GB RAM and a GeForce GTX 1070 Ti 8GB GPU to facilitate these computations. The study strikes a balance between computational complexity and the research's significance in yielding valuable insights from the obtained results.

There are several limitations in this study that need to be acknowledged. We employed EEG data derived from different brain region combinations to conduct ADHD classification against healthy control subjects using various machine learning techniques. Unlike several prior works, we deliberately refrained from employing external feature extraction techniques or feature engineering approaches to manipulate the classification features. While such comparisons could provide valuable insights, it is important to note that conducting direct performance comparisons with other studies may not yield fair or reliable results due to variations in datasets, preprocessing pipelines, features, and classifiers employed in each study. Additionally, the absence of deep learning techniques in our methodology is another limitation. Deep learning has shown promising results in various classification tasks, including medical image analysis, and its exploration could offer further improvements in the accuracy and robustness of the classification model. It is our intention to address these limitations in a future work, where we plan to include comparative analyses and explore the integration of deep learning methods to enhance the classification performance.

4. Conclusions

In this paper, we explore the application of EEG data for the classification of ADHD subjects and healthy controls. Leveraging machine learning techniques, we aimed to enhance the accuracy of ADHD diagnosis by exploiting distinctive neural patterns captured through EEG recordings. Our study uncovered several key findings that shed light on the potential of EEG-based classification in ADHD research. Through comprehensive analysis, we demonstrated the effectiveness of various classifiers, including SVM, RF, DT, AdaBoost, Naive Bayes, and LDA, in distinguishing between ADHD and healthy control groups. Notably, the Naive Bayes classifier achieved a sufficient e accuracy of 84% when considering specific combinations of brain regions. This highlights the discriminatory power of EEG data in revealing patterns associated with ADHD. Furthermore, our investigation into regional contributions revealed the significant impact of hemisphere-specific EEG data on classification performance. We observed higher AUC values in the right hemisphere, particularly with the SVM and Naive Bayes classifiers. This supports the notion that EEG data from specific brain regions can provide valuable insights into ADHD-related neural activity. The exploration of combined hemisphere data highlighted the potential of the RF classifier, which achieved a promising AUC of 72.7%. This finding underscores the importance of selecting appropriate classifiers and brain regions to maximize classification accuracy. While these classifiers have shown promising performance in previous studies and are well-suited for capturing complex relationships within the data, we acknowledge that neural networks, especially deep learning models, have gained significant attention in recent years for their capacity to learn intricate patterns in large and complex datasets. However, due to the relatively smaller size of our dataset and the interpretability offered by traditional classifiers, we opted to focus on them for our current investigation. In summary, this research advances our understanding of EEG-based ADHD classification by demonstrating the efficacy of distinct classifiers and brain region combinations. Numerous intriguing directions remain open for future explorations. Firstly, the integration of advanced machine learning techniques, such as deep learning models like convolutional neural networks (CNNs) and recurrent neural networks (RNNs), could potentially yield even more accurate and robust classification results. Additionally, the incorporation of multi-modal data, combining EEG with other physiological or behavioral measures, could provide a more comprehensive understanding of ADHD-related neural patterns. Furthermore, exploring the generalizability of our findings to larger and more diverse populations could enhance the clinical applicability of our approach. These approaches show potential for improving ADHD diagnosis methods and enhancing our knowledge of neurodevelopmental disorders.

Author Contributions: Conceptualization, N.C. and B.-J.C.; methodology, N.C.; software, N.C.; validation, N.C. and B.-J.C.; formal analysis, N.C.; investigation, N.C.; resources, B.-J.C.; data curation, N.C.; writing—original draft preparation, N.C.; writing—review and editing, N.C.; visualization, N.C.; supervision, B.-J.C.; project administration, B.-J.C.; funding acquisition, B.-J.C. All authors have read and agreed to the published version of the manuscript.

Funding: This research was supported by the Korea Institute of Advancement of Technology (KIAT) grant funded by the Korean government (MOTIE) (P0012724, HRD Program for Industrial Innovation).

Institutional Review Board Statement: Not applicable.

Informed Consent Statement: Not applicable.

Data Availability Statement: Not applicable.

Conflicts of Interest: The authors declare no conflict of interest.

References

1. American Psychiatric Association. *Diagnostic and Statistical Manual of Mental Disorders*, 5th ed.; American Psychiatric Publishing: Arlington, VA, USA, 2013.
2. Konrad, K.; Eickhoff, S.B. Is the ADHD brain wired differently? A review on structural and functional connectivity in attention deficit hyperactivity disorder. *Human Brain Mapp.* **2010**, *31*, 904–916. [CrossRef] [PubMed]
3. Smith, M. Hyperactive around the world? The history of ADHD in global perspective. *Soc. Hist. Med.* **2017**, *30*, 767–787. [CrossRef]
4. Arns, M.; Heinrich, H.; Strehl, U. Evaluation of neurofeedback in ADHD: The long and winding road. *Biol. Psychol.* **2014**, *95*, 108–115. [CrossRef] [PubMed]
5. Emerson, E.; Einfeld, S.; Stancliffe, R.J. The mental health of young children with intellectual disabilities or borderline intellectual functioning. *Soc. Psychiatry Psychiatr. Epidemiol.* **2010**, *45*, 579–587. [CrossRef] [PubMed]
6. Arns, M.; Conners, C.K.; Kraemer, H.C. A decade of EEG theta/beta ratio research in ADHD: A meta-analysis. *J. Atten. Disord.* **2013**, *17*, 374–383. [CrossRef]
7. Patidar, S.; Pachori, R.B.; Upadhyay, A.; Acharya, U.R. An integrated alcoholic index using tunable-Q wavelet transform based features extracted from EEG signals for diagnosis of alcoholism. *Appl. Soft Comput.* **2017**, *50*, 71–78. [CrossRef]
8. Micanovic, C.; Pal, S. The diagnostic utility of EEG in early-onset dementia: A systematic review of the literature with narrative analysis. *J. Neural Transm.* **2014**, *121*, 59–69. [CrossRef]
9. Daftari, C.; Shah, J.; Shah, M. Detection of Epileptic Seizure Disorder Using EEG Signals. In *Artificial Intelligence-Based Brain-Computer Interface*; Academic Press: Cambridge, MA, USA, 2022; pp. 163–188. [CrossRef]
10. Ko, D.-W.; Yang, J.-J. EEG-Based Schizophrenia Diagnosis through Time Series Image Conversion and Deep Learning. *Electronics* **2022**, *11*, 2265. [CrossRef]
11. Aljalal, M.; Aldosari, S.A.; Molinas, M.; AlSharabi, K.; Alturki, F.A. Detection of Parkinson's disease from EEG signals using discrete wavelet transform, different entropy measures, and machine learning techniques. *Sci. Rep.* **2022**, *12*, 22547. [CrossRef]
12. Wu, C.T.; Huang, H.C.; Huang, S.; Chen, I.M.; Liao, S.C.; Chen, C.K.; Lin, C.; Lee, S.H.; Chen, M.H.; Tsai, C.F.; et al. Resting-State EEG Signal for Major Depressive Disorder Detection: A Systematic Validation on a Large and Diverse Dataset. *Biosensors* **2021**, *11*, 499. [CrossRef]
13. Alba, G.; Pereda, E.; Mañas, S.; Méndez, L.D.; González, A.; González, J.J. Electroencephalography signatures of attention-deficit/hyperactivity disorder: Clinical utility. *Neuropsychiatr. Dis. Treat.* **2015**, *11*, 2755–2769. [PubMed]
14. Loo, S.K.; Makeig, S. Clinical utility of EEG in attentiondeficit/hyperactivity disorder: A research update. *Neurotherapeutics* **2012**, *9*, 569–587. [CrossRef] [PubMed]
15. Khoshnoud, S.; Nazari, M.A.; Shamsi, M. Functional brain dynamic analysis of ADHD and control children using nonlinear dynamical features of EEG signals. *J. Integr. Neurosci.* **2018**, *17*, 17–30. [CrossRef]

16. Michel, C.M.; Murray, M.M. Towards the utilization of EEG as a brain imaging tool. *NeuroImage* **2012**, *61*, 371–385. [CrossRef] [PubMed]
17. Huster, R.J. Brinkman Perspective: The potential of real-time EEG neurofeedback for optimizing anesthesia. *Front. Hum. Neurosci.* **2020**, *14*, 19.
18. Engemann, D.A.; Gramfort, A. Microstates as a tool to explore the spatiotemporal dynamics of functional connectivity in EEG. *Hum. Brain Mapp.* **2020**, *41*, 3972–3992.
19. Smith, M.E. Current advances in functional imaging of mild traumatic brain injury. *Curr. Opin. Neurol.* **2018**, *31*, 687–693.
20. Alchalabi, A.E.; Shirmohammadi, S.; Eddin, A.N.; Elsharnouby, M. FOCUS: Detecting ADHD Patients by an EEG-Based Serious Game. *IEEE Trans. Instrum. Meas.* **2018**, *67*, 1512–1520. [CrossRef]
21. Rashid, M.H.; Siddique, N.H. A novel feature extraction and selection method for classification of EEG signals using SVM. *J. Neurosci. Methods* **2017**, *284*, 48–60.
22. Wang, Z.; Li, Y.; Zhou, Z.; He, L.; Wu, S.; Guo, L. Identification of ADHD children based on ERP and machine learning methods. *IEEE Trans. Neural Syst. Rehabil. Eng.* **2020**, *28*, 3–13.
23. Chen, X.; Lai, Y.; Li, Y. Classification of ADHD children based on EEG data using neural networks. *J. Ambient Intell. Humaniz. Comput.* **2020**, *11*, 1113–1121.
24. Xie, Y.; Oniga, S. A Review of Processing Methods and Classification Algorithm for EEG Signal. *Carpathian J. Electron. Comput. Eng.* **2020**, *13*, 23–29. [CrossRef]
25. Barachant, A.; Bonnet, S.; Congedo, M.; Jutten, C. Multiclass brain–computer interface classification by Riemannian geometry. *IEEE Trans. Biomed. Eng.* **2013**, *59*, 920–928. [CrossRef] [PubMed]
26. Vapnik, V.N. An overview of statistical learning theory. *IEEE Trans. Neural Netw.* **1999**, *10*, 988–999. [CrossRef] [PubMed]
27. Breiman, L. Random forests. *Mach. Learn.* **2001**, *45*, 5–32. [CrossRef]
28. Quinlan, J.R. Induction of Decision Trees. *Mach. Learn.* **1986**, *1*, 81–106. [CrossRef]
29. Freund, Y.; Schapire, R.E. A decision-theoretic generalization of on-line learning and an application to boosting. *J. Comput. Syst. Sci.* **1997**, *55*, 119–139. [CrossRef]
30. Mitchell, T. Machine Learning. In *Chapter 6: "Bayesian Learning" Covers Naive Bayes Algorithms*; McGraw-Hill: New York, NY, USA, 1997; ISBN 0070428077.
31. Boucsein, W. *Electrodermal Activity*; Springer Science & Business Media: Berlin, Germany, 2012.
32. Nasrabadi, A.M.; Allahverdy, A.; Samavati, M.; Mohammadi, M.R. EEG Data ADHD/Control Children. *IEEE Dataport* **2020**. [CrossRef]
33. Morgan, A.E.; Hynd, G.W.; Riccio, C.A.; Hall, J. Validity of DSM-IV ADHD Predominantly Inattentive and Combined Types: Relationship to Previous DSM Diagnoses/Subtype Differences. *J. Am. Acad. Child Adolesc. Psychiatry* **1996**, *35*, 325–333. [CrossRef]
34. Onton, J.; Delorme, A.; Makeig, S. Frontal midline EEG dynamics during working memory. *NeuroImage* **2005**, *27*, 341–356. [CrossRef]
35. Larson, E.; Gramfort, A.; Engemann, D.A.; Leppakangas, J.; Brodbeck, C.; Jas, M.; Brooks, T.; Sassenhagen, J.; Luessi, M.; McCloy, D.; et al. Mne-Python. Zenodo. 23 February 2023. Available online: https://zenodo.org/record/7671973 (accessed on 8 August 2023).
36. Huang, J.; Ling, C.X. Using AUC and accuracy in evaluating learning algorithms. *IEEE Trans. Knowl. Data Eng.* **2005**, *17*, 299–310. [CrossRef]

Disclaimer/Publisher's Note: The statements, opinions and data contained in all publications are solely those of the individual author(s) and contributor(s) and not of MDPI and/or the editor(s). MDPI and/or the editor(s) disclaim responsibility for any injury to people or property resulting from any ideas, methods, instructions or products referred to in the content.

Article

Two-Stage Input-Space Image Augmentation and Interpretable Technique for Accurate and Explainable Skin Cancer Diagnosis

Catur Supriyanto [1,2,*], Abu Salam [1,2], Junta Zeniarja [1,2] and Adi Wijaya [3]

1. Faculty of Computer Science, Universitas Dian Nuswantoro, Semarang 50131, Indonesia; abu.salam@dsn.dinus.ac.id (A.S.); junta@dsn.dinus.ac.id (J.Z.)
2. Dinus Research Group for AI in Medical Science (DREAMS), Universitas Dian Nuswantoro, Semarang 50131, Indonesia
3. Department of Health Information Management, Universitas Indonesia Maju, Jakarta 12610, Indonesia; adiwjj@uima.ac.id
* Correspondence: catur.supriyanto@dsn.dinus.ac.id

Abstract: This research paper presents a deep-learning approach to early detection of skin cancer using image augmentation techniques. We introduce a two-stage image augmentation process utilizing geometric augmentation and a generative adversarial network (GAN) to differentiate skin cancer categories. The public HAM10000 dataset was used to test how well the proposed model worked. Various pre-trained convolutional neural network (CNN) models, including Xception, Inceptionv3, Resnet152v2, EfficientnetB7, InceptionresnetV2, and VGG19, were employed. Our approach demonstrates an accuracy of 96.90%, precision of 97.07%, recall of 96.87%, and F1-score of 96.97%, surpassing the performance of other state-of-the-art methods. The paper also discusses the use of Shapley Additive Explanations (SHAP), an interpretable technique for skin cancer diagnosis, which can help clinicians understand the reasoning behind the diagnosis and improve trust in the system. Overall, the proposed method presents a promising approach to automated skin cancer detection that could improve patient outcomes and reduce healthcare costs.

Keywords: deep learning; skin cancer; image augmentation; GAN; geometric augmentation; image classification; interpretable technique

1. Introduction

Skin cancer is one of the most prevalent and potentially life-threatening forms of cancer worldwide. For more effective therapy and better patient recovery, early detection and diagnosis are essential [1,2]. In recent years, the study of medical image analysis has been completely transformed by convolutional neural networks (CNNs) compared to other advanced machine learning models, supervised or unsupervised, such as k-nearest neighbor (KNN) and support vector machine (SVM), offering a promising approach for the automated detection of skin cancer [3]. CNNs have shown to be extremely effective at extracting complicated patterns and characteristics from medical images, making them an ideal tool for automating the process of skin cancer detection. This technology has the potential to assist dermatologists and healthcare professionals in identifying skin lesions and distinguishing between benign and malignant tumors.

HAM10000 and the International Skin Imaging Collaboration (ISIC) are two datasets that are widely utilized in skin cancer detection studies. HAM10000 is a comprehensive dataset containing diverse dermoscopic images of pigmented skin lesions, a common category of skin cancer [4]. An advantage of the HAM10000 dataset lies in its relatively smaller size compared to the expansive ISIC dataset. This may be beneficial for researchers facing limited computational resources or who want to focus on a specific subset of skin lesions. However, the ISIC dataset has unique advantages, including a larger scale and the inclusion of additional metadata such as lesion location and patient age. The ISIC datasets have been used for segmentation tasks, but the availability of delineated segmentation masks is limited compared to the classification tasks [5]. The choice of dataset often depends on the specific research investigation and the resources available for the study.

The issue with skin cancer detection datasets is the imbalance in the number of data samples across different classes. This imbalance is observed in both the HAM10000 and ISIC 2017–2020 datasets. In the HAM10000 dataset, which includes a total of 10,015 images, the highest number of data samples can be seen in the melanoma category, with 6705 images, while the lowest number of samples is present in the dermatofibroma category, consisting of 115 images [6]. Meanwhile, in the ISIC 2020 dataset, encompassing a total of 33,126 images, the most abundant data samples are found within the unknown (benign) category, which comprises 27,126 images, whereas the solar lentigo category contains the fewest data samples, with only 7 images [7]. Data imbalance can lead to biased results in classification because the model may be more likely to predict the overrepresented class.

Several approaches can be employed to address imbalances in the amount of data, such as geometric-transformation-based augmentation, feature-space augmentation, and GAN-based augmentation [8]. Geometric data augmentation is a technique employed in the areas of machine learning and computer vision to enhance the variability of a dataset by doing geometric modifications on the original data. The technique transforms the geometric configuration of images by moving the positions of individual pixels without modifying the values of those pixels. These transformations involve altering the position, orientation, or scale of the data while preserving their inherent characteristics. Geometric data augmentation is particularly useful for image data and is often applied to improve the performance of deep learning models. Some common geometric augmentations include rotation, scaling, translation, shearing, flipping, cropping, and zooming.

In feature-space data augmentation, there are two approaches: namely, the undersampling and oversampling approaches. In the undersampling approach, the number of samples from the majority class is reduced to create a more balanced distribution between the classes. By reducing the number of majority class samples, undersampling can help prevent the model from being biased towards the majority class and can improve its ability to recognize the minority class. However, undersampling may result in a loss of potentially valuable information, so it should be applied carefully. In the oversampling approach, additional samples from the minority class are generated to create a more balanced distribution between the classes. The goal is to increase the representation of the minority class to match the number of samples in the majority class, making the dataset more balanced. There are several oversampling methods, with one of the most commonly used techniques being SMOTE (Synthetic Minority Over-sampling Technique). In order to generate synthetic samples for the minority class, SMOTE [9] interpolates between existing data points. This helps to improve the model's ability to learn from the minority class and can lead to better classification results. However, it is important to be cautious with oversampling, as generating too many synthetic samples can lead to overfitting and reduced model generalization.

The concept of GAN-based augmentation refers to the use of generative adversarial networks (GANs) for the purpose of producing synthetic data samples that can be used to augment an existing dataset [10]. This technique is particularly useful in cases where the original dataset is small or imbalanced, as it can help to increase the size of the dataset and balance the class distribution. Augmentation by GAN has been implemented effectively in numerous domains, including medical imaging, natural language processing, and computer vision. Some examples of GAN augmentation in medical imaging include the generation of synthetic CT scans, MRI images, and X-ray images to aid in disease diagnosis and treatment.

In general, resampling approaches can be divided into two categories: namely, input-space data augmentation and feature-space data augmentation. Input-space resampling involves manipulating the original data instances themselves before any feature extraction. Meanwhile, feature-space resampling is applied after feature extraction. Geometric transformations and GAN-based augmentations are categorized as input-space data augmentation, whereas SMOTE is classified as feature-space data augmentation. The benefit of input-space data augmentation is its independence from the feature extraction method, providing greater flexibility in choosing feature extraction methods. Therefore, in this study, we propose a two-step augmentation, including geometric and GAN-based augmentation, for early detection of skin cancer. The main contributions of this research article are:

- The integration of geometric and GAN-based augmentation for skin cancer detection;
- In this study, we provide an explainable AI using SHAP to explain how the model makes decisions or predictions.

2. Related Works

The related studies in this research are categorized into three groups: studies using feature-space augmentation, geometric augmentation, and GAN-based augmentation. Augmentation or oversampling is employed to address the issues of limited data and imbalanced data. Both of these problems contribute to the reduced accuracy of the detection model. This is also observed in skin cancer detection. Several studies have been conducted regarding the use of augmentation or oversampling in skin cancer detection. Abayomi et al. [11] proposed a data augmentation strategy that entails creating a new skin melanoma dataset using dermoscopic images from the publicly available PH2 dataset. The study adopted SMOTE-conv [12], which is a variant of SMOTE. SMOTE-conv utilizes a covariance matrix to detect relationships among attributes and generate synthetic instances. The SqueezeNet deep learning network was then trained using these modified images. In the binary classification scenario, it resulted in an accuracy of 92.18%, while in the multiclass classification scenario, it achieved an accuracy of 89.2%.

SMOTE is an oversampling method used to balance the number of samples between the majority and minority classes in a dataset. SMOTE randomly selects samples from the minority class and creates new synthetic samples by combining them with those of their nearest neighbors. This helps improve the classification performance on imbalanced datasets. However, SMOTE tends to introduce noise and affect classification performance. Therefore, K-means-SMOTE [13] was developed to address SMOTE's limitations. It does this by using k-means clustering to group samples and generate synthetic samples only within clusters with fewer minority class instances. Chang et al. [14] adopted Kmeans-SMOTE to address class imbalance in the ISIC 2018 and ISIC 2019 datasets. Five pre-trained models—namely, VGG16, MELA-CNN, InceptionResNetV2, Inception V3, and the dermatologist handcrafted method—were used to extract features. The minority class data are oversampled using Kmeans-SMOTE and then classified using the Extreme Gradient Boosting (XGB) classifier. The research yielded an accuracy of 96.5%, precision of 97.4%, recall of 87.8%, AUC (Area Under the Curve) of 98.1%, and F1-score of 90.5%.

A study on a deep-learning-based skin cancer classification network (DSCC_Net) was proposed by Tahir et al. [15]; the study proposes the development of a deep learning model with multi-classification capabilities for the purpose of identifying skin cancer through the analysis of dermoscopic pictures. The model was trained and evaluated on three public datasets (HAM10000, ISIC2020, and DermIS), and the results showed that DSCC_Net outperformed other state-of-the-art models in terms of accuracy, sensitivity, specificity, and F1-score. The SMOTE Tomek [16] technique is used to balance the dataset by generating synthetic samples for the minority class and removing noisy and borderline examples from both the minority and majority classes. The DSCC_Net model demonstrates a notable level of performance, achieving an accuracy rate of 94.17%, a recall rate of 93.76%, an F1-score of 93.93%, a precision rate of 94.28%, and an AUC of 99.42%.

An alternative approach is demonstrated by Alam et al. [17], who proposed geometric augmentation in skin cancer detection. Data augmentation involved cropping the images to 256 × 256, horizontal flipping, and rotation at various angles. The study utilized the HAM10000 dataset, which initially consisted of 10,015 samples but which was increased to over 30,000 images through data augmentation. Feature extraction was performed using AlexNet, InceptionV3, and RegNetY-320. The proposed method achieved accuracy, F1, and

ROC values of 91%, 88.1%, and 0.95, respectively. A similar approach was also carried out by Sae Lim et al. [18], who proposed geometric augmentation techniques, including rotation, zooming, shifting, and flipping. Experiments were performed using MobileNet on the HAM10000 dataset, leading to performance metrics of accuracy 83.23%, specificity 87%, sensitivity 85%, and an F1-score of 82%.

Alsaidi et al. [19] demonstrated various augmentation techniques in skin cancer detection. Their research proposed the use of GAN to address imbalanced data. Several pretrained models, including EfficientNet-B0, ResNet50, ViT, and ConvNeXT, were employed. The utilization of GAN as augmentation and EfficientNet-B0 on the HAM1000 dataset yielded an accuracy rate of 96.8%, precision rate of 96.8%, recall rate of 96.9%, and F1-score of 96.8%. The development of GAN models for data augmentation was conducted by Qin et al. [20], who proposed style-based GANs. This method was tested on the ISIC 2018 dataset and achieved an accuracy of 95.2%. Using the same dataset, Ali et al. [21] proposed progressive generative adversarial networks (PGANs) and achieved an accuracy of 70.1%.

3. Materials and Methods

3.1. Dataset

HAM10000 (https://dataverse.harvard.edu/dataset.xhtml?persistentId=doi:10.7910/DVN/DBW86T, accessed on 2 July 2023) is a dataset containing clinical images of various pigmented skin lesions, including both malignant (cancerous) and benign cases. The dataset consists of 10,015 dermatoscopic images of various skin lesions. These images vary in their types and characteristics. The data are categorized into seven categories based on the type of skin lesion. These categories include melanocytic nevi (*nv*), melanoma (*mel*), benign-keratosis-like lesions (*bkl*), basal cell carcinoma (*bcc*), actinic keratoses and intraepithelial carcinoma (*akiec*), vascular lesions (*vasc*), and dermatofibroma (*df*). Figures 1 and 2 show the number and image samples of each category, respectively. Every image in the collection is accompanied by clinical metadata that includes information such as patient age and gender and the location of the skin lesion. Dermatology experts have provided annotations and diagnoses for each image in this dataset. These annotations include information about the type of lesion (whether it is malignant or benign) and its characteristics. The images in the HAM10000 dataset are of high resolution and good quality, making them suitable for in-depth analysis and diagnosis. HAM10000 is widely used by researchers and machine learning practitioners to develop and evaluate algorithms for skin cancer diagnosis. These data have played a crucial role in advancing the field of computer-aided skin cancer diagnosis. HAM10000 is a publicly available dataset, allowing researchers and developers to access and use it for non-commercial purposes.

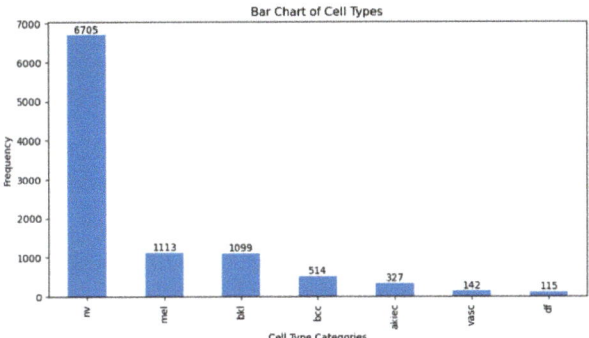

Figure 1. Category distribution of HAM10000 dataset.

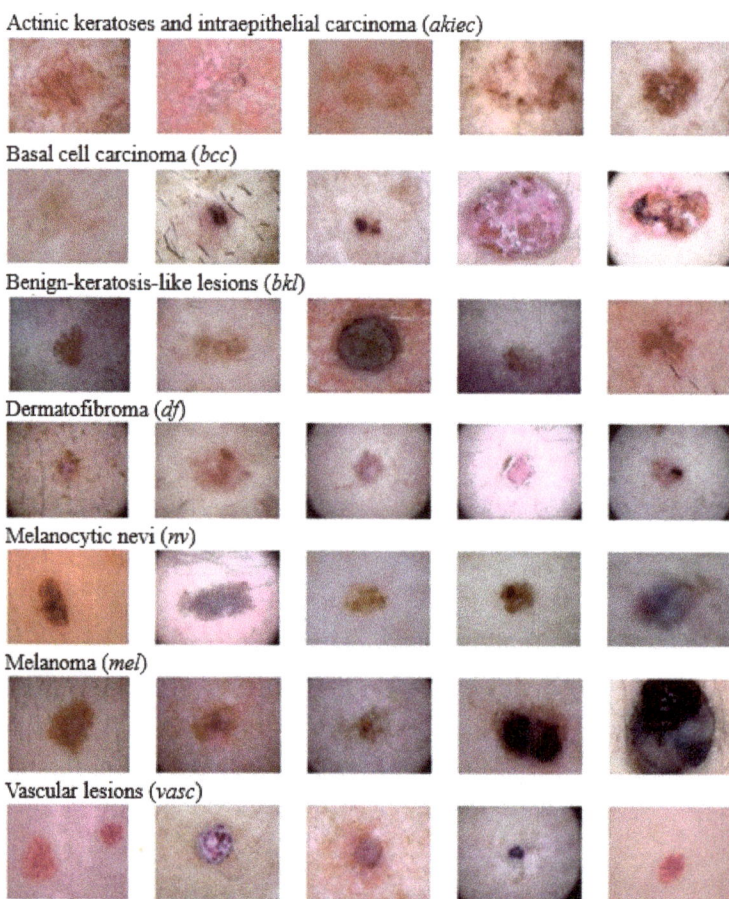

Figure 2. Example image of each class in HAM10000 dataset. From top to bottom: *akiec, bcc, bkl, df, nv, mel,* and *vasc*.

3.2. The Proposed Skin Cancer Detection Method

Transfer learning is especially beneficial when there is minimal data for the new task or when building a deep model from scratch would be computationally expensive and time-consuming. In this study, skin cancer classification employs six pre-trained CNN models, which include Xception, Inceptionv3, Resnet152v2, EfficientnetB7, InceptionresnetV2, and VGG19. In order to build a robust model, we apply augmentation techniques to categories that have a limited number of images. Two-stage input-space augmentations—namely, geometric and GAN augmentations—are proposed. Figure 3 shows the flow of skin cancer detection with the proposed augmentation.

Geometric augmentation is one of the data augmentation techniques used in computer image processing, particularly in the context of deep learning and pattern recognition. The goal of geometric augmentation is to enhance the diversity of training data by altering the geometry of the original image without changing the associated labels or class information related to that image. In this way, machine learning models can learn more general patterns and are not overly dependent on specific poses, orientations, or geometric transformations.

Some commonly used geometric augmentation techniques in deep learning include:

1. Rotation: images can be rotated by a certain angle, either clockwise or counterclockwise.
2. Translation: images can be shifted in various directions, both horizontally and vertically.
3. Scaling: images can be resized to become larger or smaller.
4. Shearing: images can undergo linear distortions, such as changing the angles.
5. Flipping: images can be flipped horizontally or vertically.
6. Cropping: parts of the image can be cut out to create variations.
7. Perspective Distortion: images can undergo perspective distortions to change the viewpoint.

By applying these geometric augmentation techniques, training data can be enriched with geometric variations, which helps machine learning models become more robust to variations in real-world images. This allows the model to perform better in pattern recognition tasks, such as object classification, object detection, or image segmentation, even when objects appear in different orientations or poses.

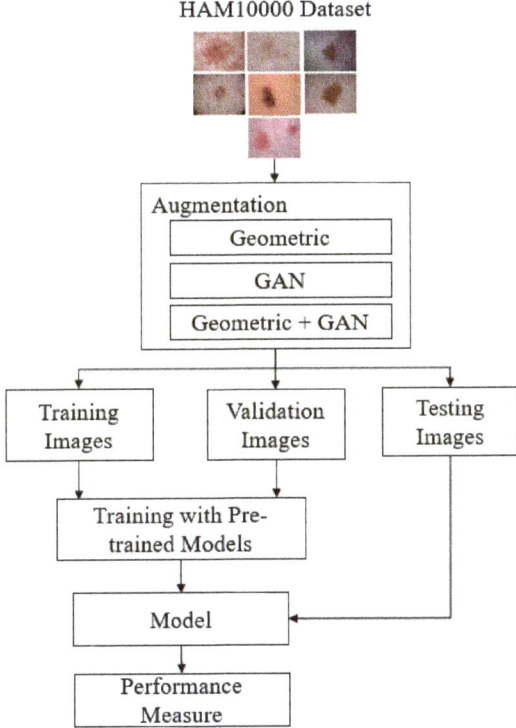

Figure 3. The proposed skin cancer detection method.

GAN [22] augmentation refers to the use of generative adversarial networks (GANs) as one of the data augmentation techniques in the context of machine learning, especially in image processing. GAN is an artificial neural network architecture consisting of two models, the generator and the discriminator, that compete in a game to improve their capabilities [23]. In the context of data augmentation, GAN augmentation involves using the GAN generator to create additional data that are similar to the existing training data. The GAN generator tries to create images that appear authentic, while the GAN discriminator attempts to distinguish between images generated by the generator and real images.

By combining the images generated by the GAN generator with the training data, the dataset can be enhanced with image variations that appear realistic. GAN augmentation has been proven effective at improving the performance of machine learning models, especially in image recognition tasks such as object classification, object detection, or image segmentation, as it can create more diverse and relevant image variations.

3.3. Design of Experiments

In the experiment, 20% of the 10,015 images, which are 2003 images, are utilized for testing, while the remaining 8012 images are split into 90% (7210) for training and 10% (802) for validation. Three methods are used to oversample the data: geometric, GAN, and geometric+GAN augmentations. Experiments are carried out using Python 3.11.5 and were run on an Nvidia DGX Station A100 with a 40 GB GPU, a 64-core CPU, and 512 GB of DDR4 RAM.

Several experimental schemes are established to achieve the best performance. In the first scheme, skin cancer detection is conducted using the original data (without augmentation). In the second, third, and fourth schemes, the original data are augmented using geometric augmentation, GAN augmentation, and geometric+GAN augmentation, respectively. This study uses rotation, shift, shear, zoom, flip, and brightness for geometric augmentation, with detailed parameter values shown in Table 1. During GAN-based augmentation, a total of 1000 epochs are run with a batch size of 64. Table 2 shows the structure of discriminator and generator networks of GAN-based augmentation. In the discriminator, we employ the Adam optimizer with a learning rate of 0.0002 along with binary cross-entropy as the loss function. LeakyReLU with $\alpha = 0.2$ is applied as the activation function for all layers except the last one, where a sigmoid activation function is utilized. A discriminator dropout with a probability of 0.2 is applied. Also in the generator network, each layer utilizes LeakyReLU with $\alpha = 0.2$ except for the final layer, which employs Tanh as the activation function. The parameter values of the training model, such as optimizer, learning rate, and epoch, are shown in Table 3. In this experiment, we also conduct trials with a custom FC layer configuration as shown in Table 4, consisting of a dense layer with 64 neurons, a dense layer with 32 neurons, and a dense layer with 7 neurons [24].

Table 1. Summary of geometric augmentation parameters.

Parameter	Value
rotation_range	20
width_shift_range	0.2
height_shift_range	0.2
shear_range	0.2
zoom_range	0.2
horizontal_flip	True
brightness_range	(0.8, 1.2)

This study also uses SHAP to explain skin cancer detection, which is a technique or approach that utilizes the concept of Shapley values to explain the contribution of each pixel or feature in an image to the model's predictions. CNNs are frequently referred to as black boxes due to the difficulty in deciphering their decision-making processes. For the purpose of understanding model behavior and building trust, SHAP assists in improving the transparency and interpretability of the CNN's decision-making. In SHAP, the concept of Shapley values is applied to measure and understand the influence of each pixel in the image on the model's output or prediction. This technique is valuable for interpreting machine learning models, including the convolutional neural network (CNN) models frequently used for image-based tasks. Positive SHAP values signify that the presence of a pixel had a positive impact on the prediction (red pixel), whereas negative values indicate the contrary (blue pixel) [25].

Table 2. Summary of GAN-based augmentation parameters.

	Layer	Activation
Discriminator	Conv2D	LeakyReLU
	Conv2D	LeakyReLU
	Conv2D	LeakyReLU
	Conv2D	LeakyReLU
	Flatten	
	Dropout	
	Dense	Sigmoid
Generator	Dense	LeakyReLU
	Conv2DTranspose	LeakyReLU
	Conv2DTranspose	LeakyReLU
	Conv2DTranspose	LeakyReLU
	Conv2D	Tanh

Table 3. Parameters of training model.

Parameter	Value
Optimizer	Adam
Learning rate	0.0001
Optimizer parameters	beta_1 = 0.9, beta_2 = 0.999
Epochs	100 (with early stopping)

Table 4. Summary of custom FC layers.

Layer	Output Shape	Activation
Dense	(None, 64)	Relu
Dense	(None, 32)	Relu
Dense	(None, 7)	Softmax

3.4. Performance Metrics

The evaluation of performance was conducted using seven metrics: accuracy (Acc), precision (Prec), recall (Rec), F1-score, SpecificityAtSensitivity, SensitivityAtSpecificity, and G-mean. Accuracy assesses the proportion of true positives and true negatives among all the images. Precision is a metric that quantifies the accuracy of a model's positive predictions. It is calculated by dividing the number of accurate positive predictions by the total number of positive predictions. Recall, also known as sensitivity, measures the ratio of true positives to all relevant elements, i.e., the true positives in the dataset. Specificity is a metric that assesses the model's capability to accurately recognize instances that are actually not part of the positive class in a classification scenario. The F1-score represents the harmonic mean of recall and precision, providing an indication of classification accuracy in imbalanced datasets. Equations (1)–(6) define these seven metrics. G-mean, short for geometric mean, is utilized to assess the effectiveness of classification models, particularly in situations where imbalanced datasets exist.

$$Accuracy = \frac{TP + TN}{TP + TN + FP + FN} \quad (1)$$

$$Precision = \frac{TP}{TP + FP} \quad (2)$$

$$Recall = Sensitivity = \frac{TP}{TP + FN} \quad (3)$$

$$F1\text{-}score = \frac{2 * Precision * Recall}{Precision + Recall} \quad (4)$$

$$Specificity = \frac{TN}{TN + FP} \quad (5)$$

$$G\text{-}mean = \sqrt{sensitivity \times specificity} \quad (6)$$

4. Results and Discussion

Before the prediction process, data augmentation is performed on the training, validation, and testing data in the HAM10000 dataset using geometric augmentation, GAN augmentation, and geometric+GAN augmentation. The limited number of images in the skin cancer class is augmented to bring it closer to the number of images in the class with the highest number of images (*nv* class). The number of images in each class before and after augmentation is shown in Table 5.

Table 6 shows the performance comparison of several pre-trained models with the proposed augmentation method. Using the original data, Resnet152v2 performed the best based on accuracy (84.12%), precision (84.77%), recall (83.67%), and F1-score (84.22%). However, when considering sensitivity, specificity, and G-mean, EfficientnetB7 achieved the best metric values with 99.49%, 94.91%, and 97.17%, respectively. Through the augmentation scheme we proposed, the accuracy of skin cancer detection can be enhanced, reaching a range of 96% to 97.95%. Overall, geometric augmentation produced the best performance based on accuracy, precision, and F1-score metrics, while geometric+GAN yielded the best metrics in terms of sensitivity, specificity, and G-mean values. SensitivityAtSpecificity, SpecificityAtSensitivity, and G-mean all approach 100% when employing geometric+GAN on a tested pre-trained model. It is clear from Table 7 that changing the FC layer makes the accuracy go up to 98.07% when EfficientnetB7 and geometric augmentation are used.

Table 5. Distribution of each skin cancer category for each augmentation scheme.

Category	Original			Geometric Aug.			GAN			Geometric Aug.+GAN		
	Train	Test	Val	Train	Test	Val	Train	Test	Val	Train	Test	Val
vasc	110	26	6	4801	1350	554	4843	1359	503	4805	1371	529
nv	4822	1347	536	4826	1316	563	4854	1302	549	4856	1300	549
mel	792	222	99	4877	1303	525	4858	1319	528	4836	1361	508
df	83	25	7	4775	1423	507	4831	1325	549	4831	1340	534
bkl	785	224	90	4887	1316	502	4813	1329	563	4831	1337	537
bcc	370	101	43	4783	1360	562	4792	1387	526	4798	1394	513
akiec	248	58	21	4844	1319	542	4802	1366	537	4836	1284	585
Num. images	7210	2003	802	33,793	9387	3755	33,793	9387	3755	33,793	9387	3755
Total images	10,015			46,935			46,935			46,935		

Figure 4 shows sample accuracy results from the training and validation of EfficientnetB7 on the original dataset and the proposed augmentation. The training and validation accuracies appear to overfit the original dataset (Figure 4a). Validation accuracy is improved by geometric augmentation (Figure 4b), thereby reducing overfitting. Training accuracy is enhanced through the use of GAN and geometric+GAN (Figure 4c,d).

The sample confusion matrices generated from the original dataset and the best-proposed model are shown in Figures 5 and 6, respectively. Both of these confusion matrices were generated using EfficientnetB7. In Figure 5, many classes are still predicted inaccurately due to imbalanced data. In Figure 6, skin cancer images in the *df* and *vasc* classes can be accurately classified with no classification errors. Only 3 images out of 1319 images in the *akiec* class were misclassified as *bcc*. Six mispredictions were observed among *bcc* samples out of 1360. Fifty-two instances of *bkl* samples were inaccurately predicted out of a comprehensive pool of 1316 samples. Out of the overall 1303 samples,

70 samples belonging to the *mel* class were predicted incorrectly. Similarly, for *nv* cases, 50 mistakes were found in 1316 samples.

Table 6. Performance of the proposed augmentation method on several pre-trained models.

Augmentation Method	Pre-Trained Model	Acc	Prec	Rec	F1	Sensitivity AtSpecificity	Specificity AtSensitivity	G-Mean	Epoch
Original Data	Xception	79.93	80.70	79.53	80.11	99.16	91.71	95.36	12
	Inceptionv3	78.88	79.51	78.48	78.99	99.31	92.31	95.75	11
	Resnet152v2	84.12	84.77	83.67	84.22	99.33	93.56	96.40	18
	EfficientnetB7	78.03	79.63	77.28	78.44	99.49	94.91	97.17	11
	InceptionresnetV2	79.63	80.20	78.68	79.44	99.28	91.76	95.45	19
	VGG19	81.73	81.83	81.63	81.73	99.12	91.26	95.11	28
Geometric	Xception	97.05	97.06	97.01	97.03	99.87	99.12	99.49	19
	Inceptionv3	97.38	97.48	97.35	97.41	99.90	99.20	99.55	31
	Resnet152v2	96.90	96.95	96.86	96.90	99.85	98.93	99.39	28
	EfficientnetB7	97.95	98.00	97.90	97.95	99.91	99.41	99.66	19
	InceptionresnetV2	97.40	97.46	97.36	97.41	99.89	99.20	99.55	28
	VGG19	97.22	97.24	97.20	97.22	99.83	98.84	99.33	32
GAN	Xception	96.08	96.35	95.96	96.16	99.86	98.70	99.28	10
	Inceptionv3	96.50	96.62	96.45	96.53	99.86	98.64	99.25	16
	Resnet152v2	96.30	96.47	96.23	96.35	99.79	98.37	99.08	20
	EfficientnetB7	96.48	96.59	96.44	96.51	99.79	98.25	99.02	20
	InceptionresnetV2	96.22	96.32	96.20	96.26	99.82	98.44	99.13	18
	VGG19	96.22	96.26	96.20	96.23	99.70	100.00	99.85	38
Geometric + GAN	Xception	96.21	96.51	96.04	96.27	99.94	99.22	99.58	9
	Inceptionv3	96.45	96.56	96.39	96.48	99.86	98.56	99.21	20
	Resnet152v2	96.59	96.75	96.45	96.60	99.86	98.87	99.36	14
	EfficientnetB7	96.50	96.61	96.43	96.52	99.89	98.92	99.40	14
	InceptionresnetV2	96.71	96.82	96.67	96.74	99.85	98.62	99.23	21
	VGG19	95.39	97.36	93.89	95.59	100.00	99.93	99.96	17

Table 7. Performance of the proposed augmentation method on the custom FC layer (three dense layers with 64 neurons, 32 neurons, and 7 neurons, respectively).

Augmentation Method	Pre-Trained Model	Acc	Prec	Rec	F1	Sensitivity AtSpecificity	Specificity AtSensitivity	G-Mean	Epoch
Geometric	EfficientnetB7	98.07	98.10	98.06	98.08	99.92	99.46	99.69	20
GAN	Inceptionv3	96.48	96.63	96.44	96.53	99.83	98.54	99.18	17
Geometric + GAN	InceptionresnetV2	96.90	97.07	96.87	96.97	99.86	98.90	99.38	22

We performed a comparative analysis to evaluate the performance of our model by comparing it to the outcomes of earlier research that utilized the use of the HAM10000 dataset, as shown in Table 8. Our proposed approach outperforms earlier findings in a number of metrics. Our limitation is mainly in terms of accuracy when compared to Gomathi et al. [4]. The accuracy rate still needs improvement, and we plan to explore other deep-learning architectures to enhance skin cancer detection. However, in terms of recall, precision, and F1, our approach outperforms the previous research. The standard deviations of accuracy, precision, and recall in our proposed methods also indicate low values, suggesting that our proposed approach demonstrates consistent performance across all three metrics. Figure 7 shows the SHAP explanations of *akiec*, *bcc*, *bkl*, *df*, *mel*, *nv*, and *vasc* samples. The explanations are displayed on a clear grey background, with the testing images on the left.

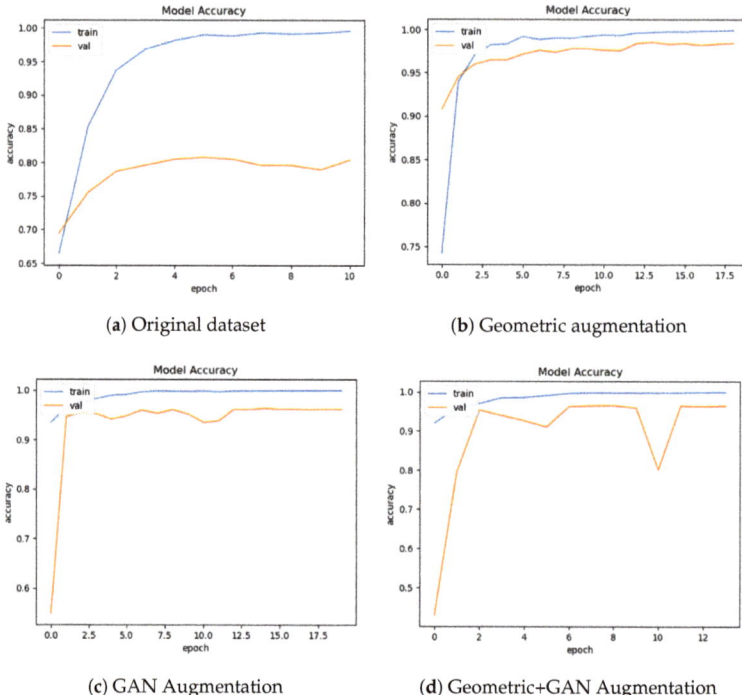

Figure 4. The samples of training and validation accuracy on EfficientnetB7.

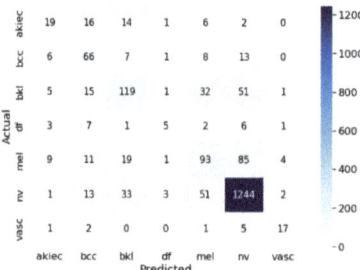

Figure 5. Confusion matrix of EfficientnetB7 on original dataset.

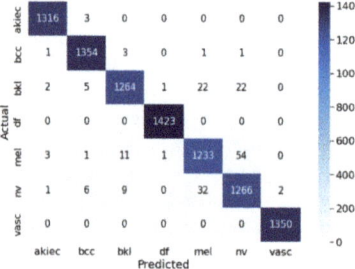

Figure 6. Confusion matrix of the best performance model (EfficientnetB7+Custom FC using geometric augmentation).

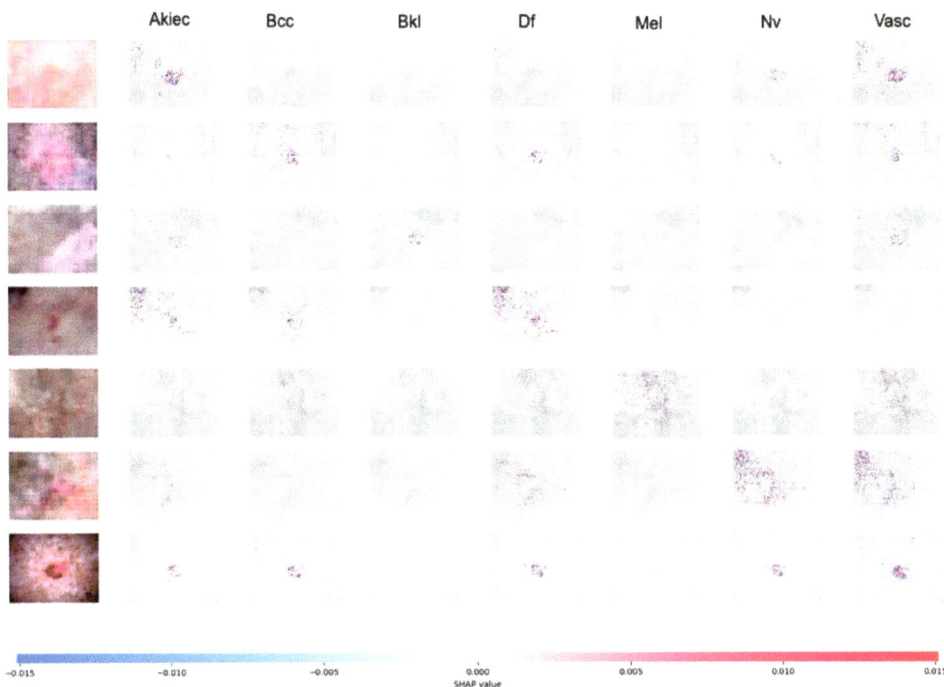

Figure 7. The results of SHAP explanation on InceptionResnetV2 using Geometric+GAN augmentation. The sample images are correctly classified as *akiec*, *bcc*, *bkl*, *df*, *mel*, *nv*, and *vasc* since the high concentrations of red pixels are located in the second, third, fourth, fifth, sixth, seventh, and eighth explanation column images, respectively.

Table 8. A comparative analysis of performance with the latest models.

Ref.	Method	Acc	Prec	Rec	F1	Stdev
Alam et al. [17]	AlexNet, InceptionV3, and RegNetY-320	91	-	-	88.1	-
Kalpana et al. [2]	ESVMKRF-HEAO	97.4	96.3	95.9	97.4	0.7767
Shan et al. [26]	AttDenseNet-121	98	91.8	85.4	85.6	6.3003
Gomathi et al. [4]	DODL net	98.76	96.02	95.37	94.32	1.7992
Alwakid et al. [27]	InceptionResnet-V2	91.26	91	91	91	0.1501
Sae-Lim et al. [18]	Modified MobileNet	83.23	-	85	82	-
Ameri [28]	AlexNet	84	-	-	-	-
Chaturvedi et al. [6]	ResNeXt101	93.2	88	88	-	3.0022
Shahin Ali et al. [29]	DCNN	91.43	96.57	93.66	95.09	2.5775
Sevli et al. [30]	Custom CNN architecture	91.51	-	-	-	-
Fraiwan et al. [31]	DenseNet201	82.9	78.5	73.6	74.4	4.6522
Balambigai et al. [32]	Grid search ensemble	77.17	-	-	-	-
Shaheen et al. [33]	PSOCNN	97.82	-	-	98	-
This study	Geometric+EfficientnetB7+Custom FC	98.07	98.10	98.06	98.08	0.0002
	GAN+InceptionV3	96.50	96.62	96.45	96.53	0.0009
	Geometric+GAN+InceptionresnetV2+Custom FC	96.90	97.07	96.87	96.97	0.0011

5. Conclusions

This study provides valuable insights into a deep-learning approach for the early detection of skin cancer using image augmentation techniques. The proposed two-stage image augmentation technique, involving both geometric augmentation and GAN augmentation, demonstrated high performance. The proposed model achieves an accuracy of

96.90%, precision of 97.07%, recall of 96.87%, and F1-score of 96.97%. The other metrics, such as sensitivity, specificity, and G-mean, of the proposed augmentation method also achieve better performance compared to the results from the original dataset. The use of an interpretable technique for skin cancer diagnosis is also a significant contribution to the field, as it can help clinicians understand the reasoning behind the diagnosis and improve trust in the system. Overall, this research paper presents a promising approach to automated skin cancer detection that could have a significant impact on patient outcomes and healthcare costs. For future research, we will include another dataset, namely ISIC 2020, to validate the results of the next experiments.

Author Contributions: Conceptualization, C.S.; Data curation, A.S. and J.Z.; Formal analysis, A.W.; Investigation, C.S. and A.W.; Methodology, C.S.; Resources, A.S.; Software, A.S. and J.Z.; Supervision, A.W.; Validation, C.S. and A.W.; Writing—original draft, C.S.; Writing—review and editing, C.S. All authors have read and agreed to the published version of the manuscript.

Funding: This research was funded by DRTPM-DIKTI: 065/A38-04/UDN-09/VII/2023 for research funding in 2023.

Data Availability Statement: Dataset is publicly available at: https://dataverse.harvard.edu/dataset.xhtml?persistentId=doi:10.7910/DVN/DBW86T (accessed on 2 July 2023)

Conflicts of Interest: The authors declare no conflict of interest.

References

1. Bozkurt, F. Skin lesion classification on dermatoscopic images using effective data augmentation and pre-trained deep learning approach. *Multimed. Tools Appl.* **2023**, *82*, 18985–19003. [CrossRef]
2. Kalpana, B.; Reshmy, A.; Senthil Pandi, S.; Dhanasekaran, S. OESV-KRF: Optimal ensemble support vector kernel random forest based early detection and classification of skin diseases. *Biomed. Signal Process. Control* **2023**, *85*, 104779. [CrossRef]
3. Girdhar, N.; Sinha, A.; Gupta, S. DenseNet-II: An improved deep convolutional neural network for melanoma cancer detection. *Soft Comput.* **2023**, *27*, 13285–13304. [CrossRef]
4. Gomathi, E.; Jayasheela, M.; Thamarai, M.; Geetha, M. Skin cancer detection using dual optimization based deep learning network. *Biomed. Signal Process. Control* **2023**, *84*, 104968. [CrossRef]
5. Cassidy, B.; Kendrick, C.; Brodzicki, A.; Jaworek-Korjakowska, J.; Yap, M.H. Analysis of the ISIC image datasets: Usage, benchmarks and recommendations. *Med. Image Anal.* **2022**, *75*, 102305. [CrossRef] [PubMed]
6. Chaturvedi, S.S.; Tembhurne, J.V.; Diwan, T. A multi-class skin Cancer classification using deep convolutional neural networks. *Multimed. Tools Appl.* **2020**, *79*, 28477–28498. [CrossRef]
7. Alsahafi, Y.S.; Kassem, M.A.; Hosny, K.M. Skin-Net: A novel deep residual network for skin lesions classification using multilevel feature extraction and cross-channel correlation with detection of outlier. *J. Big Data* **2023**, *10*, 105. [CrossRef]
8. Mumuni, A.; Mumuni, F. Data augmentation: A comprehensive survey of modern approaches. *Array* **2022**, *16*, 100258. [CrossRef]
9. Chawla, N.V.; Bowyer, K.W.; Hall, L.O.; Kegelmeyer, W.P. SMOTE: Synthetic Minority Over-sampling Technique. *J. Artif. Intell. Res.* **2002**, *16*, 321–357. [CrossRef]
10. Zhang, Y.; Wang, Z.; Zhang, Z.; Liu, J.; Feng, Y.; Wee, L.; Dekker, A.; Chen, Q.; Traverso, A. GAN-based one dimensional medical data augmentation. *Soft Comput.* **2023**, *27*, 10481–10491. [CrossRef]
11. Abayomi-Alli, O.O.; Damaševičius, R.; Misra, S.; Maskeliūnas, R.; Abayomi-Alli, A. Malignant skin melanoma detection using image augmentation by oversampling in nonlinear lower-dimensional embedding manifold. *Turk. J. Electr. Eng. Comput. Sci.* **2021**, *29*, 2600–2614. [CrossRef]
12. Leguen-deVarona, I.; Madera, J.; Martínez-López, Y.; Hernández-Nieto, J.C. SMOTE-Cov: A New Oversampling Method Based on the Covariance Matrix. In *Data Analysis and Optimization for Engineering and Computing Problems: Proceedings of the 3rd EAI International Conference on Computer Science and Engineering and Health Services, Mexico City, Mexico, 28–29 November 2019*; Vasant, P., Litvinchev, I., Marmolejo-Saucedo, J.A., Rodriguez-Aguilar, R., Martinez-Rios, F., Eds.; Springer International Publishing: Cham, Switzerland, 2020; pp. 207–215.
13. Douzas, G.; Bacao, F.; Last, F. Improving imbalanced learning through a heuristic oversampling method based on k-means and SMOTE. *Inf. Sci.* **2018**, *465*, 1–20. [CrossRef]
14. Chang, C.C.; Li, Y.Z.; Wu, H.C.; Tseng, M.H. Melanoma Detection Using XGB Classifier Combined with Feature Extraction and K-Means SMOTE Techniques. *Diagnostics* **2022**, *12*, 1747. [CrossRef] [PubMed]
15. Tahir, M.; Naeem, A.; Malik, H.; Tanveer, J.; Naqvi, R.A.; Lee, S.W. DSCC_Net: Multi-Classification Deep Learning Models for Diagnosing of Skin Cancer Using Dermoscopic Images. *Cancers* **2023**, *15*, 2179. [CrossRef] [PubMed]
16. Batista, G.E.A.P.A.; Bazzan, A.L.C.; Monard, M.C. Balancing Training Data for Automated Annotation of Keywords: A Case Study. *WOB* **2003**, *3*, 10–18.

17. Alam, T.M.; Shaukat, K.; Khan, W.A.; Hameed, I.A.; Almuqren, L.A.; Raza, M.A.; Aslam, M.; Luo, S. An Efficient Deep Learning-Based Skin Cancer Classifier for an Imbalanced Dataset. *Diagnostics* **2022**, *12*, 2115. [CrossRef] [PubMed]
18. Sae-Lim, W.; Wettayaprasit, W.; Aiyarak, P. Convolutional Neural Networks Using MobileNet for Skin Lesion Classification. In Proceedings of the 2019 16th International Joint Conference on Computer Science and Software Engineering (JCSSE), Chonburi, Thailand, 10–12 July 2019; pp. 242–247. [CrossRef]
19. Alsaidi, M.; Jan, M.T.; Altaher, A.; Zhuang, H.; Zhu, X. Tackling the class imbalanced dermoscopic image classification using data augmentation and GAN. *Multimed. Tools Appl.* **2023**. [CrossRef]
20. Qin, Z.; Liu, Z.; Zhu, P.; Xue, Y. A GAN-based image synthesis method for skin lesion classification. *Comput. Methods Programs Biomed.* **2020**, *195*, 105568. [CrossRef]
21. Ali, I.S.; Mohamed, M.F.; Mahdy, Y.B. Data Augmentation for Skin Lesion using Self-Attention based Progressive Generative Adversarial Network. *arXiv* **2019**, arXiv:1910.11960.
22. Goodfellow, I.; Pouget-Abadie, J.; Mirza, M.; Xu, B.; Warde-Farley, D.; Ozair, S.; Courville, A.; Bengio, Y. Generative Adversarial Nets. In Proceedings of the Advances in Neural Information Processing Systems, Montreal, QC, Canada, 8–13 December 2014; Curran Associates, Inc.: Red Hook, NY, USA, 2014; Volume 27.
23. Shamsolmoali, P.; Zareapoor, M.; Shen, L.; Sadka, A.H.; Yang, J. Imbalanced data learning by minority class augmentation using capsule adversarial networks. *Neurocomputing* **2021**, *459*, 481–493. [CrossRef]
24. Shahin, M.; Chen, F.F.; Hosseinzadeh, A.; Khodadadi Koodiani, H.; Shahin, A.; Ali Nafi, O. A smartphone-based application for an early skin disease prognosis: Towards a lean healthcare system via computer-based vision. *Adv. Eng. Inform.* **2023**, *57*, 102036. [CrossRef]
25. Bhandari, M.; Shahi, T.B.; Neupane, A. Evaluating Retinal Disease Diagnosis with an Interpretable Lightweight CNN Model Resistant to Adversarial Attacks. *J. Imaging* **2023**, *9*, 219. [CrossRef] [PubMed]
26. Shan, P.; Chen, J.; Fu, C.; Cao, L.; Tie, M.; Sham, C.W. Automatic skin lesion classification using a novel densely connected convolutional network integrated with an attention module. *J. Ambient. Intell. Humaniz. Comput.* **2023**, *14*, 8943–8956. [CrossRef]
27. Alwakid, G.; Gouda, W.; Humayun, M.; Jhanjhi, N.Z. Diagnosing Melanomas in Dermoscopy Images Using Deep Learning. *Diagnostics* **2023**, *13*, 1815. [CrossRef] [PubMed]
28. Ameri, A. A Deep Learning Approach to Skin Cancer Detection in Dermoscopy Images. *J. Biomed. Phys. Eng.* **2020**, *10*, 801–806. [CrossRef] [PubMed]
29. Ali, M.S.; Miah, M.S.; Haque, J.; Rahman, M.M.; Islam, M.K. An enhanced technique of skin cancer classification using deep convolutional neural network with transfer learning models. *Mach. Learn. Appl.* **2021**, *5*, 100036. [CrossRef]
30. Sevli, O. A deep convolutional neural network-based pigmented skin lesion classification application and experts evaluation. *Neural Comput. Appl.* **2021**, *33*, 12039–12050. [CrossRef]
31. Fraiwan, M.; Faouri, E. On the Automatic Detection and Classification of Skin Cancer Using Deep Transfer Learning. *Sensors* **2022**, *22*, 4963. [CrossRef]
32. Balambigai, S.; Elavarasi, K.; Abarna, M.; Abinaya, R.; Vignesh, N.A. Detection and optimization of skin cancer using deep learning. *J. Phys. Conf. Ser.* **2022**, *2318*, 012040. [CrossRef]
33. Shaheen, H.; Singhn, M.P. Multiclass skin cancer classification using particle swarm optimization and convolutional neural network with information security. *J. Electron. Imaging* **2022**, *32*, 042102. [CrossRef]

Disclaimer/Publisher's Note: The statements, opinions and data contained in all publications are solely those of the individual author(s) and contributor(s) and not of MDPI and/or the editor(s). MDPI and/or the editor(s) disclaim responsibility for any injury to people or property resulting from any ideas, methods, instructions or products referred to in the content.

Article

Predicting Time-to-Healing from a Digital Wound Image: A Hybrid Neural Network and Decision Tree Approach Improves Performance

Aravind Kolli [1,†], Qi Wei [2,†] and Stephen A. Ramsey [1,3,*]

1 School of Electrical Engineering and Computer Science, Oregon State University, Corvallis, OR 97331, USA; kollia@oregonstate.edu
2 Institute for Systems Biology, 401 Terry Ave N, Seattle, WA 98109, USA
3 Department of Biomedical Sciences, Oregon State University, Corvallis, OR 97331, USA
* Correspondence: ramseyst@oregonstate.edu
† These authors contributed equally to this work.

Simple Summary: In this work, we explored computational methods for analyzing a color digital image of a wound and predicting (from the analyzed image) the number of days it will take for the wound to fully heal. We used a hybrid computational approach combining deep neural networks and decision trees, and within this hybrid approach, we explored (and compared the accuracies of) different types of models for predicting the time to heal. More specifically, we explored different models for finding the outline of the wound within the wound image and we proposed a model for computing the proportions of different types of tissues within the wound bed (e.g., fibrin slough, granulation, or necrotic tissue). Our work clarifies what type of model should be used for the computational prediction of wound time-to-healing and establishes that, in order to predict time-to-healing accurately, it is important to incorporate (into the model) data on the proportions of different types in the wound bed.

Abstract: Despite the societal burden of chronic wounds and despite advances in image processing, automated image-based prediction of wound prognosis is not yet in routine clinical practice. While specific tissue types are known to be positive or negative prognostic indicators, image-based wound healing prediction systems that have been demonstrated to date do not (1) use information about the proportions of tissue types within the wound and (2) predict time-to-healing (most predict categorical clinical labels). In this work, we analyzed a unique dataset of time-series images of healing wounds from a controlled study in dogs, as well as human wound images that are annotated for the tissue type composition. In the context of a hybrid-learning approach (neural network segmentation and decision tree regression) for the image-based prediction of time-to-healing, we tested whether explicitly incorporating tissue type-derived features into the model would improve the accuracy for time-to-healing prediction versus not including such features. We tested four deep convolutional encoder–decoder neural network models for wound image segmentation and identified, in the context of both original wound images and an augmented wound image-set, that a SegNet-type network trained on an augmented image set has best segmentation performance. Furthermore, using three different regression algorithms, we evaluated models for predicting wound time-to-healing using features extracted from the four best-performing segmentation models. We found that XGBoost regression using features that are (i) extracted from a SegNet-type network and (ii) reduced using principal components analysis performed the best for time-to-healing prediction. We demonstrated that a neural network model can classify the regions of a wound image as one of four tissue types, and demonstrated that adding features derived from the superpixel classifier improves the performance for healing-time prediction.

Keywords: wound monitoring; computer vision; hybrid learning; image segmentation; superpixel; regression

Citation: Kolli, A.; Wei, Q.; Ramsey, S.A. Predicting Time-to-Healing from a Digital Wound Image: A Hybrid Neural Network and Decision Tree Approach Improves Performance. Computation 2024, 12, 42. https://doi.org/10.3390/computation12030042

Academic Editor: Anando Sen

Received: 23 January 2024
Revised: 20 February 2024
Accepted: 23 February 2024
Published: 28 February 2024

Copyright: © 2024 by the authors. Licensee MDPI, Basel, Switzerland. This article is an open access article distributed under the terms and conditions of the Creative Commons Attribution (CC BY) license (https://creativecommons.org/licenses/by/4.0/).

1. Introduction

1.1. Motivation

Chronic wounds affect 6.5 million Americans [1,2], reduce quality of life, and lead to USD 25 billion per year in healthcare costs in the United States [3]. Proper care and the clinical monitoring of the wound are critical to improving outcomes [4]. Clinicians are trained to recognize the prognostically useful visual characteristics of the wound, such as red granulation tissue, yellow fibrin slough, and black necrosis [5]. However, the cost and distance limit the frequency with which patients can visit a clinic for wound examination, necessitating self-monitoring and wound care in the home setting [6]. Many patients lack the knowledge and tools to do so effectively, which increases the likelihood of (1) delayed healing and (2) poor clinical outcomes [6]. Given the well-recognized need for improved home wound monitoring [1], recent advances in informatics have stimulated interest in developing smart in-home monitoring solutions that would analyze a patient's self-acquired image of the wound [7–9]. With the increasing availability of the public-domain sets of wound images with useful metadata [10] as well as image augmentation methods [11], the use of deep learning methods has become feasible for developing computational systems for image-based wound assessment [12]. A critical consideration in the development of wound image analysis methods is that the predicted variable should be clinically useful. In the context of wound care, one of the key prediction tasks that an image-based machine learning model (i.e., a computer vision model) can be reasonably trained for is the regression problem of predicting the number of days it will take for a wound to heal [13]. Conducting well-controlled studies of healing of standardized, surgically induced wounds is very difficult in humans due to ethical challenges, difficulties in obtaining cohorts with standardized and comparable wounds, and due to the varied conditions that necessitate surgical intervention [14]. Dogs have therefore been used on many occasions as an animal model for the controlled studies of wound healing (see Ref. [15] and references therein).

1.2. Previous Efforts

1.2.1. Traditional Computer Vision Methods Using Wound Images

Prior to the extensive use of deep neural networks for semantic pixel-wise segmentation—assigning a label to each pixel in an image [16]—traditional computer vision methods utilized manually engineered image features. Notably, Gupta et al. [17] used depth information for object boundary detection and hierarchical grouping for category segmentation, and Silberman et al. [18] combined color- and depth-based cues. For the specific application of machine-learning for image-based wound assessment, previous advances include the following Song and Sacan [19], who (1) extracted features using edge-detection, thresholding, and region growing, and (2) used a multilayer perceptron neural network; Hettiarachchi et al. [20], who used active contour models for identifying wound borders irrespective of coloration and shape; and Fauzi et al. [21], who used a four-dimensional color probability map to guide the segmentation process, enabling the handling of different tissue types observed in a wound. While the Fauzi et al. study introduced tissue-type-specific segmentation in the context of image-based wound assessment, the relatively simple region-growing segmentation method that was used limited the resulting tissue-type classification accuracy to approximately 75%.

1.2.2. Deep Learning Models Using Wound Images

The application of deep convolutional neural networks (CNNs) in computer vision has led to significant advances in the area of semantic segmentation. By learning to decode low-resolution image representations to pixel-wise predictions, CNNs eliminate the need for manually engineered features and integrate feature extraction and decision making. For example, Cui et al. [22] developed a CNN-based method for wound region segmentation that outperformed traditional segmentation methods [21]. Fully convolutional networks (FCNs) [23] are another example, allowing for arbitrary input sizes and preserving spatial information. Several FCN-based methods have been proposed for wound segmentation: Wang et al. [7] used an FCN to estimate wound areas and predict the wound healing progress using Gaussian process regression (GPR); Yuan et al. [24] used a deep convolutional encoder–decoder neural network for skin lesion segmentation without relying on prior data knowledge; Milletari et al. [25] proposed the "V-Net" model for 3D medical image segmentation; Goyal et al. [26] applied a two-tier transfer learning approach based on the FCN-16 architecture to segment wound images; Liu et al. [27] presented a modified FCN model replacing the classic FCN decoder with a skip-layer concatenation upsampled with bilinear interpolation; Wang et al. [28] proposed an efficient framework based on MobileNetsV2 [29] to automatically segment wound regions; and, in a key foundation for this paper (see Section 2.2.1), Blanco et al. [9] pioneered the use of the multiclass superpixel [30] classification to map different tissue types within the wound bed. In summary, deep learning and CNNs have shown promising results in the field of semantic segmentation, outperforming traditional methods and paving the way for applications in wound monitoring.

1.3. Our Approach

In this study, using both unlabeled and time-series-labeled images from both humans and dogs (the dog data are from a controlled study with wound images taken every 48 h [31]), we investigated the utility of a hybrid model—using both deep artificial neural networks for feature extraction and using decision trees or Gaussian processes for regression—for predicting how long it will take for a wound to fully heal based on a color digital image of the wound. In the context of regression using features extracted from a deep neural encoder–decoder network segmentation model, we investigated the performance of three regression algorithms (Gaussian process regression (GPR) [32], random forest regression (RFR) [33], and XGBoost regression [34]); two different types of segmentation network architectures (SegNet [35] and U-Net [36]); and two different image sets (original images and a geometrically augmented image-set). Furthermore, we investigated whether the performance of the best such hybrid model could be improved by adding the features derived from a multilayer network trained to categorize the sub-regions as one of four tissue types relevant to wound healing (not wound, necrotic, granulation, or fibrin, a tissue type classification originally proposed by Blanco et al. [9]). From these studies, we obtained the best performance using a feature-set combining two different types of features, which we call Phase 1 and Phase 2 features, as described below.

Principal Contribution of this Work

Our main aim in this work was to improve the prediction of time-to-healing from a color image of the wound. Our work's key contribution to the field of computer vision for wound assessment and monitoring is that it demonstrates that (1) the decomposition of the wound image into tissue type sub-regions (Section 2.1.2) provides features that substantially improve the prediction of the wound time-to-healing; and (2) XGBoost regression provides a superior performance for this regression task over alternative regression models. Our work further clarifies the relative contributions of tissue sub-region proportion data (versus image segmentation-derived features) and of the image augmentation and segmentation model type to performance in predicting wound time-to-healing.

2. Materials and Methods

2.1. Overview of Our Computational Approach

2.1.1. Our Approach for Obtaining Phase 1 Features

In Phase 1 of our approach (Figure 1, bottom), we (1) segment the high-resolution wound image at the pixel level (into "wound" and "not wound"); (2) extract high-dimensional feature information from an inner layer of the encoder–decoder segmentation neural network. This phase has two steps:

Step 1: In this step, we use an encoder–decoder neural network model to carry out a pixel-level binary segmentation (classifying pixels as inside or outside the wound bed) to extract the inner layer's states as a feature encoding of the wound image. Furthermore, we use the segmentation model to extract the wound area and wound percentage area, which are included as features in the regression model (and which are also used in computing the dependent variable for the ground-truth-labeled image-set for the regression task; see Section 2.9). For the neural network model for segmentation, we used deep convolutional network approaches. Specifically, in this work, we evaluated the segmentation performance of two network architectures each for two architecture classes, SegNet and U-Net, using the pixel-level overlap between predicted and ground-truth-segmented images. For each of the four network architectures (two SegNet and two U-Net architectures), we evaluated the performance when the model is trained using original images and when it is trained using an augmented image-set. From the eight models, we used independent labeled images (not used in training and tuning) to select the four best-performing segmentation models and used those to extract features to use in regression.

Step 2: In this step, for two of the network architectures whose inner layers were high-dimensional, we used principal component analysis (PCA) to reduce the dimension of the inner layer-level image encoding, to obtain suitable feature vectors for regression.

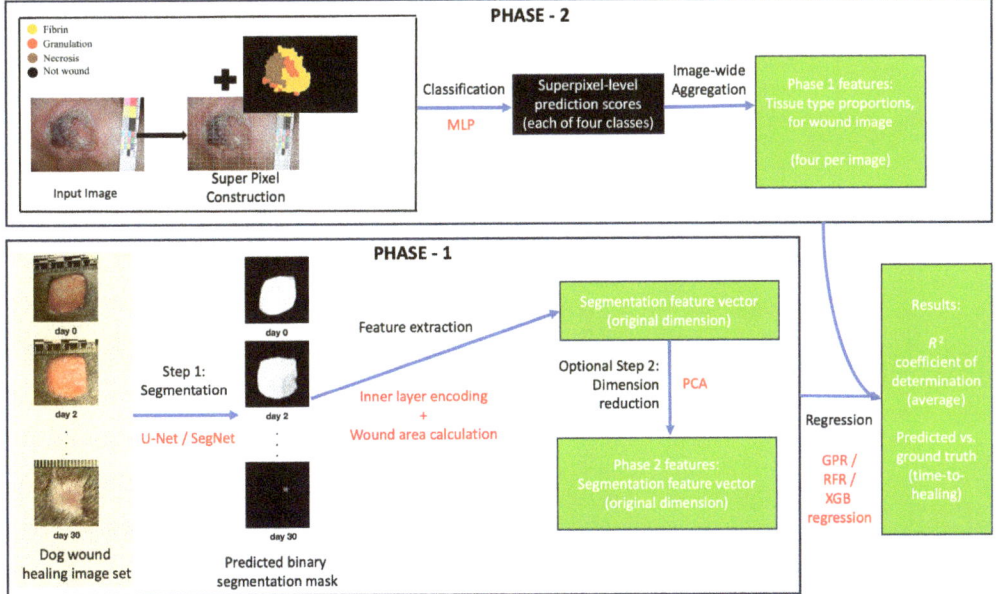

Figure 1. Overview of the two-phase approach that we developed for computationally predicting the time-to-healing from a wound image. (In the Phase 1 section, the inset wound images are the courtesy of Dr. Bryden J. Stanley (see Acknowledgements). In the Phase 2 section, the inset wound images are reprinted from the Blanco et al. study [9], © 2020 with permission from Elsevier (Amsterdam, The Netherlands)).

2.1.2. Approach for Obtaining Phase 2 Features

In this phase (Figure 1, top), our approach classifies the sub-regions of the wound image into the four tissue-type categories (Section 1.3). To do this, we use a multilayer perceptron (MLP) model (Section 2.4) that we trained on labeled 70×70 pixel (px) wound sub-images ("superpixels") from a public dataset [9]. Our approach splits the image into superpixels and then generates a prediction score for each of the four classes, for each superpixel. Four features are then extracted from the superpixel predictions by summing class-specific scores across all superpixels of the image.

2.1.3. Regression

In our approach, we use a regression model to predict, based on features extracted from a wound image in Phase 1 and Phase 2, the number of days it will take for a wound to heal. For the training image set, we used time-series wound images from a study of wound healing in dogs [31]; these images were labeled for the number of days until the wound was fully healed (Section 2.9). We explored the performance of the ensemble decision tree and GPR models to predict wound the time-to-healing, first using the features derived from segmentation alone (Phase 1) and then using the features derived from both the superpixel classifier and from the segmentation model (i.e., Phase 1 and Phase 2 features combined).

For this work, we ran all analyses in Python version 3.5.5 under Ubuntu 16.04 on a Dell XPS 8700 computer (x86_64 architecture) equipped with an NVIDIA Titan RTX GPU (24 GiB GDDR6 memory). We implemented the classification (superpixel and segmentation) and regression model pipelines, including cross-validation and performance evaluation, using the Python software packages Tensorflow (ver. 2.9.1) and scikit-learn (ver. 1.0.2).

2.2. Wound Image Datasets

2.2.1. Ulcer Wound Superpixel Data Set

In Phase 2 of our approach (see Figure 1, top), in order to train a neural network model that can classify the 70 × 70 "superpixel" sub-images of wound images into four tissue types (not-wound, fibrin, granulation, and necrotic), we used the publicly available ULCER_SET images from the Blanco et al. study [9] (see Data Availability Statement). This set comprises 44,893 expert-labeled 70 × 70 px color (red-green-blue) superpixels derived from 40 human lower-limb ulcerous wound images (82.8% not-wound, 8.9% fibrin, 7.3% granulation, and 1.0% necrotic).

2.2.2. Dog Wound Healing Image Set

To train the regression model that predicts the time-to-healing from a digital wound image, we used a previously published [31] set of 136 color images (4000 × 6000 px; acquired every other day over 32 days and labeled by date) of ten 2 × 2 cm^2 dog cutaneous surgical wounds (full-thickness surgical wounds in the trunk; see Ref. [31] for details). The ten dogs were male adult beagles (13–18 weeks of age). The wound images included standard rulers which enabled conversion of mm^2 to px^2 (Section 2.9). Given the size of the segmentation models (Sections 2.6 and 2.7), to fit a reasonably sized batch of training images into the GPU memory, we resized and cropped the raw wound images to 224 × 224 px, to prepare them for feature extraction using the previously trained binary segmentation models. We augmented the 224 × 224 px wound-bed images as described in Section 2.3.1. To enable the use of the dog wound images for training the segmentation model, we manually segmented the images as described in Section 2.3.2 (Figure 2).

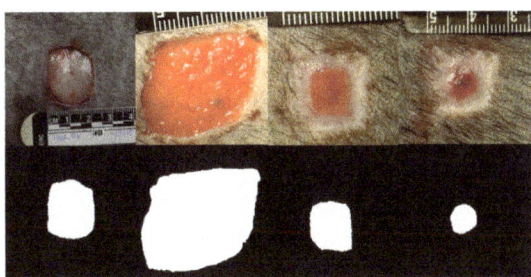

Figure 2. Example wound images (**top row**) and corresponding human-segmented images (**bottom row**) from the dog wound image dataset [31] used for training the healing-time model. Top-row images are cropped from original images that were provided courtesy of Dr. B. Stanley (see acknowledgments).

2.3. Image Augmentation and Annotation

2.3.1. Augmentor

To geometrically augment (e.g., rotate or flip) the sets of wound images used in this study, we used the Augmentor [11] tool as described in the Supplementary Materials Table S1, yielding four augmented images for each dog wound image. In this study, we compared the performances of different wound segmentation algorithms trained on original images (without augmentation) as well as algorithms trained on augmented image sets (see Section 3.2).

2.3.2. Pixel Annotation

To binary-segment wound images at the pixel level as "in-wound" our "not in wound" with human guidance, we used the PixelAnnotationTool software tool (ver. 0.14.0), which uses the marker-based variant of the watershed segmentation algorithm [37] from the OpenCV software library [38]. The manually annotated wound image masks were used as labeled data for training the segmentation algorithms (Section 2.11.2).

2.4. Superpixel Classifier Model Architecture

To classify the wound image superpixels by tissue type (not wound, fibrin, granulation, and necrotic), we implemented a six-layer perceptron [39] with rectified linear unit (ReLU) activation in each intermediate hidden layer and softmax activation with four classes in the output layer. For each of the four classes, we measured the model's prediction performance by computing the area under the receiver-operating characteristic (AUROC) curve for the class's prediction scores using the one vs. rest strategy [40] for comparison.

2.5. Segmentation

In our two-classe segmentation model architectures, SegNet (Section 2.6) and U-Net (Section 2.7), we do not use recurrent edges, whose use in segmentation has been advocated [41] in applications requiring multiscale object recognition (which is not an issue in our application). The two main classes of segmentation model architectures are described in the following two subsections.

2.6. SegNet Model Architecture

Of the two classes of neural network architectures that we used for image segmentation, the first is SegNet [35], a ten-layer convolutional network. SegNet is built on the FCN [23] architecture, which consists of an encoder network that computes a set of compact feature maps on high-resolution images, a decoder network that upsamples the feature maps, and a pixel-wise classification layer that outputs the full-size segmentation masks. The main difference separating a SegNet model from a common FCN model is that the decoder layers of a SegNet model directly use the pooling indices computed in the corresponding encoder layers' max-pooling step. In this way, when a decoder layer performs the nonlinear upsampling of its lower-resolution input feature map from the previous layer, there is no need to learn the weights of the decoder part to upsample again. The advantages of reusing max-pooling indices include drastically reducing the number of parameters needed in the training process, improved on boundary delineation, and minimal modification required to implement upsampling. The encoder network is constructed by stacking basic computation blocks like convolution, nonlinear transformation using ReLU activation function, spatial pooling, and local response normalization [42]. To produce probability maps, a softmax layer is appended to the end of the network. To mitigate the downscaling effects of the convolution and pooling layers, the decoder network of the SegNet model is constructed by a stack of layers with upsampling operations. From the pixel-level probabilities, a threshold of 0.5 is used to produce the segmentation mask.

2.7. U-Net Model Architecture

The second class of the segmentation network architecture that we evaluated is U-Net [36], a U-shaped, 23-layer FCN. To localize, high-resolution features from the contracting path are combined with the upsampled output; a successive convolution layer can then learn to assemble a more precise output based on this more detailed information. U-Net has many feature channels in the upsampling part, which allow the network to propagate context information to higher resolution layers. U-Net does not have any fully connected layers and only uses the valid part of each convolution, allowing the segmentation of arbitrarily large images by an overlap-tile strategy (which also enables training on high-resolution images). For border region prediction, the missing context is extrapolated by mirroring the input image. At each downsampling step, the number of feature filters is doubled. The contracting path has repeated two 3×3 convolution layers (unpadded convolutions), each followed by a ReLU activation layer and a 2×2 max pooling operation with stride 2 for downsampling. The expanding path has an upsampling of the feature map followed by a 2×2 convolution layer ("up-convolution") that halves the number of feature filters, a concatenation with the correspondingly cropped feature map from the contracting path, and two 3×3 convolutions, each followed by a ReLU activation layer. A 1×1 convolution is used to map each 64-component feature vector (each component is a feature filter) to the desired number of classes at the final layer. The U-Net has been reported to work well for segmentation applications with small training sets [28].

2.8. Feature Engineering and Extraction

We extracted features from both the Phase 2 (superpixel tissue-type classifier) model and from the Phase 1, Step 1 (pixel-level binary segmentation) models, as described below.

2.8.1. Feature Extraction from a Superpixel Model

For each of the four wound-tissue types (Section 2.2.1), we estimated the proportion of the wound image of that type adding up the prediction value for that type's softmax class output across all superpixels. This procedure generated four features per image.

2.8.2. Feature Extraction from Segmentation Models

We extracted features from the segmentation models (Sections 2.6 and 2.7) in two ways—using the network's inner layer states as a vector encoding of the image, and by calculating summary statistics on the pixel-level binary segmentation (which were appended to the encoding vector).

Inner Layer Encoding

For the SegNet architecture (Section 2.6), we used the output of the intermediate layer "Conv5" as a feature vector. For the SegNet-1 network, the "Conv5" layer's dimension is 6272, and thus, we reduced it using PCA (see Section 2.8.3) to 404 principal components.

For the SegNet-2 network, the "Conv5" layer's dimension is 1568 and we directly used that layer's values as the feature vector. For the U-Net architecture in Section 2.7, we used the output of intermediate layer "Conv5" as a feature vector. For the U-Net-1 network, the "Conv5" layer's dimension is 50,176, and thus, we reduced it using PCA to 342 principal components. From the U-Net-2 network, the "Conv5" layer's dimension is 3136 and we used that layer's values directly as the feature vector.

Wound Area Calculation

For each segmented wound image, from the pixel-level segmentation mask, we computed two summary-level features, the overall pixel wound area (see also Section 2.9), and the percentage of pixels of the image that are in the wound area. These two features were added to the "Conv5"-derived features to generate the complete Phase 1 feature-set for use in the regression.

2.8.3. PCA Reduction in Segmentation Feature Vectors

We carried out PCA using the function "PCA" from the sklearn.decomposition package, with parameter svd_solver set to "full" and parameter n_components set to 0.95 (which selects the number of principal components so that it explains at least 95% of the variance in the feature vector).

2.9. How We Obtained the Dependent Variable for Regression Training

Using the output from the segmentation models (Section 2.8.2), we first determined the length scale "ratio" of each image by calculating the pixel length of 1 cm space on the ruler in the image (manually counted by visual image inspection). We obtained the predicted wound area by counting the pixel area on the segmentation mask annotations and converted it to area in cm^2 using the empirically determined linear pixel density per cm of each image. With the predicted wound area for each image, we obtained the remaining proportion of wound area as a feature for each image by dividing the predicted wound area of the current day with the predicted wound area of day zero. In the healing status prediction task, we did not use the number of days post-injury as a feature; this is because we used the image date to determine the dependent variable (i.e., the number of days to full healing) for the regression.

2.10. Regression Model Training

We evaluated three general-purpose regression algorithms (which are well described in the literature and highly versatile) that are well suited to our problem from the standpoints of sample-size of our labeled image-set, the high dimensionality of the feature-space, and the fact that the features are continuous: random forest regression, Gaussian process regression, and XGBoost.

2.11. Model Training, Tuning, and Evaluation

We implemented cross-validation using RandomizedSearchCV and GridSearchCV from the package sklearn.model_selection, as described below. For all prediction evaluation metrics, we used functions from the sklearn.metrics package as described below.

2.11.1. Superpixel Classifier

For the Phase 2 classifier (Sections 2.1.2 and 2.4), we used stratified ten-fold cross-validation (25 epochs) to obtain performance measurements (sample-averaged categorical cross-entropy loss) on the training-set images for hyperparameter tuning. We calculated the sample-averaged categorical cross-entropy loss as follows:

$$L = -\frac{1}{N}\sum_{i,j} y_{ij} \log \hat{y}_{ij},$$

where j ranges over the four possible class labels, i ranges over the N samples, y_{ij} is the ground-truth (one-hot encoded) class label for class label j for sample i, and $\hat{y}_{ij}$ is the prediction score for the class label j for sample i. We ultimately trained the model with a batch size of 1024 and for 25 epochs to using stochastic gradient descent [43] with a learning rate of 0.005, to minimize the categorical cross-entropy loss. For measuring AUROC on the test set of superpixels, we used the function "roc_auc_score" with the parameter "average" set to "weighted".

2.11.2. Binary Pixel-Level Segmentation

For the image segmentation models (Sections 2.6 and 2.7), we used five-fold cross-validation to obtain unbiased performance measures (precision, recall, and Dice overlap) on the training images for hyperparameter tuning. The hyperparameters were the number of epochs, batch size, and learning rate. We ultimately trained both SegNet models and both U-Net models with a batch size of two and for 2000 epochs (with early stopping) using stochastic gradient descent with a learning rate of 10^{-4} to minimize cross-entropy loss.

2.11.3. Regression Using Decision Trees or Gaussian Process

For Gaussian process regression, we used the class GaussianProcessRegressor from the package sklearn.gaussian_process; for random forest regression, we used the class RandomForestRegressor from the sklearn.ensemble package; and for XGBoost regression, we used the class XGBoostRegressor from the same package. For Random Forest and XGBoost, we first tuned model hyperparameters using RandomizedSearchCV using a larger set and range of hyperparameters (six for random forest, and four for XGBoost); for the Gaussian process model, we did not use the random-search hyperparameter tuning. Then, for all three models, we carried out exhaustive grid search hyperparameter tuning using the GridSearchCV function from sklearn.model_selection (four hyperparameters for Gaussian process, six for random forest, and four for XGBoost); for both, we used five-fold cross-validation (see Ref. [42] for details). For XGBoost regression model with the Phase 2 features only, to avoid overfitting, the hyperparameter grid-search space was reduced as follows: max_depth $\in \{1, 2, 3\}$, learning_rate $\in \{0.0005, 0.001, 0.005, 0.01, 0.05\}$, and n_estimators $\in \{10, 20, 30\}$.

2.11.4. Regression Using a Deep Neural Network

Using the Phase 1 (Section 2.1.1) and Phase 2 (Section 2.1.2) features as inputs and days until healing as the dependent variable (Section 2.9), we trained a five-layer fully connected deep neural network with ReLU activation functions and with the following numbers of neurons at each layer: 256, 128, 64, 32, and 1.

2.11.5. Confidence Interval Estimation

For the regression task, we estimated $\pm 1\,\sigma$ confidence intervals for the test-set coefficient of variation (R^2) using bootstrap resampling [44] with 1000 iterations.

3. Results

3.1. Superpixel Tissue-Type Classification Performance

Under the premise that for predicting time-to-healing, the utility of features extracted from a tissue-type multiclass superpixel classifier depends on the classifier's performance, we investigated (Section 2.4) the MLP model's performance for annotating non-overlapping 70×70 px wound image superpixels. For this analysis, we used the ULCER_SET (44,893 superpixels; see Section 2.2.1) in which each superpixel was expert-labeled with one of four tissue types [42].

We randomly separated the superpixels into training/validation and test sets (80% and 20%, respectively), tuned the classifier hyperparameters as described in Section 2.11.1, and obtained categorical cross-entropy loss on both the training/validate and test sets of superpixels and AUROC model performance (each class against all others) on the test set of superpixels. The prediction error on the test set was comparable to the training set (Figure 3), indicating that the model was not overfitted. The single-class-versus-others AUROC prediction performance exceeded 0.8 for all classes except class 3 (necrotic) (Figure 4); performance was best for discriminating class 2 (granulation) superpixels from the other superpixel classes (AUROC 0.94). The overall accuracy of the model for superpixel type prediction was 86.4% (which is comparable to the MLP results in Blanco et al. [9]).

Having measured the accuracy of the wound image superpixel classifier, we calculated, for each image, four summary features representing the total sum (across all superpixels of the image) of the class-specific prediction scores (Section 2.1.2); these four features are the "Phase 2" features, whose relative utility for the healing-time prediction regression task we evaluate in Section 3.4.

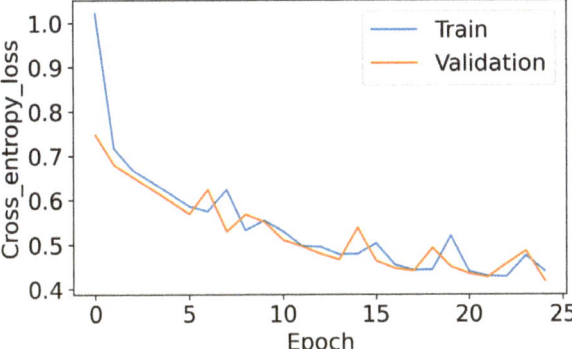

Figure 3. Average categorical cross-entropy loss of the MLP model for each of 25 training epochs, evaluated on the train/validate superpixels and on the test-set superpixels.

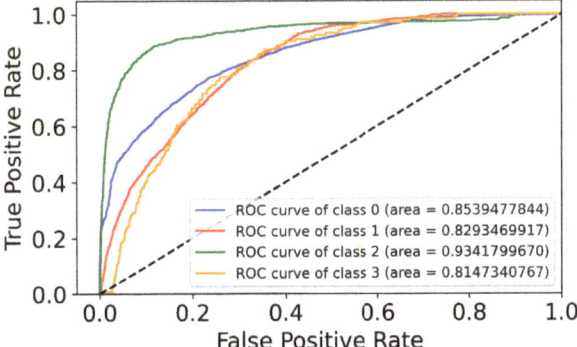

Figure 4. Test-set performance for classifying superpixels as class 0 (not-wound), class 1 (fibrin), class 2 (granulation), or class 3 (necrotic). Each receiver operating characteristic (ROC) curve represents the test-set performance on the binary task of predicting whether a superpixel is of the indicated class or not. The class sample counts as follows: Class 0, 7437 samples; Class 1, 794 samples; Class 2, 656 samples; and Class 3, 89 samples. "Area" denotes AUROC. Each ROC curve shows a relationship between sensitivity (the vertical axis) and the false positive error rate (the horizontal axis), for correctly recognizing superpixels that are members of one class versus mis-predictions of the other three class types. Each ROC curve rises steeply as easy cases are discriminated and then saturates as more borderline cases require increasingly permissive thresholds for making a positive prediction.

3.2. Binary Segmentation (Phase 1, Step 2) Model Performance to Find the Best Four Models

Under the premise that, for predicting wound time-to-healing, the utility of features derived from the pixel-level segmentation model will depend on the segmentation model's performance, we investigated (Sections 2.6 and 2.7) the performance of four neural network architectures (SegNet-1, SegNet-2, U-Net-1, and U-Net-2) for binary pixel-level segmentation. For this model-selection task, used a set of 136 high-resolution dog wound images [31] that had been pixel-wise labeled (Section 2.2.2) by human annotators as "within wound" or "outside wound". Of the 136 dog wound images, we set aside 26 images (images from two wounds every 48 h) as a test-set of labeled images for evaluating the performance of the trained regression models (Section 3.3); we used the other 110 images for both segmentation model selection and for training the regression models. Given previous reports that image-set geometric augmentation improves the learning performance in some computer-vision tasks [45] and does not improve performance in others [46], we investigated the performance of the four network architectures using the original dog wound image set ("OIS": 74 images) and using an augmented (Section 2.3.1) image set ("AIS": 945 images).

We randomly partitioned the 110 images into 74 "OIS" images (images from five wounds) for training/validation and 36 images (images from three wounds) for model testing; we augmented 945 "AIS" images from 74 "OIS" images for training/validation and the same 36 original (unaugmented) images for testing. Using the training/validation set, we tuned the hyperparameters as described in Section 2.11.2. Then, for each of the eight combinations of the network architecture and image set ("AIS" or "OIS"), we trained on the 74 images of "OIS" and 945 images of "AIS" and measured the average performance on the test image-set by precision, recall, and Dice overlap. SegNet-1/AIS has the highest recall (0.953) and Dice coefficient (0.921) values (Table 1), whereas U-Net-1/AIS had the highest average precision (0.962).

Table 1. Test-set performance (averaged over 36 images) of each of eight combinations of network architecture and training-set performance (original, i.e., OIS; or augmented, i.e., AIS) on the segmentation task of classifying the wound image pixels as "within wound" or "not within wound". OIS: original image set (without augmentation); AIS: augmented image set.

Architecture	Set	Precision	Recall	Dice
SegNet-1	AIS	0.955	**0.953**	**0.921**
SegNet-1	OIS	**0.962**	0.710	0.773
SegNet-2	AIS	0.933	0.906	0.887
SegNet-2	OIS	0.740	0.920	0.787
U-Net-1	AIS	**0.962**	0.950	0.916
U-Net-1	OIS	0.946	0.883	0.878
U-Net-2	AIS	0.930	0.937	0.890
U-Net-2	OIS	0.952	0.948	0.919

For all architectures except U-Net-2, the performance was higher when trained on AIS than when trained on OIS; on average, augmentation improved performance by 7.7%. We selected the best model within each architecture type, to take forward to use for extracting segmentation-based features for time-to-healing prediction: SegNet-1/AIS, SegNet-2/OIS, U-Net-1/AIS, and U-Net-2/OIS. We then extracted image-level features from each of the four models using intermediate-layer neuron values (Section 2.8); due to the high dimensionality of these layers in SegNet-1 and U-Net-1, we reduced the dimensions for the SegNet-1 (Section 2.6) and U-Net-1 (Section 2.7) derived features using PCA. We combined the segmentation-derived features with the superpixel classification-derived features (Section 2.1.2).

3.3. Time-to-Healing Prediction Performance without Phase 2 Features

Next, we turned to the regression task of predicting, from features extracted from the wound image segmentation (i.e., not including the superpixel-based features), the number of days remaining for the wound to heal. We used the features extracted from one of the four binary segmentation models in Phase 1, Step 2 (with PCA used to reduce the dimension of the segmentation vectors extracted from SegNet-1 and U-Net-1, as described in Section 3.2), yielding a total of 404 features per image for SegNet-1 and 342 features per image for U-Net-1. We then added two summary-level features derived from the segmentation, consisting of the wound area and wound percentage area (see Section 2.8.2). Next, for the labeled training examples, we used a collection of 544 augmented time-series images of ten dog wounds (Section 2.2.2) from a controlled study of wound healing [31], for which we estimated (Section 2.8) the time-until-complete-healing (i.e., the dependent variable for the regression) for each of the images. We sought to evaluate three different regression algorithms, (Gaussian process regression, random forest regression, and XGBoost regression) separately against the feature-sets from the four different segmentation models for a set of 12 models. Using the set of 440 augmented (110 unaugmented images) dog images (images of eight different wounds), we tuned (Section 2.11.3) the regression algorithms' hyperparameters and then trained each of the 12 models. Using either the 104 augmented (or 26 original) dog images (from two wounds) that were withheld as a test set, we obtained the average coefficient of variation (R^2) performance measurements, for each of the 12 models. The combination of XGBoost with the SegNet-1/AIS-derived features had the best regression performance, with $R^2 = 0.839$ (Table 2) (SegNet-1/AIS also yielded the best segmentation performance; Table 1).

Table 2. Average prediction performance of three regression models—each with input features from one of four different segmentation models—on a test set of 104 images (26 original images of two dog wounds; four-fold augmented as described in Section 2.3.1), as measured by R^2 coefficient of variation. The models shown here were trained with Phase 1 features only; they did not include any superpixel-derived (i.e., Phase 2) features. Column abbreviations as follows: Arch., segmentation network architecture; PCA, indicates whether or not that segmentation model's image encoding was PCA-reduced; CV R^2, cross-validation average R^2 on the set of 440 images used for training the regression model; Test R^2, average R^2 on the test set of images; L.C.I., lower confidence interval (1 σ) on the test R^2; U.C.I., upper confidence interval (1 σ) on test R^2.

Model	Arch.	PCA?	Feat.	CV R^2	Test R^2	L.C.I.	U.C.I.
GPR	U-Net-1	Yes	344	0.916	0.773	0.751	0.799
GPR	U-Net-2	No	3138	0.791	0.762	0.739	0.781
GPR	SegNet-1	Yes	406	0.904	0.778	0.749	0.797
XGBoost	U-Net-1	Yes	344	0.919	0.766	0.749	0.775
XGBoost	U-Net-2	No	3138	0.867	0.795	0.774	0.821
XGBoost	SegNet-1	Yes	406	**0.922**	**0.839**	0.825	0.852
XGBoost	SegNet-2	no	1570	0.890	0.811	0.799	0.826
RFR	U-Net-1	yes	344	0.914	0.770	0.765	0.785
RFR	U-Net-2	no	3138	0.848	0.782	0.769	0.803
RFR	SegNet-1	yes	406	0.916	0.831	0.817	0.851
RFR	SegNet-2	no	1570	0.893	0.797	0.781	0.818

While the SegNet-1/AIS/XGBoost performance is only slightly better (0.839 vs. 0.831) than SegNet-1/AIS/Random-Forest, XGBoost trained faster than Random Forest (63 min vs. 118 min) than the RFR model when using the SegNet-1-PCA feature vector. Overall, across the four feature sets, XGBoost regression models had a 3.8% better performance than GPR models and 1.1% better than the random forest.

3.4. Time-to-Healing Prediction Performance Including Phase 2 Features

Having established (Section 3.3) that the combination of using the SegNet-1 model (Section 2.6) for segmentation (trained with image augmentation) and using XGBoost for regression has the best performance among models tested for predicting wound time-to-healing, we next investigated whether adding four features derived (Section 2.8.1) from the superpixel tissue-type classifier (Section 2.4; i.e., Phase 2 of our method, as shown in Figure 1), which would improve the performance for the time-to-healing prediction. On the same train/validation and test sets of images as used for the regression model using only the segmentation (Phase 1) features (Section 3.3), we measured the average R^2 performance on the training/validation and test image sets, for two XGBoost regression models: a "Phase 1 model" trained with only the 406 SegNet-1 segmentation-derived features, and a "Phase 1 + 2 model" trained with all of those features plus four superpixel tissue type classifier-derived features (for a total of 410 features). We found that the test-set performance was higher for the "Phase 1 + 2 model" (0.863) than for the "Phase 1" model (0.839) (Table 3).

Finally, to compare the performance of the hybrid approach (consisting of deep learning-derived features and decision tree-based regression, as shown in Figure 1) with a fully neural network approach, we implemented an alternative fully neural-network-based approach in which the same Phase 1 and Phase 2 features were used as inputs to a deep neural network regression model (see Section 2.11.4 for details). The fully neural network approach's R^2 performance (0.823 on CV, and 0.813 on the test set) was significantly lower than with the XGBoost-based hybrid approach (0.966 on CV, and 0.863 on the test set).

Table 3. Average prediction performance of XGBoost regression on a test set of 104 images (images from two different dog wounds; 4× augmented from 26 original wound images), as measured by R^2 coefficient of variation, for models trained with 406 features (from SegNet-1 segmentation only) and/or with the four superpixel-derived tissue classification features (Section 2.8.1). Column abbreviations are as follows: Arch., segmentation network architecture; Phase, indicates which feature-sets were included; CV R^2, cross-validation average R^2 on the augmented set of 440 images used for regression training; Test R^2, average R^2 on the test set of images; L.C.I., lower-confidence interval (1 σ) on test R^2; U.C.I., upper-confidence interval (1 σ) on test R^2. Performance on the Phase 2 only feature-set was sufficiently low that no C.I. permutation analysis was performed.

Model	Arch.	Phase	Feat.	CV R^2	Test R^2	L.C.I.	U.C.I.
XGBoost	SegNet-1	1 and 2	410	0.966	0.863	0.851	0.875
XGBoost	SegNet-1	1	406	0.922	0.839	0.825	0.852
XGBoost	SegNet-1	2	4	0.603	0.042		

4. Discussion

Our first point of discussion concerns the biological rationale for including image-wide tissue type (superpixel)-derived features in the regression model. The results of Section 3.1 indicate that the MLP model can accurately predict wound image superpixels' tissue types among four classes (Section 2.2.1). Biologically, at the hemostasis (blood clotting) stage, platelets in the blood adhere to the injured site [47]. Platelet activation leads to the activation of fibrin, which autopolymerizes and further promotes platelet aggregation. Thus, a high proportion of "fibrin"-labeled tissue in the wound would be expected to indicate the healing process is in the early stage. Similarly, at the proliferative stage, granulation tissue can be noted from the healthy wound buds that protrude from the wound base [5]; thus, the proportion of "granulation"-labeled tissue would be expected to correlate with the healthy healing process, whereas the proportion of "necrosis"—(i.e., tissue death)-labeled tissue would be expected to be anti-correlated with time-to-healing. The size of superpixels represents a balance between the potential resolution for mapping tissue types in the wound-bed and accuracy for detecting tissue types based on color and texture patterns within the superpixel; in our case, in the dataset of labeled superpixels that we had access to, the superpixel size was chosen to be 70 × 70 px to balance those two priorities.

A second point of discussion concerns the (currently manual) step of counting pixels per cm in order to assess the wound area (which was necessary in order to estimate the days until healing, as explained in Section 2.9). Approaches toward automating this step could include (1) using the Hough transform [48] for detecting the ruler line; (2) using a separate neural network specifically for ruler detection and length calculation; or (3) using image metadata regarding image distance and field size to calculate the field's physical dimensions.

While the results of Table 3 represent a significant new finding in terms of the types of features that are useful for predicting the wound time-to-healing, the context of the dataset is relevant to interpreting the model's absolute performance. The ground-truth set of labeled images used for the regression model in this work are from a series of controlled images (acquired with a single digital camera at fixed distance) from a controlled study [31] of surgical wounds. Thus, we would expect that attaining equivalent absolute predictive performance on wound images from uncontrolled data acquisition settings and from a broader array of wound conditions (e.g., burns, ulcers, etc.) would likely require retraining the regression model with a substantially larger and more diverse set of images. However, the key findings from this work (Tables 1–3) are based on the relative regression performance on a feature set including superpixel tissue-derived features vs. without them. Although the superpixel tissue-derived features (Phase 2 features) by themselves have relatively poor performance on the regression task (see Table 3), when combined with the segmentation-derived features (Phase 1 features) they significantly improve the regression performance.

As far as we are aware, this work represents the first effort to leverage wound images from a controlled study of surgical wound healing for the purpose of regression model selection and feature-set selection for computationally predicting wound time-to-healing. Our long-term goal is to apply these results to develop models for predicting healing time in humans. Furthermore, a natural progression of the work would be to integrate wound images with other measurements such as C-reactive protein (CRP) [49] and immunoglobulin G [50], as well as relevant clinical comorbidities and demographic/anthropometric parameters (e.g., diabetes, nutritional status, age, body mass index, and smoking status) [13] to accurately predict the time-to-healing in human clinical settings and to flag wounds requiring intervention. While these types of demographic, anthropometric, and comorbidity data were not needed for the specific questions that we focused on in this controlled study of canine wound healing (i.e., Can a tissue type-augmented hybrid approach improve time-to-healing prediction? and Which regression model leveraging a hybrid tissue-type and segmentation-derived feature-set gives best performance for time-to-healing prediction?), it is expected that such data would be required in order to maximize accuracy in clinical settings.

5. Conclusions

For three of the four image segmentation models, using the augmented wound image set led to a better image segmentation performance than the models learned using original images without augmentation. Among the four segmentation models, the SegNet-1-PCA model using augmented images had the highest test-set Dice performance (0.921). Using the segmentation-derived features, out of the three different regression models that we studied, XGBoost regression outperformed both Gaussian process regression and random forest regression, reaching a test-set R^2 of 0.839 (95% confidence range of 0.825–0.852). We further found that, given the high-resolution wound images without tissue type labels, the SegNet-PCA model is a powerful tool for extracting low-dimensional feature vectors while generating reasonable wound segmentation masks; this yields a mask that retains the wound area information and features that retain biological patterns that enable the improved image-based prediction of wound time-to-healing. Finally, we demonstrated that incorporating tissue-type superpixel-derived features into the regression model significantly improves the prediction of wound time-to-healing, versus using features derived only from the image segmentation model.

Supplementary Materials: The following supporting information can be downloaded at: https://www.mdpi.com/article/10.3390/computation12030042/s1, Table S1: Dog wound image augmentation transformations used.

Author Contributions: Overall study design: S.A.R., Q.W. and A.K.; implementation of computational approach: Q.W. and A.K.; Data analysis: A.K., Q.W. and S.A.R.; manuscript writing: S.A.R., A.K. and Q.W. All authors have read and agreed to the published version of the manuscript.

Funding: This work was supported by the National Institutes of Health [grant number R01EB028104].

Institutional Review Board Statement: Not applicable.

Informed Consent Statement: Not applicable.

Data Availability Statement: The Python and Tensorflow source code for our wound image analysis method is available on GitHub at https://github.com/ramseylab/wound-analysis (accessed on 26 February 2024). under an open source software license. The ULCER_SET images from the Blanco et al. study [9] are freely and publicly available on GitHub at https://github.com/gu-blanco/qtdu (accessed on 26 February 2024). The dog wound images that were used in the study were previously published by Kurach et al. [31] and were provided to us courtesy of one of that study's co-authors (Bryden Stanley at Michigan State University); the images are available from Stanley upon request.

Acknowledgments: We thank Elain Fu, Matt Johnston, and Arun Natarajan for their ideas and feedback on the project. We thank Bryden J. Stanley and Milan Milovancev for kindly providing the wound images from Ref. [31].

Conflicts of Interest: The authors declare no conflicts of interest. The funders had no role in the design of the study; in the collection, analyses, or interpretation of data; in the writing of the manuscript; or in the decision to publish the results.

Abbreviations

The following abbreviations are used in this manuscript:

AIS	augmented image set
AUROC	area under the receiver operating characteristic curve
CV	cross-validation
FCN	fully convolutional network
MLP	multi-layer perceptron
OIS	original image set
PCA	principal components analysis
px	pixel
R^2	coefficient of determination
ReLU	rectified linear unit

References

1. Branski, L.K.; Gauglitz, G.G.; Herndon, D.N.; Jeschke, M.G. A review of gene and stem cell therapy in cutaneous wound healing. *Burns* **2009**, *35*, 171–180. [CrossRef]
2. McLister, A.; McHugh, J.; Cundell, J.; Davis, J. New Developments in Smart Bandage Technologies for Wound Diagnostics. *Adv. Mater.* **2016**, *28*, 5732–5737. [CrossRef]
3. Sen, C.K.; Gordillo, G.M.; Roy, S.; Kirsner, R.; Lambert, L.; Hunt, T.K.; Gottrup, F.; Gurtner, G.C.; Longaker, M.T. Human skin wounds: A major and snowballing threat to public health and the economy. *Wound Repair Regen.* **2009**, *17*, 763–771. [CrossRef]
4. Falcone, M.; Angelis, B.D.; Pea, F.; Scalise, A.; Stefani, S.; Tasinato, R.; Zanetti, O.; Paola, L.D. Challenges in the management of chronic wound infections. *J. Glob. Antimicrob. Resist.* **2021**, *26*, 140–147. [CrossRef]
5. Grey, J.E.; Enoch, S.; Harding, K.G. Wound assessment. *BMJ* **2006**, *332*, 285–288. [CrossRef]
6. Seaman, M.; Lammers, R. Inability of patients to self-diagnose wound infections. *J. Emerg. Med.* **1991**, *9*, 215–219. [CrossRef]
7. Wang, C.; Yan, X.; Smith, M.; Kochhar, K.; Rubin, M.; Warren, S.M.; Wrobel, J.; Lee, H. A unified framework for automatic wound segmentation and analysis with deep convolutional neural networks. In Proceedings of the 37th Annual International Conference of the IEEE Engineering in Medicine and Biology Society, Milan, Italy, 25–29 August 2015; pp. 2415–2418. [CrossRef]
8. Veredas, F.J.; Luque-Baena, R.M.; Martín-Santos, F.J.; Morilla-Herrera, J.C.; Morente, L. Wound image evaluation with machine learning. *Neurocomputing* **2015**, *164*, 112–122. [CrossRef]

9. Blanco, G.; Traina, A.J.M.; Traina, C., Jr.; Azevedo-Marques, P.M.; Jorge, A.E.S.; de Oliveira, D.; Bedo, M.V. A superpixel-driven deep learning approach for the analysis of dermatological wounds. *Comput. Methods Programs Biomed.* **2020**, *183*, 105079. [CrossRef] [PubMed]
10. Anisuzzaman, D.M.; Patel, Y.; Rostami, B.; Niezgoda, J.; Gopalakrishnan, S.; Yu, Z. Multi-modal wound classification using wound image and location by deep neural network. *Sci. Rep.* **2022**, *12*, 20057. [CrossRef]
11. Bloice, M.D.; Roth, P.M.; Holzinger, A. Biomedical image augmentation using Augmentor. *Bioinformatics* **2019**, *35*, 4522–4524. [CrossRef] [PubMed]
12. Zhang, R.; Tian, D.; Xu, D.; Qian, W.; Yao, Y. A Survey of Wound Image Analysis Using Deep Learning: Classification, Detection, and Segmentation. *IEEE Access* **2022**, *10*, 79502–79515. [CrossRef]
13. Berezo, M.; Budman, J.; Deutscher, D.; Hess, C.T.; Smith, K.; Hayes, D. Predicting Chronic Wound Healing Time Using Machine Learning. *Adv. Wound Care* **2022**, *11*, 281–296.
14. Sullivan, T.P.; Eaglstein, W.H.; Davis, S.C.; Mertz, P. The pig as a model for wound healing. *Wound Repair Regen.* **2001**, *9*, 66–76. [CrossRef] [PubMed]
15. Volk, S.W.; Bohling, M.W. Comparative wound healing—Are the small animal veterinarian's clinical patients an improved translational model for human wound healing research? *Wound Repair Regen.* **2013**, *21*, 372–381. [CrossRef] [PubMed]
16. Stockman, G.; Shapiro, L.G. *Computer Vision*, 1st ed.; Prentice Hall: Hoboken, NJ, USA, 2001.
17. Gupta, S.; Arbelaez, P.; Malik, J. Perceptual Organization and Recognition of Indoor Scenes from RGB-D Images. In Proceedings of the IEEE Conference on Computer Vision and Pattern Recognition, Portland, OR, USA, 23–28 June 2013; pp. 564–571. [CrossRef]
18. Silberman, N.; Hoiem, D.; Kohli, P.; Fergus, R. Indoor Segmentation and Support Inference from RGBD Images. In Proceedings of the Computer Vision–ECCV 2012, Florence, Italy, 7–13 October2012; Fitzgibbon, A., Lazebnik, S., Perona, P., Sato, Y., Schmid, C., Eds.; Springer: Berlin/Heidelberg, Germany, 2012; pp. 746–760.
19. Song, B.; Sacan, A. Automated wound identification system based on image segmentation and Artificial Neural Networks. In Proceedings of the International Conference on Bioinformatics and Biomedicine, Philadelphia, PA, USA, 4–7 October 2012; pp. 1–4. [CrossRef]
20. Hettiarachchi, N.; Mahindaratne, R.; Mendis, G.; Nanayakkara, H.; Nanayakkara, N.D. Mobile based wound measurement. In Proceedings of the Point-of-Care Healthcare Technologies, Bangalore, India, 16–18 January 2013; pp. 298–301.
21. Fauzi, M.F.A.; Khansa, I.; Catignani, K.; Gordillo, G.; Sen, C.K.; Gurcan, M.N. Computerized segmentation and measurement of chronic wound images. *Comput. Biol. Med.* **2015**, *60*, 74–85. [CrossRef] [PubMed]
22. Cui, C.; Thurnhofer-Hemsi, K.; Soroushmehr, R.; Mishra, A.; Gryak, J.; Dominguez, E.; Najarian, K.; Lopez-Rubio, E. Diabetic Wound Segmentation using Convolutional Neural Networks. In Proceedings of the 41st Annual International Conference of the IEEE Engineering in Medicine and Biology Society, Berlin, Germany, 23–27 July 2019; pp. 1002–1005. [CrossRef]
23. Long, J.; Shelhamer, E.; Darrell, T. Fully convolutional networks for semantic segmentation. In Proceedings of the Conference on Computer Vision and Pattern Recognition, Boston, MA, USA, 7–12 June 2015; pp. 3431–3440.
24. Yuan, Y.; Chao, M.; Lo, Y.C. Automatic Skin Lesion Segmentation Using Deep Fully Convolutional Networks With Jaccard Distance. *IEEE Trans. Med. Imaging* **2017**, *36*, 1876–1886. [CrossRef] [PubMed]
25. Milletari, F.; Navab, N.; Ahmadi, S.A. V-net: Fully convolutional neural networks for volumetric medical image segmentation. In Proceedings of the 4th International Conference on 3D Vision, Stanford, CA, USA, 25–28 October 2016; pp. 565–571.
26. Goyal, M.; Yap, M.H.; Reeves, N.D.; Rajbhandari, S.; Spragg, J. Fully convolutional networks for diabetic foot ulcer segmentation. In Proceedings of the International Conference on Systems, Man, and Cybernetics, Banff, AB, Canada, 5–8 October 2017; pp. 618–623. [CrossRef]
27. Liu, X.; Wang, C.; Li, F.; Zhao, X.; Zhu, E.; Peng, Y. A framework of wound segmentation based on deep convolutional networks. In Proceedings of the 10th International Congress on Image and Signal Processing, Biomedical Engineering and Informatics, Shanghai, China, 14–16 October 2017; pp. 1–7.
28. Wang, C.; Anisuzzaman, D.M.; Williamson, V.; Dhar, M.K.; Rostami, B.; Niezgoda, J.; Gopalakrishnan, S.; Yu, Z. Fully automatic wound segmentation with deep convolutional neural networks. *Sci. Rep.* **2020**, *10*, 21897. [CrossRef] [PubMed]
29. Sandler, M.; Howard, A.; Zhu, M.; Zhmoginov, A.; Chen, L.C. Mobilenetv2: Inverted residuals and linear bottlenecks. In Proceedings of the Conference on Computer Vision and Pattern Recognition, Salt Lake City, UT, USA, 18–23 June 2018; pp. 4510–4520.
30. Achanta, R.; Shaji, A.; Smith, K.; Lucchi, A.; Fua, P.; Süsstrunk, S. SLIC Superpixels Compared to State-of-the-Art Superpixel Methods. *IEEE Trans. Pattern Anal. Mach. Intell.* **2012**, *34*, 2274–2282. [CrossRef] [PubMed]
31. Kurach, L.M.; Stanley, B.J.; Gazzola, K.M.; Fritz, M.C.; Steficek, B.A.; Hauptman, J.G.; Seymour, K.J. The Effect of Low-Level Laser Therapy on the Healing of Open Wounds in Dogs. *Vet. Surg.* **2015**, *44*, 988–996. [CrossRef] [PubMed]
32. Rasmussen, C.E.; Williams, C.K.I. *Gaussian Processes for Machine Learning*; MIT Press: Cambridge, MA, USA, 2005. [CrossRef]
33. Breiman, L. Random Forests. *Mach. Learn.* **2001**, *45*, 5–32. [CrossRef]
34. Chen, T.; Guestrin, C. XGBoost. In Proceedings of the 22nd ACM SIGKDD International Conference on Knowledge Discovery and Data Mining, San Francisco, CA, USA, 13–17 August 2016; pp. 785–794. [CrossRef]
35. Badrinarayanan, V.; Kendall, A.; Cipolla, R. SegNet: A deep convolutional encoder-decoder architecture for image segmentation. *IEEE Trans. Pattern Anal. Mach. Intell.* **2017**, *39*, 2481–2495. [CrossRef] [PubMed]

36. Ronneberger, O.; Fischer, P.; Brox, T. U-Net: Convolutional Networks for Biomedical Image Segmentation. In Proceedings of the Medical Image Computing and Computer-Assisted Intervention, Munich, Germany, 5–9 October 2015; Navab, N., Hornegger, J., Wells, W.M., Frangi, A.F., Eds.; Springer: Berlin/Heidelberg, Germany, 2015; pp. 234–241.
37. Beucher, S. The Watershed Transformation Applied To Image Segmentation. *Scanning Microsc.* **1992**, *6*, 299–314.
38. Bradski, G.; Kaehler, A. OpenCV. *Dr. Dobb's J.* **2000**, *3*, 122–125.
39. McCulloch, W.S.; Pitts, W. A logical calculus of the ideas immanent in nervous activity. *Bull. Math. Biophys.* **1943**, *5*, 115–133. [CrossRef]
40. Stiller, C.; Lappe, D. Gain/cost controlled displacement-estimation for image sequence coding. In Proceedings of the International Conference on Acoustics, Speech, and Signal Processing, Toronto, ON, Canada, 14–17 May 1991; pp. 2729–2730. [CrossRef]
41. Jiang, D.; Qu, H.; Zhao, J.; Zhao, J.; Liang, W. Multi-level graph convolutional recurrent neural network for semantic image segmentation. *Telecommun. Syst.* **2021**, *77*, 563–576. [CrossRef]
42. Wei, Q. Machine Learning for Precision Medicine: Application to Cancer Chemotherapy Response Prediction and Wound Healing Status Assessment. Ph.D. Thesis, Oregon State University, Corvallis, OR, USA, 2021.
43. Ruder, S. An overview of gradient descent optimization algorithms. *arXiv* **2016**, arXiv:1609.04747.
44. Hastie, T.; Tibshirani, R.; Friedman, J. *The Elements of Statistical Learning*; Springer Series in Statistics; Springer: New York, NY, USA, 2001.
45. Anwar, T.; Zakir, S. Effect of Image Augmentation on ECG Image Classification Using Deep Learning. In Proceedings of the International Conference on Artificial Intelligence, Islamabad, Pakistan, 5–7 April 2021; pp. 182–186. [CrossRef]
46. Elgendi, M.; Nasir, M.U.; Tang, Q.; Smith, D.; Grenier, J.P.; Batte, C.; Spieler, B.; Leslie, W.D.; Menon, C.; Fletcher, R.R.; et al. The Effectiveness of Image Augmentation in Deep Learning Networks for Detecting COVID-19: A Geometric Transformation Perspective. *Front. Med.* **2021**, *8*, 629134. [CrossRef]
47. Kehrel, B.E. Blood platelets: Biochemistry and physiology. *Hamostaseologie* **2003**, *23*, 149–158.
48. Duda, R.O.; Hart, P.E. Use of the Hough Transformation to Detect Lines and Curves in Pictures. *Commun. Assoc. Comput. Mach.* **1972**, *15*, 11–15. [CrossRef]
49. Pepys, M.B.; Hirschfield, G.M. C-reactive protein: A critical update. *J. Clin. Investig.* **2003**, *111*, 1805–1812. [CrossRef]
50. Vidarsson, G.; Dekkers, G.; Rispens, T. IgG Subclasses and Allotypes: From Structure to Effector Functions. *Front. Immunol.* **2014**, *5*, 520. [CrossRef] [PubMed]

Disclaimer/Publisher's Note: The statements, opinions and data contained in all publications are solely those of the individual author(s) and contributor(s) and not of MDPI and/or the editor(s). MDPI and/or the editor(s) disclaim responsibility for any injury to people or property resulting from any ideas, methods, instructions or products referred to in the content.

Article

Creation of a Simulated Sequence of Dynamic Susceptibility Contrast—Magnetic Resonance Imaging Brain Scans as a Tool to Verify the Quality of Methods for Diagnosing Diseases Affecting Brain Tissue Perfusion

Seweryn Lipiński

Department of Electrical Engineering, Power Engineering, Electronics and Automation, Faculty of Technical Sciences, University of Warmia and Mazury in Olsztyn, 10-036 Olsztyn, Poland; seweryn.lipinski@uwm.edu.pl

Abstract: DSC-MRI examination is one of the best methods of diagnosis for brain diseases. For this purpose, the so-called perfusion parameters are defined, of which the most used are CBF, CBV, and MTT. There are many approaches to determining these parameters, but regardless of the approach, there is a problem with the quality assessment of methods. To solve this problem, this article proposes virtual DSC-MRI brain examination, which consists of two steps. The first step is to create curves that are typical for DSC-MRI studies and characteristic of different brain regions, i.e., the gray and white matter, and blood vessels. Using perfusion descriptors, the curves are classified into three sets, which give us the model curves for each of the three regions. The curves corresponding to the perfusion of different regions of the brain in a suitable arrangement (consistent with human anatomy) form a model of the DSC-MRI examination. In the created model, one knows in advance the values of the complex perfusion parameters, as well as basic perfusion descriptors. The shown model study can be disturbed in a controlled manner—not only by adding noise, but also by determining the location of disturbances that are characteristic of specific brain diseases.

Keywords: virtual DSC-MRI examination; perfusion descriptors; tracer concentration curves; brain model; pathology simulation

Citation: Lipiński, S. Creation of a Simulated Sequence of Dynamic Susceptibility Contrast—Magnetic Resonance Imaging Brain Scans as a Tool to Verify the Quality of Methods for Diagnosing Diseases Affecting Brain Tissue Perfusion. *Computation* **2024**, *12*, 54. https://doi.org/10.3390/computation12030054

Academic Editor: Anando Sen

Received: 25 January 2024
Revised: 5 March 2024
Accepted: 6 March 2024
Published: 8 March 2024

Copyright: © 2024 by the author. Licensee MDPI, Basel, Switzerland. This article is an open access article distributed under the terms and conditions of the Creative Commons Attribution (CC BY) license (https:// creativecommons.org/licenses/by/ 4.0/).

1. Introduction

DSC-MRI (Dynamic Susceptibility Contrast—Magnetic Resonance Imaging) is one of the most modern brain diagnostic methods. It allows for imaging perfusion, i.e., assessing the degree of blood supply and blood flow through tissues. As the perfusion changes in pathological lesions, DSC-MRI brain imaging allows for early diagnosis and the indication of the location of brain tissues that put patients at risk of pathologies such as cancerous tumors, damage resulting from a stroke, epilepsy, migraine headaches, dementia, Moyamoya disease, and many more [1–5].

In a DSC-MRI study, the response of the examined brain area is observed over time in the form of a sequence of MRI scans [6–9], after prior injection of a paramagnetic tracer (e.g., gadolinium-based chelates) into the bloodstream [8,9]. A tracer passes successively from the injection site, through the circulatory system and cerebral artery, and to the examined area (Region of Interest—ROI). Tracer flow causes changes in the measured MRI signal [8–10]. From the temporal sequence of MRI scans obtained during the examination, a temporal course of changes in the MRI signal is created for each pixel of the examined brain cross-section [10]. Its shape corresponds to changes in tracer concentration in the ROI. On this basis, the so-called perfusion parameters, containing diagnostic information, are calculated [11,12].

The most frequently used perfusion parameters are Cerebral Blood Volume (CBV), Cerebral Blood Flow (CBF), and Mean Transit Time (MTT). CBV can be calculated by assessing the area under the concentration–time curve. The most used approach to obtaining CBF is to utilize singular value decomposition (SVD) to estimate CBF through deconvolution of the arterial input function (AIF, i.e., the function that is an excitation for the ROI), while MTT is calculated by dividing CBV by CBF [2,9,11,12]. The basis of the diagnosis are so-called parametric images, which are maps of the values of perfusion parameters in the examined brain cross-section [13,14].

There are many approaches to determining perfusion parameters [12,13,15–17]. The most frequently used is the non-parametric approach, in which no internal structure of the system under study is assumed, and a specific regression function is fitted to the measurement data [12,16]. In the case of the parametric approach, a hypothesis is put forward regarding the functioning of the system under study, in the form of its model, which allows the parameters of such a model to be given a physical interpretation [17,18]. However, regardless of the approach used, there is a problem with assessing its quality and its comparison with others. This problem results from the small number of available measurement samples and the low quality of DSC-MRI data [19,20]. For this reason, tracer concentration curves are often simulated using various regression functions (e.g., gamma variate). Tracer concentration patterns in blood are usually created by selecting the parameters of the regression curve so that the perfusion parameters calculated on its basis agree with the literature values [16,21]. Consequently, the choice of a specific regression function affects the subsequent simulation results, favoring those methods in which perfusion parameters are estimated based on the same regression function.

Generally, in simulation-based DSC-MRI studies, a statistically significant set of curves is created, and then, affected in a specific way [19,21]. In most studies, complex perfusion parameters are determined, i.e., the above-mentioned CBF, CBV, and MTT. Meanwhile, in many cases, also related to the fact that DSC-MRI is used in an increasing range of brain diseases [1–7], it turns out that the basic perfusion descriptors may become more diagnostically useful [22,23]. The basic perfusion descriptors are those that can be determined directly from the tracer concentration curve, i.e., BAT (Bolus Arrival Time—the time of appearance of the tracer in the ROI), MPC (Maximum Peak Concentration—the maximum amplitude of the tracer concentration in the ROI), TTP (Time to Peak—the time taken to reach the maximum amplitude), and FWHM (Full Width at Half Maximum—the width of the tracer concentration curve at the height of half the maximum of the curve) [16,22]. This is significant because, especially in relation to time descriptors (BAT and TTP), the diagnosis depends not only on the descriptor value itself, but also on the connection of the measurement value with the location in the brain cross-section [24].

For brain tissue, there are two basic shapes of tracer concentration curves, i.e., for white and gray brain matter. In addition to signals from both tissues, signals from blood vessels are measured [11,13,14]. So, the aim of the first part of the article is to obtain three DSC-MRI curves characteristic of different brain regions, created not using the standard approach of simulating characteristic curves in the form of arbitrarily chosen regression functions, but from a set of measurement data from a clinical DSC-MRI study. The obtained characteristic curves will be then used to create a virtual DSC-MRI brain examination.

It will be possible to introduce known disturbances into the obtained model brain study (e.g., in the form of signals typical for pathologies such as tumor, stroke, etc.), and then, examine the effectiveness of their identification for various methods and computational algorithms. This will enable research to be conducted for various dimensions and shapes of pathological disorders and for various amounts of measurement disturbances.

To create such a virtual sequence of DSC-MRI scans covering the entire cross-section of the brain, model tracer concentration curves with a shape corresponding to perfusion in various brain regions, after conversion into a DSC-MRI measurement signal, must be appropriately arranged (corresponding to the anatomy of a specific brain cross-section).

The key motivation for creating a virtual DSC-MRI study in this form is the fact that, unlike the above-mentioned approach, in which we only simulate curves and not the entire sequence of images, in the proposed approach, we will know in advance the values of both complex perfusion parameters, as well as simple perfusion descriptors. Thanks to the possibility of introducing curves with a shape characteristic for various types of pathological lesions in a known place, their detectability can be evaluated. The model examination can be disturbed in a controlled way—noise with assumed characteristics can be introduced into the MRI signals [14,16,21]. The advantage of creating a sequence covering the entire cross-section of the brain is that we can also influence the entire sequence by disrupting the images created, not just the signals that form them. It is also possible to test the reliability of the methods when introducing pathologies of different diameters and profiles into the examination, which is impossible in the case of the classical approach, while the size of the detected pathology often implies the method of treatment (including the decision on surgical intervention), e.g., allows for the distinction between reversible and irreversible ischemia [23,25].

2. Materials

To maintain the uniformity of the input data, all clinical DSC-MRI measurements used for the creation of the virtual brain DSC-MRI examination were performed on a GE scanner with the following parameters of the sequence of scans: B = 1.5T, SE-EPI, 12 layers of size (slice thickness) 5–10 mm, 60 measurement points, TR (Repetition Time) = 1250–1610 ms, TE (Time Echo Delay) = 32–53 ms, Ts (sampling interval) = 1.43 s.

Gadopentetic acid (Gd-DTPA) was used as a paramagnetic contrast agent.

To maintain the anonymity of the subjects, 24 DSC-MRI sequences from adults assumed to be healthy (age 23–68) of both sexes were randomly selected from the available ones, which gave 1200 MRI images for analysis.

It should be noted that TR values, although varying, were considered appropriate for the model study, which results from the fact that since a short TR was below 700 ms, while a long TR was above 2000 ms [26], the range of 1250–1610 ms could be considered without deviations on either side.

3. Basic Perfusion Descriptors

Figure 1 shows the typical shape of the tracer concentration curve in the ROI [16,18].

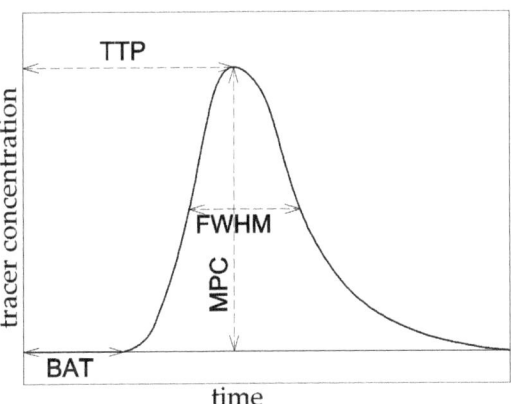

Figure 1. Tracer concentration curve with selected basic perfusion descriptors.

Figure 1 also shows the basic perfusion descriptors that can be determined directly from this curve [14,16,18]:
- The time of appearance of the tracer in the ROI—BAT (Bolus Arrival Time);
- The maximum amplitude of the tracer concentration in the ROI—MPC (Maximum Peak Concentration);
- The time taken to reach the maximum amplitude by the curve—TTP (Time to Peak);

- The width of the tracer concentration curve at the height of half the maximum of the curve—FWHM (Full Width at Half Maximum).

The values of the descriptors characterizing the tracer concentration curve are different in different areas of the brain. Based on the literature study, the following were selected for further consideration: TTP, MPC, and FWHM [18,27]. The analysis of the values of these descriptors for the entire cross-section of the brain allows us to distinguish three subsets corresponding to three distinguished brain areas.

The following relationships exist between individual perfusion descriptors [16,28–30]:
- $MPC_A > MPC_{GM} > MPC_{WM}$;
- $TTP_{WM} > TTP_{GM} > TTP_A$;
- $FWHM_{WM} > FWHM_{GM} > FWHM_A$.

In the above relationships, *WM* represents white matter, *GM* gray matter, and *A* arteries.

The literature [29–31] presents classification methods using, among others, the above relationships to also assign each pixel of the brain slice to one of the three sets (*WM*, *GM*, or *A*). The research presented in this paper has a different goal, which is to obtain characteristic tracer concentration curves for three distinguished areas. Curves with a shape deviating from the characteristic shape (as shown in Figure 1) were identified as anomalous due to inappropriate perfusion parameters or due to excessive noise and were not included as components of the characteristic curves.

4. Method of Calculating Selected Perfusion Descriptors

In the three-compartment model of the process of the tracer passing through the circulatory system to the examined area in the brain, presented, e.g., in [17,18], AIF measurements, as well as measurements made in the ROI, basic regression functions were fitted, two- and three-exponential, respectively:

$$f_{regrAIF}(t) = p_1 \cdot e^{-p_2 \cdot t} + p_3 \cdot e^{-p_4 \cdot t}, \tag{1}$$

$$f_{regrROI}(t) = p_5 \cdot e^{-p_6 \cdot t} + p_7 \cdot e^{-p_8 \cdot t} + p_9 \cdot e^{-p_{10} \cdot t}, \tag{2}$$

where $p_1 \div p_{10}$ are simply the parameters of the above regression functions. These regression function parameters were estimated based on DSC-MRI measurements. The AIF parameters, i.e., $p_1 \div p_4$, were the same for the whole sequence, while the parameters $p_5 \div p_{10}$ were calculated for each ROI separately, i.e., in accordance with the contrast agent concentration in a particular ROI.

The basic features of the parametric approach to calculating the most diagnostically important perfusion parameters (i.e., CBF, CBV, and MTT) are the avoidance of numerical deconvolution and the possibility of using stochastic filtering to improve the noise properties of the analyzed data. Both features are unique to the parametric approach [17,18,32]. Moreover, in the case of calculating the perfusion descriptors from Figure 1 using regression functions (1) and (2), these descriptors can be calculated directly using estimates of the parameters of the regression function.

The subject of interest in this work are the signals in the ROI (i.e., not the arterial input function). Therefore, a regression function in the form of Equation (2) was used, and the parameters were estimated. The regression function was fitted to first-pass samples, i.e., those corresponding to the first passage of the marker through the ROI [17,32]. For this purpose, the LS method and the Marquardt–Levenberg (M-L) algorithm were used.

After determining the estimates of the parameters of the regression function (3) for each considered curve, the perfusion descriptors were calculated as follows:

TTP and MPC

TTP, as the time in which the concentration curve reaches its maximum, is calculated by solving the following equation:

$$\frac{df_{regROI}(t)}{dt} = 0, \tag{3}$$

so, using Equation (2) as $f_{regROI}(t)$, we obtain

$$\left(p_5 \cdot e^{-p_6 \cdot t} + p_7 \cdot e^{-p_8 \cdot t} + p_9 \cdot e^{-p_{10} \cdot t}\right)' = 0, \tag{4}$$

$$-p_5 \cdot p_6 \cdot e^{-p_6 \cdot t} - p_7 \cdot p_8 \cdot e^{-p_8 \cdot t} - p_9 \cdot p_{10} \cdot e^{-p_{10} \cdot t} = 0. \tag{5}$$

The solution to Equation (5) is t_{MPC} = TTP. According to Figure 1, the TTP descriptor clearly indicates the maximum MPC, and therefore,

$$MPC = f_{regROI}(t_{MPC}). \tag{6}$$

FWHM

The FWHM descriptor value is calculated using the previously obtained MPC and TTP values. The value of the FWHM descriptor is defined as

$$FWHM = t_2 - t_1, \tag{7}$$

where t_1 and t_2 are determined from the two following relationships:

$$f_{regROI}(t_1) = \frac{f_{regROI}(t_{MPC})}{2}, \ t < t_{MPC}, \tag{8}$$

$$f_{regROI}(t_2) = \frac{f_{regROI}(t_{MPC})}{2}, \ t > t_{MPC}. \tag{9}$$

In other words, we calculate the time taken for the regression function to reach half of its maximum before reaching it (t_1), as well as the time taken for the value of this function to fall back to half of its maximum after reaching it (t_2). According to the definition of FWHM, the difference in these times gives us the exact value of this descriptor.

5. Creation and Verification of Model Curves

In the next step, each measurement curve was assigned a vector of previously selected descriptors characterizing it, related to the brain area based on previously shown relationships existing between individual perfusion descriptors, as given in [16,28–30]: $D_n = [MPC_n, TTP_n, FWHM_n], n = 1, 2, \ldots, N$, where N is the number of all curves.

These vectors were assigned to appropriate sets, in this case, four. Three of them correspond to three brain areas, while the fourth is intended for non-standard curves that do not correspond to any of the three brain sets. The curves in the fourth set were too noisy, or they came from areas affected by the disease. Curves from this set were not components of characteristic signals—they were rejected, as they could influence the quality of the resulting model DSC-MRI study.

The unsupervised clustering method was used to divide the set of all curves. Generally, clustering is the task of dividing a multidimensional set of data (in this case, N vectors) characterized by a feature vector (in this case, a D_n vector of descriptors) into subsets in such a way that the elements of each subset are similar to each other while being as different as possible from elements belonging to other groups. For this purpose, this work uses the k-means algorithm, in which the data set is initially divided into a predetermined number of classes (in this case, equal to 4). Then, the obtained division is iteratively improved in such a way that some elements are transferred to other classes until the minimum variance within the obtained classes is obtained [33,34].

An unsupervised clustering algorithm was chosen so that the resulting sets of curves would not be subject to any arbitrarily selected thresholds or other factors that could influence the selection and, consequently, mismatch of the model curves to the brain region they are supposed to represent.

After clustering, characteristic curves were created by averaging each of the three obtained sets. Figure 2 shows the average curves from the first three sets, i.e., those containing measurements from the white and gray matter of the brain, and from blood vessels (arteries). These are the tracer concentration curves characteristic of particular brain regions.

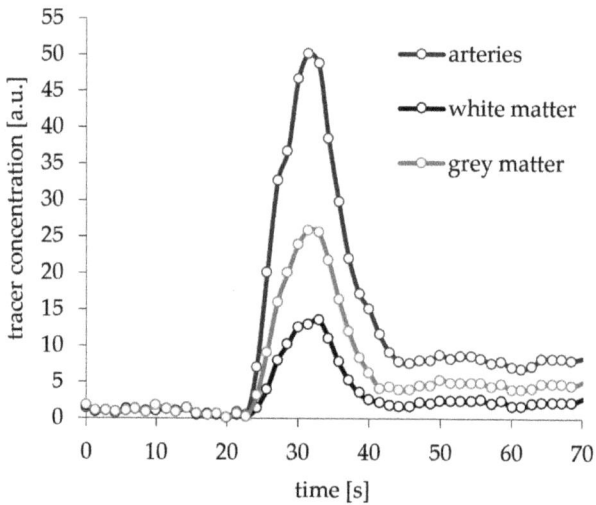

Figure 2. Contrast agent concentration curves characteristic of three brain regions. On these curves, we can identify baseline (0–24 s), first-pass of contrast agent (24–40 s), and contract agent recirculation (40–60 s).

Visual assessment of the averaged curves from Figure 2 allows us to initially conclude that the selected classification method works properly. The differences between them correspond to those known from the literature [11,13,14,28,29].

Better verification of the quality of the obtained curves can be achieved by comparing the CBV parameter values calculated for each of the obtained characteristic curves with the values known from the literature. The choice of the CBV for this purpose was dictated by the fact that of the three complex perfusion parameters (i.e., CBF, CBV, and MTT), this one is the most computationally explicit and, in some cases, it is possible to use a regional relative description of CBV without knowing the arterial input function [35]. CBV is, by definition, given as [11,13,14,36]

$$CBV = \frac{\int_0^\infty C_{ROI}(t)dt}{\int_0^\infty C_{AIF}(t)dt}, \tag{10}$$

where $C_{ROI}(t)dt$ and $C_{AIF}(t)dt$ are the tracer concentration curves for the ROI and for the AIF, respectively.

As follows from Equation (10), the absolute value of *CBV* depends on the arterial input function. The value of the denominator (i.e., integral of the arterial input function) is the same for each pixel of the brain cross-section, so to verify the quality of the obtained characteristic curves, the *CBV* ratios for the three individual brain regions can be used. This will make the obtained results independent of the possible impact of the arterial input function on their quality.

So, CBV_A/CBV_{GM} and CBV_{GM}/CBV_{WM} were calculated, and the obtained results were compared with values obtained from six different clinical studies [29,31,36–39]. The results are shown in Table 1.

Table 1. Comparison of results obtained based on curves characteristic of three brain regions obtained in this study with published parameter values obtained from clinical studies.

	CBV_A/CBV_{GM}	CBV_{GM}/CBV_{WM}
Based on the Model Curves Presented in This Work	1.97	2.19
Artzi et al. [29]	1.60–2.10	2–2.4
Bjornerud and Emblem [31]	(not investigated)	1.60–1.98 or 1.74–2.18 (depending on the calculation method)
Ibaraki et al. [36]	(not investigated)	1.60–2.40 or 2.30–2.50 (depending on the ROI)
Schreiber et al. [37]	(not investigated)	1.90–2.30
Wenz et al. [38]	(not investigated)	1.60–2.60
Fuss et al. [39]	(not investigated)	1.50–2.80

The results presented in Table 1 show that the proposed approach enables very good compliance of the perfusion parameters with those published in the literature. This means that the curves shown in Figure 2 can be used as tracer concentration curves specific to the white and gray matter of the brain and to blood vessels. It is worth noting that only the authors of [29] present the CBV_A/CBV_{GM} ratio. In the same paper, Artzi et al. [29] point out that it is impossible to compare the result obtained by them with the literature values, as such values are not published; however, a comparison of the highest curve in Figure 2 with the curves obtained in arteries and presented, for example, in the works [11,13,14,28,29] shows very good compliance between the shapes of these curves. The lack of other CBV_A values in the literature can probably be explained by the fact that, in general, determining the position of arteries in a cross-section of the brain is a difficult task, and in fact, it is most often only used to determine curves that can be candidates for AIF [18,27].

However, the obtained value of the CBV_{GM}/CBV_{WM} ratio is consistent with five of the six clinical trial results shown. The result from [31] differs from others, which may be due to the use of a different calculation method or the fact that in many studies, the ROI selection is performed manually. This may lead to the over- or under-estimation of values due to the inaccurate marking of gray- and white-matter regions, or by incorrectly including measurements from vessels in these regions.

The tracer concentration curves shown in Figure 2 can be used to evaluate and compare methods for calculating complex perfusion parameters. Their main advantage is the fact that they were obtained based on an actual DSC-MRI examination of the brain, and not using the standard approach of simulating regression curves. However, this is not their only possible application. The next part of this article will show other possibilities of using the obtained characteristic curves, i.e., creating a simulated DSC-MRI brain examination.

6. Creation of DSC-MRI Measurement Curves and a Brain Anatomy Model

From the $c(t)$ tracer concentration curve in the blood, the DSC-MRI measurement signal $S(t)$ is obtained from the following relationship [40]:

$$S(t) = S_0 \cdot e^{-\kappa \cdot c(t) \cdot TE} \tag{11}$$

where S_0 is the amplitude of the measurement signal before contrast administration, and κ is the proportionality coefficient (depending on the properties of the tissue and measurement conditions resulting from the device used).

Figure 3 shows model DSC-MRI signals calculated using Equation (11) and based on the previously obtained model tracer concentration curves for individual brain regions.

The DSC-MRI measurements used to create the model curves consisted of 50 scans, so this was the resolution of the model study.

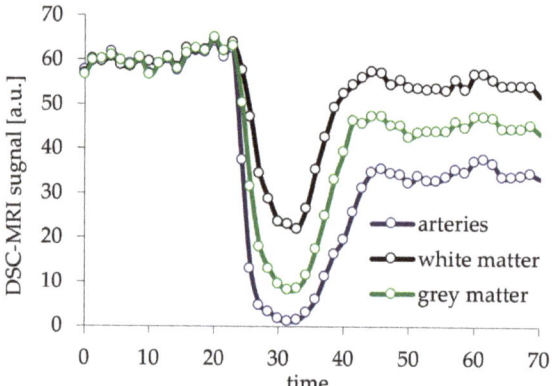

Figure 3. DSC-MRI signals derived from model contrast agent concentration curves for each of three brain regions: gray matter, white matter, and arteries.

To create the model study, a physiological base was needed, i.e., the arrangement of individual tissues in a cross-section of the brain. The BrainWeb database was used for this purpose [41–44]. This database provides, among others, 20 virtual static brain MRI images [43]. The MRI simulator described in [44] includes blood vessels, which, combined with the model DSC-MRI measurement signals from Figure 3, makes it possible to use it to create a model series of dynamic brain images including blood vessels, without the need to obtain an anatomical base from the segmentation of brain areas from other imaging studies.

It should be noted that at this stage, changes in the brain resulting from aging (like the WM/GM ratio) or sex differences should be considered if necessary [45–47].

Figure 4 shows where the gray (b) and white (c) matter of the brain and blood vessels (d) are located in two selected brain cross-sections (a). The maps shown in Figure 4b–d allow for the appropriate location of the model DSC-MRI signals, and thus, the creation of a simulation of a DSC-MRI study, consisting of a sequence of scans with the same parameters as the actual studies constituting the basis for obtaining the model DSC-MRI signals.

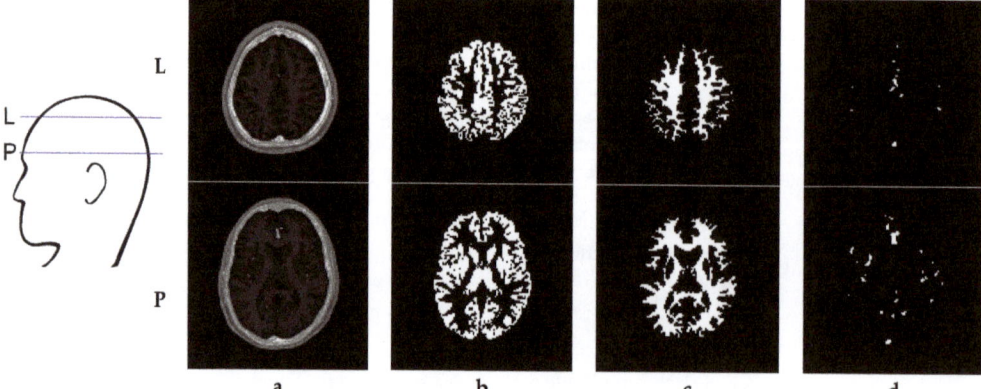

Figure 4. Brain cross-sections serving as bases of two models of DSC-MRI examinations, (a) and localization of gray matter (b), white matter (c), and arteries (d) on these cross-sections.

7. Results—Exemplary DSC-MRI Study Models

Figure 5 shows five selected exemplary scans from a model DSC-MRI study created based on cross-section P of Figure 4: one scan from before the tracer appeared (t_1), one when the tracer reached its maximum value (t_3), two adjacent to it (t_2 and t_4), and one after tracer passing, i.e., during the recirculation (t_5). One study (row a) contains no noise, and two others (rows b and c) are artificially noised. The DSC-MRI signals were distorted using Gaussian white noise at levels of SNR of 25 dB (b) and 20 dB (c).

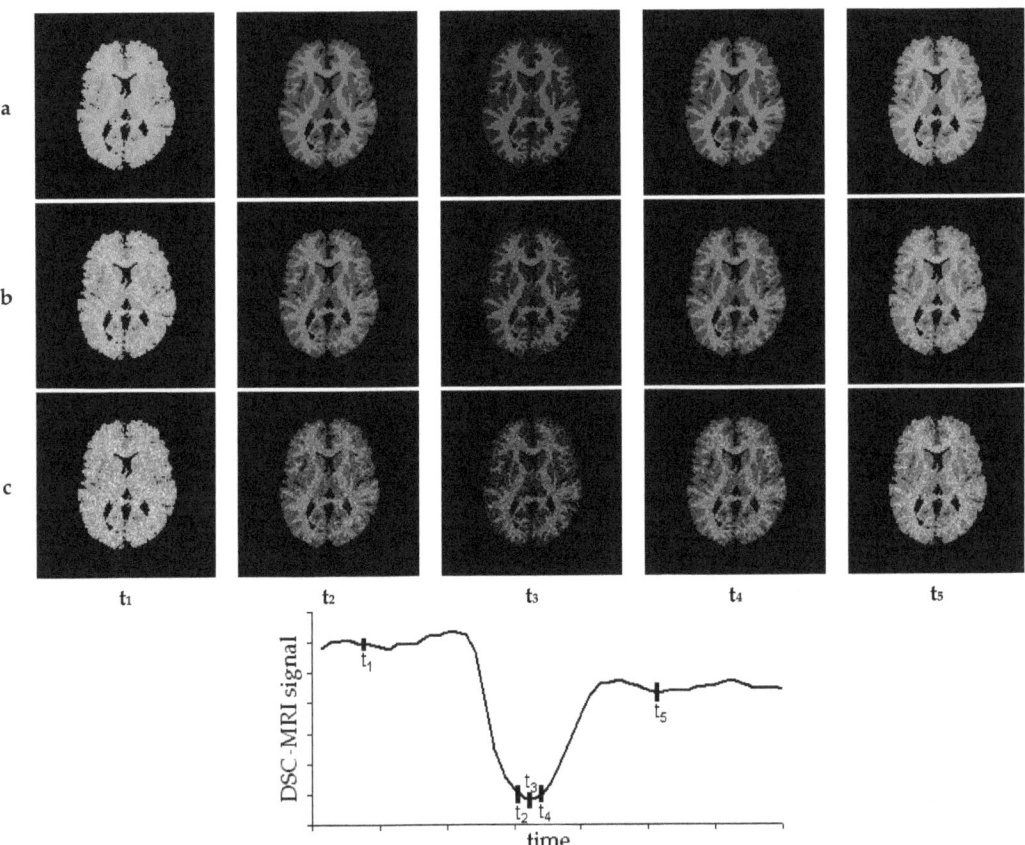

Figure 5. Representative scans from exemplary model DSC-MRI examination: sequence of scans for healthy brain (**a**) and the same sequence with low (**b**) and high (**c**) noise content.

DSC-MRI examination is used for the diagnosis of an increasing range of brain diseases. Each disease (as well as its stage) has more or less specific characteristics and, in a peculiar way, affects simple and complex perfusion parameters [1,14,15,22,48]. Therefore, the introduction of a pathology with a specific diameter and profile into a model sequence can show how effective each approach to calculating perfusion parameters is in terms of detecting a specific pathology. Figure 6 shows the location of two pathologies with different dimensions introduced into the model DSC-MRI examination.

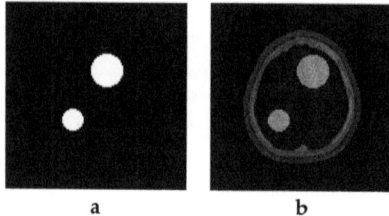

Figure 6. Shapes (**a**) and localizations (**b**) of the pathologies introduced into the model research.

A pathology with the shape shown in Figure 6 was introduced into a model study based on the brain cross-section from row L of Figure 4. The model DSC-MRI signals were modified based on the thresholds imposed on the perfusion descriptors proposed by Grandin et al. [23] to detect ischemic cerebral infarction. The DSC-MRI examination created in this way is shown in Figure 7. The scans were selected analogously to those from Figure 5. The first sequence (a) does not contain noise or a pathology, the second one (b) has a pathology introduced as described above, and the third one (c) contains a pathology and is also disturbed by noise at an SNR level of 25 dB.

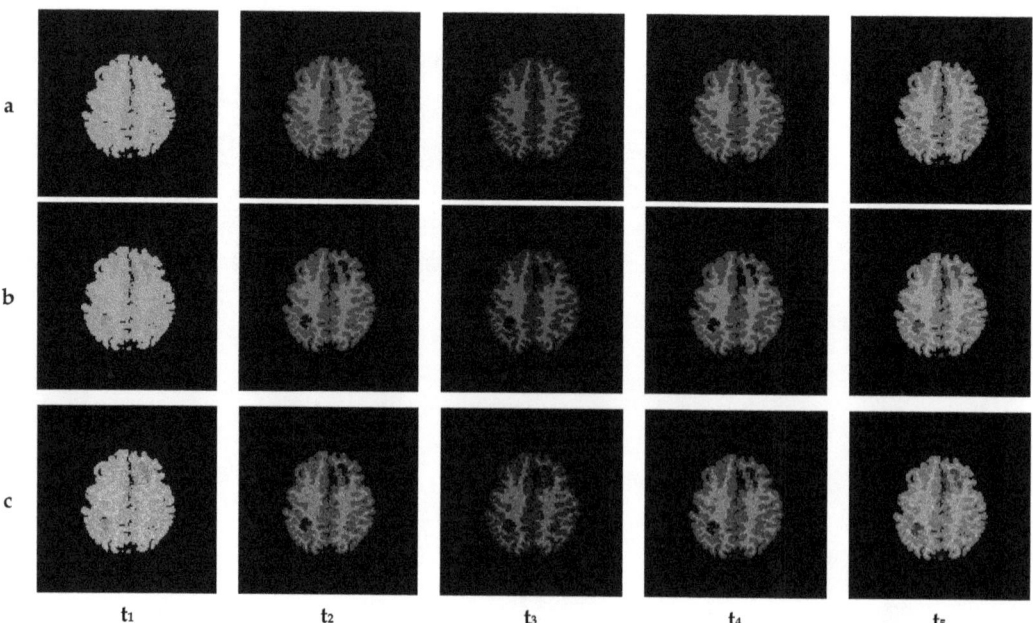

Figure 7. Representative scans from different exemplary model DSC-MRI examinations: healthy brain (**a**), brain with inserted pathologies (**b**), and brain with two pathologies and noise content (**c**).

Figure 8 shows maps of the three most important perfusion parameters, i.e., CBV, CBF, and MTT, calculated based on the sequence shown in Figure 7c.

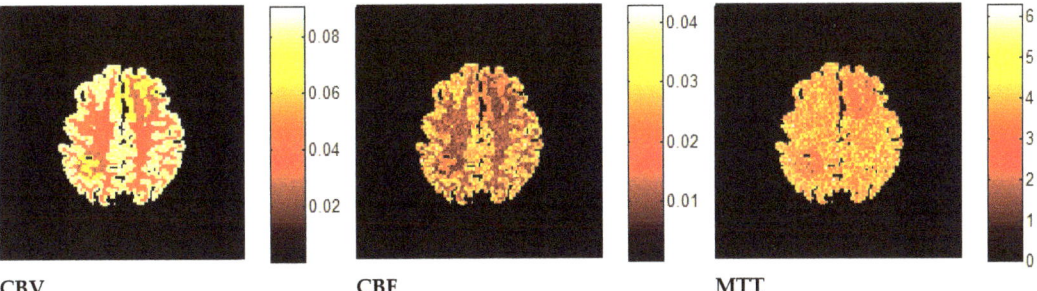

Figure 8. CBV, CBF, and MTT perfusion maps obtained based on the model sequence shown in Figure 7c.

As can be seen in Figure 8, the introduced pathologies in the parametric images are clearly observable and consistent with the shape of the pathologies shown in Figure 6. In the case shown, both pathologies are relatively obvious, and they also have a large diameter; therefore, for a diagnostician, it would be hard not to notice the impact of the disease on the parameter map. However, the pathological changes appearing in the brain cross-section may have a completely different shape (irregular) and a much smaller diameter, and their impact on the tracer concentration curve may also be much more subtle. In such cases, it is extremely desirable that the perfusion parameter map assessed by the doctor is as reliable as possible. The proposed virtual study can be used as a tool for this purpose. With its help, it is possible to objectively compare different approaches to calculating perfusion parameters by checking how the obtained parametric maps correlate with a known pathological change introduced in a specific location.

8. Discussion and Conclusions

This article proposes the creation of a simulated DSC-MRI examination. Curves corresponding to the perfusion of various brain regions in an appropriate arrangement (corresponding to the anatomy of a specific brain cross-section) create model DSC-MRI studies. Using sequences created in this way, we know in advance the values of both complex perfusion parameters and their simple descriptors. We can disrupt the model study in a controlled way—not only by introducing disturbances of the assumed size and characteristics to the DSC-MRI signals, but also by influencing their location (and distribution) in the brain cross-section(s). Other types of interference that can be introduced into such sequences, typical for image data, are blurring, geometric shifts between subsequent images in the sequence, etc. Assessment of the resistance of the methods for determining perfusion parameters to this type of interference is impossible when the interference is introduced only to DSC-MRI signals, i.e., as in the approach classically proposed in the literature. The described research model will allow us to reliably and objectively, while considering the type of brain tissue (white and gray matter), assess the quality of various approaches to calculating perfusion parameters.

Additionally, disturbed signals characteristic of specific pathologies can be introduced into such a sequence. Then, an additional criterion for assessing the approach to determining perfusion parameters is to check the threshold (this threshold may be, for example, the diameter of the introduced disorder) for detecting perfusion disorders in the presence of various types of disturbances, either automatically or indirectly, through the assessment of the diagnosing doctor based on a map of specific parameters.

The one limitation of the proposed approach is the fact that reliable data are needed on the impact of each specific disease on the flow of the tracer through the brain tissue, so that the disorders introduced into the virtual study reflect reality.

Another limitation of the proposed approach is the fact that if an analogous model study for long or short TR was to be created, a different set of DSC-MRI studies must be used for the creation of model curves.

It is also worth emphasizing that this study focuses on a method that can be considered invasive, as it involves tracer injection, while in recent years, non-invasive methods of brain perfusion imaging have been gaining popularity, with particular emphasis on Arterial Spin Labeling (ASL) perfusion MRI [49–51]. Since the use of this method is based on a different principle, the described approach cannot be used to create an analogous research model.

Funding: This research received no external funding.

Data Availability Statement: The dataset is available on request from the author.

Conflicts of Interest: The author declares no conflicts of interest.

References

1. Forsting, M.; Weber, J. MR perfusion imaging: A tool for more than a stroke. *Eur. Radiol. Suppl.* **2004**, *14* (Suppl. 5), M2–M7.
2. Lanzman, B.; Heit, J. Advanced MRI measures of cerebral perfusion and their clinical applications. *Top. Magn. Reson. Imaging* **2017**, *26*, 83–90. [CrossRef] [PubMed]
3. Lee, H.; Fu, J.F.; Gaudet, K.; Bryant, A.G.; Price, J.C.; Bennett, R.E.; Johnson, K.A.; Hyman, B.T.; Hedden, T.; Salat, D.H.; et al. Aberrant vascular architecture in the hippocampus correlates with tau burden in mild cognitive impairment and Alzheimer's disease. *J. Cereb. Blood Flow Metab.* **2023**, 0271678X231216144. [CrossRef]
4. Petrella, J.R.; Provenzale, J.M. MR Perfusion Imaging of the Brain: Techniques and Applications. *Am. J. Roentgenol.* **2000**, *175*, 207–219. [CrossRef] [PubMed]
5. Zuloaga, K.L.; Zhang, W.; Yeiser, L.A.; Stewart, B.; Kukino, A.; Nie, X.; Roese, N.E.; Grafe, M.R.; Pike, M.M.; Raber, J.; et al. Neurobehavioral and imaging correlates of hippocampal atrophy in a mouse model of vascular cognitive impairment. *Transl. Stroke Res.* **2015**, *6*, 390–398. [CrossRef] [PubMed]
6. Bakhtiari, A.; Vestergaard, M.B.; Benedek, K.; Fagerlund, B.; Mortensen, E.L.; Osler, M.; Lauritzen, M.; Larsson, H.B.W.; Lindberg, U. Changes in hippocampal volume during a preceding 10-year period do not correlate with cognitive performance and hippocampal blood–brain barrier permeability in cognitively normal late-middle-aged men. *GeroScience* **2023**, *45*, 1161–1175. [CrossRef]
7. Choi, J.D.; Moon, Y.; Kim, H.J.; Yim, Y.; Lee, S.; Moon, W.J. Choroid plexus volume and permeability at brain MRI within the Alzheimer disease clinical spectrum. *Radiology* **2022**, *304*, 635–645. [CrossRef]
8. Gordon, Y.; Partovi, S.; Müller-Eschner, M.; Amarteifio, E.; Bäuerle, T.; Weber, M.A.; Kauczor, H.-U.; Rengier, F. Dynamic contrast-enhanced magnetic resonance imaging: Fundamentals and application to the evaluation of the peripheral perfusion. *Cardiovasc. Diagn. Ther.* **2014**, *4*, 147.
9. Koh, T.S.; Bisdas, S.; Koh, D.M.; Thng, C.H. Fundamentals of tracer kinetics for dynamic contrast-enhanced MRI. *J. Magn. Reson. Imaging* **2011**, *34*, 1262–1276. [CrossRef]
10. Kalicka, R.; Lipiński, S. A fast method of separation of the noisy background from the head-cross section in the sequence of MRI scans. *Biocybern. Biomed. Eng.* **2010**, *30*, 15–27.
11. Calamante, F.; Thomas, D.L.; Pell, G.S.; Wiersma, J.; Turner, R. Measuring Cerebral Blood Flow Using Magnetic Resonance Imaging Techniques. *J. Cereb. Blood Flow. Metab.* **1999**, *19*, 701–735. [CrossRef] [PubMed]
12. Jackson, D. Analysis of dynamic contrast enhanced MRI. *Br. J. Radiol.* **2004**, *77*, S154–S166. [CrossRef] [PubMed]
13. van Osch, T. *Evaluation of Cerebral Hemodynamics by Quantitative Perfusion MRI*; PrintPartners Ipskamp: Enschede, The Netherlands, 2002.
14. Sorensen, A.G.; Reimer, P. *Cerebral MR Perfusion Imaging. Principles and Current Applications*; Georg Thieme Verlag: Stuttgart, Germany, 2000.
15. Kane, I.; Carpenter, T.; Chappell, F.; Rivers, C.; Armitage, P.; Sandercock, P.; Wardlaw, J. Comparison of 10 Different Magnetic Resonance Perfusion Imaging Processing Methods in Acute Ischemic Stroke. *Stroke* **2007**, *38*, 3158–3164. [CrossRef]
16. Perthen, J.E.; Calamante, F.; Gadian, D.G.; Connelly, A. Is Quantification of Bolus Tracking MRI Reliable Without Deconvolution? *Magn. Reson. Med.* **2002**, *47*, 61–67. [CrossRef] [PubMed]
17. Kalicka, R.; Pietrenko-Dąbrowska, A. Parametric Modeling of DSC-MRI Data with Stochastic Filtration and Optimal Input Design Versus Non-Parametric Modeling. *Ann. Biomed. Eng.* **2007**, *35*, 453–464. [CrossRef] [PubMed]
18. Lipiński, S.; Kalicka, R. Automatic selection of arterial input function in DSC-MRI measurements for calculation of brain perfusion parameters using parametric modelling. *Math. Model. Nat. Phenom.* **2018**, *13*, 58. [CrossRef]
19. Salluzzi, M.; Frayne, R.; Smith, M.R. Is correction necessary when clinically determining quantitative cerebral perfusion parameters from multi-slice dynamic susceptibility contrast MR studies? *Phys. Med. Biol.* **2006**, *51*, 407–424. [CrossRef]
20. Smith, M.R.; Lu, H.; Frayne, R. Signal-to-Noise Ratio Effects in Quantitative Cerebral perfusion Using Dynamic Susceptibility Contrast Agents. *Magn. Reson. Med.* **2003**, *49*, 122–128. [CrossRef]

21. Knutsson, L.; Stahlberg, F.; Wirestam, R. Aspects on the accuracy of cerebral perfusion parameters obtained by dynamic susceptibility contrast MRI: A simulation study. *Magn. Reson. Imaging* **2004**, *22*, 789–798. [CrossRef]
22. Calamante, F.; Ganesan, V.; Kirkham, F.J.; Chir, B.; Jan, W.; Chong, W.K.; Gadian, D.G.; Connelly, A. MR Perfusion Imaging in Moyamoya Syndrome. *Stroke* **2001**, *32*, 2810–2816. [CrossRef]
23. Grandin, C.B.; Duprez, T.P.; Smith, A.M.; Oppenheim, C.; Peeters, A.; Robert, A.R.; Cosnard, G. Which MR-derived perfusion parameters are the best predictors of infarct growth in hyperacute stroke? Comparative study between relative and quantitative measurements. *Radiology* **2002**, *223*, 361–370. [CrossRef] [PubMed]
24. Akella, N.S.; Twieg, D.B.; Mikkelsen, T.; Hochberg, F.H.; Grossman, S.; Cloud, G.A.; Nabors, L.B. Assessment of Brain Tumor Angiogenesis Inhibitors Using Perfusion Magnetic Resonance Imaging: Quality and Analysis Results of a Phase I Trial. *J. Magn. Reson. Imaging* **2004**, *20*, 913–922. [CrossRef]
25. Wintermark, M.; Sesay, M.; Barbier, E.; Borbély, K.; Dillon, W.P.; Eastwood, J.D.; Glenn, T.C.; Grandin, C.B.; Pedraza, S.; Soustiel, J.-F.; et al. Comparative Overview of Brain Perfusion Imaging Techniques. *Stroke* **2005**, *36*, e83–e99. [CrossRef]
26. Bitar, R.; Leung, G.; Perng, R.; Tadros, S.; Moody, A.R.; Sarrazin, J.; McGregor, C.; Christakis, M.; Symons, S.; Nelson, A.; et al. MR pulse sequences: What every radiologist wants to know but is afraid to ask. *Radiographics* **2006**, *26*, 513–537. [CrossRef]
27. Mouridsen, K.; Christensen, S.; Gyldensted, L.; Ostergaard, L. Automatic Selection of Arterial Input Function Using Cluster Analysis. *Magn. Reson. Med.* **2006**, *55*, 524–531. [CrossRef] [PubMed]
28. Koshimoto, Y.; Yamada, H.; Kimura, H.; Maeda, M.; Tsuchida, C.; Kawamura, Y.; Ishii, Y. Quantitative Analysis of Cerebral Microvascular Hemodynamics with T2-Weighted Dynamic MR Imaging. *J. Magn. Reson. Imaging* **1999**, *9*, 462–467. [CrossRef]
29. Artzi, M.; Aizenstein, O.; Hendler, T.; Bashat, D.B. Unsupervised multiparametric classification of dynamic susceptibility contrast imaging: Study of a healthy brain. *Neuroimage* **2011**, *56*, 858–864. [CrossRef] [PubMed]
30. Kao, Y.-H.; Guo, W.-Y.; Wu, Y.-T.; Liu, K.-C.; Chai, W.-Y.; Lin, C.-Y.; Hwang, Y.-S.; Liou, A.J.-K.; Wu, H.-M.; Cheng, H.-C.; et al. Hemodynamic Segmentation of MR Brain Perfusion Images Using Independent Component Analysis, Thresholding, and Bayesian Estimation. *Magn. Reson. Med.* **2003**, *49*, 885–894. [CrossRef]
31. Bjornerud, A.; Emblem, K.E. A fully automated method for quantitative cerebral hemodynamic analysis using DSC-MRI. *J. Cereb. Blood Flow. Metab.* **2010**, *30*, 1066–1078. [CrossRef]
32. Kalicka, R.; Lipiński, S. Valuation of usefulness of Kalman filtration to improve noise properties of DSC-MRI brain research data. *Meas. Automat. Monit.* **2008**, *54*, 118–121.
33. Theodoridis, S.; Pikrakis, A.; Kotroumbas, K.; Cavouras, D. *An Introduction to Pattern Recognition: A MATLAB Approach*; Elsevier Academic Press: Cambridge, MA, USA, 2010.
34. Jain, A.K.; Duin, R.P.W.; Mao, J. Statistical Pattern Recognition: A review. *IEEE Trans. Pattern Anal.* **2000**, *22*, 4–37. [CrossRef]
35. Calamante, F.; Gadian, D.G.; Connelly, A. Quantification of Perfusion Using Bolus Tracking Magnetic Resonance Imaging in Stroke: Assumptions, Limitations, and Potential Implications for Clinical Use. *Stroke* **2002**, *33*, 1146–1151. [CrossRef]
36. Ibaraki, M.; Ito, H.; Shimosegawa, E.; Toyoshima, H.; Ishigame, K.; Takahashi, K.; Kanno, I.; Miura, S. Cerebral vascular mean transit time in healthy humans: A comparative study with PET and dynamic susceptibility contrast-enhanced MRI. *J. Cereb. Blood Flow. Metab.* **2007**, *27*, 404–413. [CrossRef] [PubMed]
37. Schreiber, W.G.; Guckel, F.; Stritzke, P.; Schmiedek, P.; Schwartz, A.; Brix, G. Cerebral Blood Flow and Cerebrovascular Reserve Capacity: Estimation by Dynamic Magnetic Resonance Imaging. *J. Cereb. Blood Flow. Metab.* **1998**, *18*, 1143–1156. [CrossRef]
38. Wenz, F.; Rempp, K.; Brix, G.; Knopp, M.V.; Gückel, F.; Hess, T.; van Kaick, G. Age dependency of the regional cerebral blood volume (rCBV) measured with dynamic susceptibility contrast MR imaging (DSC). *Magn. Reson. Imaging* **1996**, *14*, 157–162. [CrossRef]
39. Fuss, M.; Wenz, F.; Scholdei, R.; Essig, M.; Debus, J.; Knopp, M.V.; Wannenmacher, M. Radiation-induced regional cerebral blood volume (rCBV) changes in normal brain and low-grade astrocytomas: Quantification and time and dose-dependent occurrence. *Int. J. Radiat. Oncol. Biol. Phys.* **2000**, *48*, 53–58. [CrossRef]
40. Ostergaard, L.; Weisskoff, R.M.; Chesler, D.A.; Gyldensted, C.; Rosen, B.R. High resolution Measurement of Cerebral Blood Flow using Intravascular Tracer Bolus Passages. Part I: Mathematical Approach and Statistical Analysis. *Magn. Reson. Med.* **1996**, *36*, 715–725. [CrossRef] [PubMed]
41. Cocosco, C.A.; Kollokian, V.; Kwan, R.K.-S.; Evans, A.C. BrainWeb: Online Interface to a 3D MRI Simulated Brain Database. *NeuroImage* **1997**, *5*, 1996.
42. Kwan, R.K.-S.; Evans, A.C.; Pike, G.B. An Extensible MRI Simulator for Post-Processing Evaluation. *Lect. Notes Comput. Sci.* **1996**, *1131*, 135–140.
43. Aubert-Broche, B.; Griffin, M.; Pike, G.B.; Evans, A.C.; Collins, D.L. Twenty new digital brain phantoms for creation of validation image data bases. *IEEE Trans. Med. Imaging* **2006**, *25*, 1410–1416. [CrossRef]
44. Aubert-Broche, B.; Evans, A.C.; Collins, D.L. A new improved version of the realistic digital brain phantom. *NeuroImage* **2006**, *32*, 138–145. [CrossRef]
45. Fujita, S.; Mori, S.; Onda, K.; Hanaoka, S.; Nomura, Y.; Nakao, T.; Yoshikawa, T.; Takao, H.; Hayashi, N.; Abe, O. Characterization of brain volume changes in aging individuals with normal cognition using serial magnetic resonance imaging. *JAMA Netw. Open* **2023**, *6*, e2318153. [CrossRef] [PubMed]
46. Gómez-Ramírez, J.; Fernández-Blázquez, M.A.; González-Rosa, J.J. A causal analysis of the effect of age and sex differences on brain atrophy in the elderly brain. *Life* **2022**, *12*, 1586. [CrossRef] [PubMed]

47. Usui, K.; Yoshimura, T.; Tang, M.; Sugimori, H. Age Estimation from Brain Magnetic Resonance Images Using Deep Learning Techniques in Extensive Age Range. *Appl. Sci.* **2023**, *13*, 1753. [CrossRef]
48. Tofts, P. (Ed.) *Quantitative MRI of the Brain. Measuring Changes Caused by Disease*; John Wiley and Sons: Hoboken, NJ, USA, 2004.
49. Bambach, S.; Smith, M.; Morris, P.P.; Campeau, N.G.; Ho, M.L. Arterial spin labeling applications in pediatric and adult neurologic disorders. *J. Magn. Reson. Imaging* **2022**, *55*, 698–719. [CrossRef] [PubMed]
50. Iutaka, T.; de Freitas, M.B.; Omar, S.S.; Scortegagna, F.A.; Nael, K.; Nunes, R.H.; Pacheco, F.T.; Maia Júnior, A.C.M.; do Amaral, L.L.F.; da Rocha, A.J. Arterial spin labeling: Techniques, clinical applications, and interpretation. *Radiographics* **2022**, *43*, e220088. [CrossRef] [PubMed]
51. Qin, Q.; Alsop, D.C.; Bolar, D.S.; Hernandez-Garcia, L.; Meakin, J.; Liu, D.; Nayak, K.S.; Schmid, S.; van Osch, M.J.P.; Wong, E.C.; et al. Velocity-selective arterial spin labeling perfusion MRI: A review of the state of the art and recommendations for clinical implementation. *Magn. Reson. Med.* **2022**, *88*, 1528–1547. [CrossRef]

Disclaimer/Publisher's Note: The statements, opinions and data contained in all publications are solely those of the individual author(s) and contributor(s) and not of MDPI and/or the editor(s). MDPI and/or the editor(s) disclaim responsibility for any injury to people or property resulting from any ideas, methods, instructions or products referred to in the content.

Article

COVID-19 Image Classification: A Comparative Performance Analysis of Hand-Crafted vs. Deep Features

Sadiq Alinsaif

College of Computer Science and Engineering, University of Hafr Al Batin, Hafar Al Batin 39524, Saudi Arabia; alinsaif@uhb.edu.sa

Abstract: This study investigates techniques for medical image classification, specifically focusing on COVID-19 scans obtained through computer tomography (CT). Firstly, handcrafted methods based on feature engineering are explored due to their suitability for training traditional machine learning (TML) classifiers (e.g., Support Vector Machine (SVM)) when faced with limited medical image datasets. In this context, I comprehensively evaluate and compare 27 descriptor sets. More recently, deep learning (DL) models have successfully analyzed and classified natural and medical images. However, the scarcity of well-annotated medical images, particularly those related to COVID-19, presents challenges for training DL models from scratch. Consequently, I leverage deep features extracted from 12 pre-trained DL models for classification tasks. This work presents a comprehensive comparative analysis between TML and DL approaches in COVID-19 image classification.

Keywords: machine learning; deep learning; convolutional neural networks; deep features; COVID-19; classification; CT scan

1. Introduction

Coronavirus disease 2019 (COVID-19) poses a significant global health threat due to its highly contagious nature, primarily transmitted through respiratory droplets expelled during coughing, sneezing, or speaking [1,2]. This respiratory illness represents one of the most lethal infectious diseases of our time [3], often leading to a substantial decline in the quality of life for afflicted individuals [4]. While the standard diagnostic tool, reverse transcriptase-polymerase chain reaction (RT-PCR), is widely employed, its limitations include a non-negligible rate of false negative results [5]. Therefore, developing and exploring alternative methodologies for accurate COVID-19 diagnosis is crucial.

Chest computed tomography (CT) [6] has emerged as a valuable adjunct to RT-PCR testing in the context of COVID-19 screening and diagnosis. Studies such as those by Fang et al. [7] and Ai et al. [8] have demonstrated the efficacy of CT scans in identifying COVID-19 patients with high sensitivity, even in cases where initial RT-PCR results were negative. This suggests the potential benefit of utilizing CT scans, particularly for patients exhibiting suggestive clinical symptoms despite negative RT-PCR findings [7].

Manual analysis of COVID-19 chest CT scans by radiologists presents a time-consuming burden, especially in emergency settings with high patient volumes. This necessitates the development of robust computer-aided diagnosis (CAD) systems capable of leveraging the rich information embedded within digital CT scans. Machine learning (ML) frameworks, in conjunction with image processing techniques, offer promising avenues for the construction of such CAD systems [9]. Given their potential to expedite and improve COVID-19 identification, ultimately facilitating timely and appropriate treatment interventions, CAD systems hold significant clinical value. Existing approaches for COVID-19 classification from CT scans can be broadly classified based on the type of feature descriptors extracted: traditional ML and deep learning (DL) techniques [10,11].

Conventional ML approaches for COVID-19 identification rely on meticulously crafted feature descriptors [12,13]. Conversely, DL models, specifically convolutional neural

Citation: Alinsaif, S. COVID-19 Image Classification: A Comparative Performance Analysis of Hand-Crafted vs. Deep Features. *Computation* **2024**, *12*, 66. https://doi.org/10.3390/computation12040066

Academic Editor: Anando Sen

Received: 4 March 2024
Revised: 28 March 2024
Accepted: 28 March 2024
Published: 30 March 2024

Copyright: © 2024 by the author. Licensee MDPI, Basel, Switzerland. This article is an open access article distributed under the terms and conditions of the Creative Commons Attribution (CC BY) license (https://creativecommons.org/licenses/by/4.0/).

networks (CNNs), offer the unique capability of end-to-end training [14,15]. However, their data-intensive nature necessitates substantial labeled data samples for training from scratch [16]. To circumvent this limitation, pre-trained CNN models can be fine-tuned and deployed as feature extractors, effectively capturing the salient information within medical images like CT scans [17].

Accurate and timely COVID-19 diagnosis presents a significant challenge in clinical settings. Robust automated detection methods can significantly aid medical professionals in making treatment decisions upon confirmation of the disease. This study, therefore, investigates a comprehensive range of feature descriptors for the classification of COVID-19 chest CT images. The primary objective is to identify a robust and efficient set of descriptors that accurately classify COVID-19 versus non-COVID-19 cases.

In the following, I provide a literature review of classification techniques for chest CT scans of patients afflicted with COVID-19 and outline this work's contributions.

2. Related Work

The automatic classification of COVID-19 CT images has garnered significant attention in recent research, as evidenced by a plethora of contributions documented in the literature [18–20]. In this section, I provide a review of these computer-aided system techniques, categorizing them into two distinct paradigms: traditional machine learning (TML) and deep learning (DL)-based approaches.

2.1. Traditional ML-Based Techniques

TML approaches for COVID-19 image classification typically adopt a pipelined structure. This workflow encompasses three key stages: (1) feature extraction, where relevant image characteristics (e.g., shape, color, texture) are isolated; (2) feature selection, which involves choosing a subset of informative features; and (3) classification model construction, where a model is trained to distinguish between COVID-19 and non-COVID-19 images. The ultimate goal is to achieve a robust classifier with minimal classification error. For instance, Hussain et al. [21] employed texture and morphological features to train various supervised classifiers for COVID-19 classification. Similarly, Chen et al. [22] utilized texture features derived from the Gray-Level Co-Occurrence Matrix (GLCM) to train a support vector machine (SVM) classifier within a 10-fold cross-validation (CV) framework.

Other studies have explored the utility of statistical moments for differentiating COVID-19 from non-COVID-19 images. Elaziz et al. [12] proposed the extraction of Fractional Multichannel Exponent Moments (FEMs) as features for classifier training. Their approach was evaluated on two independent datasets, achieving accuracies of 96.09% and 98.09%, respectively.

Ismael and Şengür [13] explored the efficacy of various multiresolution analysis techniques, namely wavelet, shearlet, and contourlet transforms, for COVID-19 detection in X-ray images. Following image decomposition, they extracted entropy and normalized entropy as features from the resulting subbands. These feature vectors were subsequently employed to train extreme learning machines (ELMs) for classification. The study utilized an imbalanced dataset comprising 200 healthy control samples and 361 COVID-19 X-ray images. The authors compared their proposed traditional method with the performance of DL-derived features. Interestingly, they concluded that traditional methods retain relevance and do not necessarily yield inferior results compared to DL approaches. This observation is supported by their finding that shearlet-based descriptors achieved an accuracy of 99.28%.

In a similar tendency to previous TML works, this research proposes to delve into a comprehensive exploration of hand-engineered descriptors. To mitigate the introduction of extraneous biases, the training and testing protocols will be held constant, while leveraging the identical COVID-19 dataset employed throughout the study.

2.2. DL-Based Techniques

In contrast to TML approaches, DL models offer the capability of end-to-end training directly on raw COVID-19 image data. The efficacy of DL architectures stems from their inherent capacity to autonomously acquire and unveil multi-tiered representations from data. Initial strata within the network typically concentrate on the extraction of fundamental characteristics, such as chromatic properties and boundaries [23]. Subsequently, higher strata progressively abstract these features, culminating in the formation of semantically significant representations of the input data. As an example, Ismael and Şengür [24] explored both fine-tuning pre-trained CNNs and training a CNN from scratch for COVID-19 detection in chest X-ray images. The utilized dataset comprised 180 COVID-19 and 200 healthy control X-ray images. The study evaluated various pre-trained models, including ResNet50, ResNet101, VGG16, and VGG19. Extracted deep features from these models were subsequently fed to an SVM for classification. The authors reported accuracies of 94.7%, 92.6%, and 91.6% for utilizing unsupervised deep features extraction, fine-tuning pre-trained models, and training from scratch, respectively.

Mirroring the approach of Ismael and Şengür [24], Haque et al. [25] investigated the utility of DL for COVID-19 detection in chest X-ray images. They explored both a custom-designed CNN model and fine-tuned pre-trained models (ResNet50, VGG-16, and VGG-19). Their proposed CNN architecture achieved an accuracy of 98.3% and a precision of 96.72%.

Furthermore, Jain et al. [26] investigated the use of X-ray images for COVID-19 detection through a DL model trained with data augmentation techniques. While their model was validated using a 5-fold CV scheme, it achieved an accuracy of 98.93%.

Saiz and Barandiaran [27] proposed an object detection DL architecture, which was trained and tested using publicly available datasets of 1500 images of normal and abnormal COVID-19 patients. The authors' primary goal was to classify the patients as infected or non-infected with COVID-19. The reported sensitivity and specificity were 94.92% and 92%, respectively.

Sahin et al. [28] investigated the application of DL methodologies for COVID-19 diagnosis utilizing CT imagery. Their approach leveraged Faster R-CNN and Mask R-CNN architectures for the classification of patients with COVID-19 and pneumonia. The study conducted a comparative analysis employing VGG-16 as the backbone for the Faster R-CNN model, while ResNet-50 and ResNet-101 backbones were utilized for the Mask R-CNN model. The implemented Faster R-CNN model achieved an accuracy of 93.86%. The Mask R-CNN model, employing ResNet-50 and ResNet-101 backbones, yielded mean average precision (mAP) values of 97.72% and 95.65%, respectively.

Avola et al. [29] investigated the effectiveness of twelve pre-trained DL models to differentiate between chest X-ray images from healthy individuals, those exhibiting signs of viral pneumonia (encompassing both generic and SARS-CoV-2 strains), and those with bacterial pneumonia. The experiment employed a dataset consisting of 6330 images, subdivided into training, validation, and testing sets. Standard classification metrics, such as precision and F1 scores, were computed for all models. The findings revealed that many of the implemented architectures achieved an average F1 score of up to 84.46% when distinguishing between the four designated classes.

Kathamuthu et al. [30] explored the efficacy of various deep transfer learning-based CNN architectures for the detection of COVID-19 in chest CT imagery. The investigation leverages pre-trained models including VGG16, VGG19, Densenet121, InceptionV3, Xception, and Resnet50 as foundational elements. The results demonstrate that the VGG16 model achieves superior performance within this study, attaining an accuracy of 98.00%.

Analogous to prior investigations documented within the literature, this study leverages established DL models; however, these models are employed solely for unsupervised feature extraction, eschewing fine-tuning. My approach focuses on utilizing the terminal layer within the network hierarchy, situated immediately before the classification layer. This selection is predicated on the assumption that these deep features encapsulate a semantic representation of the input data. Notably, Nanni et al. [31] proposed a system that exploits features learned by CNNs across multiple levels. Their system advocates for the fusion of these learned features, subsequently leveraging them for various image classification tasks.

2.3. Contribution

Many computational techniques have been developed for the identification of COVID-19 using traditional and DL approaches. Many of these techniques lack standardized training and testing approaches. The importance of this research can be comprehended by answering the following questions:

- Why COVID-19 Detection is Still Important?
    - Long-Term Effects: COVID-19 can cause lingering health problems even after recovery. Thus, early detection potentially helps in managing these effectively.
    - Variants and Future Outbreaks: New variants can emerge, and having robust detection systems is indispensable for future outbreaks.
    - Improved Healthcare Systems: Coherent detection tools can minimize unnecessary hospitalizations and allocate resources better.
- Why a Comparative Study for COVID-19 is important?
    - Benchmarking Progress: Contrasting different techniques allows us to identify the well-performing models and track advancements in the field.
    - Understanding Best-Performing Methods: Knowing best-performing methods guides future development to generalize and adapt for other classification tasks in biomedicine.
    - Focus on Improving Techniques: Even if the overall trend of COVID-19 is decreasing, a comparative analysis of various techniques could potentially identify and improve robust techniques for COVID-19 detection.

Thus, I compute the performances while utilizing 27 descriptors on one popular COVID-19 dataset with the same experimental setting. Moreover, I compare the results achieved by handcrafted features with the results obtained by the state-of-the-art deep features. As such, a comparative experimental study was conducted on how well-advanced deep CNNs trained on ImageNet. To this end, I experimented with 12 deep networks that have different architectural designs and varying depths. These models are utilized as unsupervised feature extractors. As an advantage of using CNNs as unsupervised feature extractors, I avoid training and fine-tuning the models, and thus, fewer computational resources are needed. I also evaluate the robustness of both hand-crafted and deep features with an SVM, which is trained in the context of a 5-fold nested CV.

3. Methods

First and foremost, the progress and improvement in developing techniques related to the classification of COVID-19 chest CT scans are due to the public availability of such datasets. For instance, Angelov and Soares [32] is a highly cited paper that collected 2482 images of COVID-19 samples that I use in my investigation to evaluate and compare traditional and deep features techniques.

Machine learning [33] refers to the field of computer science where algorithms are trained to learn and solve problems from examples rather than being explicitly programmed. In the context of medical image analysis, particularly COVID-19, this involves building mathematical models based on datasets to achieve the task of differentiating between healthy vs diseased patients. These data-driven algorithms are constantly optimized through various optimization algorithms [34] to achieve high accuracy and efficiency in their performance. Ultimately, the goal is to develop a generalizable ML model that can accurately predict outcomes even for unseen data, meaning new medical images that are not included in the training dataset.

The construction of robust classifiers within a TML paradigm for COVID-19 image analysis hinges on the extraction of informative features from the data. Commonly utilized features, as documented in the literature, include morphological descriptors, textural descriptors, and those derived from spectral methods. These extracted feature vectors subsequently serve as input to the classification model. As an alternative, DL approaches offer the distinct advantage of directly learning features from the raw medical images in an end-to-end manner, avoiding the feature extraction step. However, a notable limitation of DL techniques lies in their data-intensive nature, often requiring substantial labeled data samples for effective training from scratch.

A review of both traditional and DL-based techniques applied to COVID-19 patient datasets, as presented in the prior section, highlights their capability to achieve impressive classification performance. This suggests the potential of ML as a pre-screening tool to support radiologists in clinical settings. Notably, the literature indicates the efficacy of spectral methods, such as shearlet [35] and contourlet [36], coupled with statistical analysis for image analysis. The multiresolution and multi-scale nature of sub-bands obtained through image decomposition facilitates in-depth exploration. Notably, Ismael and Şengür [13] extracted shearlet coefficients and utilized entropy and normalized entropy as features, demonstrating the continued relevance of traditional methods.

My study addresses the challenge of limited medical data samples by evaluating and comparing various methods capable of mitigating this issue. I explore the performance of twenty-seven traditionally hand-crafted features. While DL methods have demonstrated promising classification results on diverse datasets, including those pertaining to COVID-19, they generally require substantial training data. To circumvent this limitation, a common practice in DL, particularly with scarce image samples, is to fine-tune a pre-trained model alongside data augmentation techniques to achieve optimal classification performance and avoid overfitting. In contrast, my approach investigates the use of DL models trained on non-medical image datasets (i.e., ImageNet [37]) as unsupervised feature extractors. My technique is implemented using MATLAB® 2021b. The experimental platform consisted of a computer system equipped with an Intel Core i7-9700 central processing unit (CPU) operating at a clock speed of 3.00 GHz. Additionally, the system was outfitted with an NVIDIA GeForce RTX 2080 graphics processing unit (GPU) possessing 8 GB of dedicated video memory.

3.1. Handcrafted Descriptors for COVID-19 Image Classification

Many conventional feature extraction methods aspire to detect a region of interest in images by computing geometric and appearance features [38], subsequently, these features are utilized to train traditional ML algorithms. Geometric features are computed based on the shape, locality of features, and salient points [39]. On the other hand, appearance-based attributes are based on texture information. In this study, I examine a set of 27 descriptors (MATLAB ToolboxDESC contains the implementation of 27 sets of descriptors that can be accessed via https://github.com/cigdemturan/ToolboxDESC, accessed on 1 January 2024) (as shown in Table 1). The same descriptors are utilized by Turan and Lam [38] to study facial expression recognition, but I examine these features to classify COVID-19 images.

Table 1. Details about the feature vector length of utilized hand-crafted features.

ID	Method	Abbreviation	Dimension
1	Binary Pattern of Phase Congruency [40]	BPPC	1062
2	Gradient Directional Pattern [41]	GDP	256
3	Gradient Direction Pattern [42]	GDP2	8
4	Gradient Local Ternary Pattern [43]	GLTeP	512
5	Improved Weber Binary Coding [44]	IWBC	2048
6	Local Arc Pattern [45]	LAP	272
7	Local Binary Pattern [46]	LBP	59
8	Local Directional Pattern [47]	LDiP	56
9	Local Directional Pattern Variance [48]	LDiPv	56
10	Local Directional Number Pattern [49]	LDN	56

Table 1. *Cont.*

ID	Method	Abbreviation	Dimension
11	Local Directional Texture Pattern [50]	LDTP	72
12	Local Frequency Descriptor [51]	LFD	512
13	Local Gabor Binary Pattern Histogram Sequence [52]	LGBPHS	256
14	Local Gabor Directional Pattern [53]	LGDiP	280
15	Local Gradient Increasing Pattern [54]	LGIP	37
16	Local Gradient Pattern [55]	LGP	7
17	Local Gabor Transitional Pattern [56]	LGTrP	256
18	Local Monotonic Pattern [57]	LMP	256
19	Local Phase Quantization [58]	LPQ	256
20	Local Ternary Pattern [59]	LTeP	512
21	Local Transitional Pattern [60]	LTrP	256
22	Monogenic Binary Coding [61]	MBC	3072
23	Median Binary Pattern [59]	MBP	256
24	Median Robust Extended Local Binary Pattern [62]	MRELBP	800
25	Median Ternary Pattern [59]	MTP	512
26	Pyramid of Histogram of Oriented Gradients [63]	PHOG	168
27	Weber Local Descriptor [64]	WLD	32

3.2. Deep Models for COVID-19 Images

Rather than training from scratch, I leverage pre-trained models capable of extracting meaningful features, i.e., these models are trained on vast datasets of non-medical images like ImageNet. These pre-trained models act as powerful but unsupervised feature extractors, generating deep features for COVID-19 image classification. Subsequently, I compare the performance of an SVM model trained and tested solely on these extracted features under a 5-fold nested CV scheme.

This study evaluates the capability of different CNN architectures to capture valuable information from COVID-19 images. I investigate both lightweight models like SqueezeNet 1.1 [65] and MobileNet v2 [66] for their efficiency, and larger models like ResNet-18 [67] and DenseNet-201 [68] for their potential in capturing richer details. Particularly, SqueezeNet utilizes diverse filter sizes to potentially extract both fine-grained and broader features from the images. MobileNet v2 boasts superior speed compared to other efficient models like ShuffleNet [69] and NASNet [70]. Notably, DenseNet-201 leverages feature reuse, where previously learned features are incorporated into subsequent layers, potentially enriching the information available for processing. In contrast, ResNet-18 employs element-wise addition to combine feature maps, offering a different approach to information flow.

My investigation encompassed the exploration of alternative DL models with comparable structures. Notably, Inception [71] exhibits similarities to DenseNet in its utilization of skip connections for depth-wise feature map concatenation. However, Inception's wider building block, constructed using diverse kernel sizes, resulted in subpar performance for COVID-19 image classification compared to DenseNet. Conversely, older models devoid of skip connections, such as VGG architectures [72], are susceptible to vanishing gradients and potentially slower training times. Nonetheless, My experimentation revealed promising results when applying these models to COVID-19 image classification.

A brief description of the CNN models that are used in this study is as follows.

- GoogLeNet (Inception) [73]: architecture relies on LeNet and AlexNet CNN models, but with the modification of depth and width of the layers. This model consists of 22 layers. It employs a parallel structure to significantly lessen the training time. As such, the model is designed to avoid patch-alignment problems by applying filter sizes of 1×1, 3×3, and 5×5.
- Inception-ResNet-v2 [74]: model consists of 164 layers. This model relies on the family of Inception, but instead comprises residual connections. As such, this model replaces the filter concatenation step of the Inception CNN model.

- Inception-v3 [71]: comprising 48 layers, tackles the challenge of positional variance in salient image features by employing a multi-branch architecture. This architecture allows the network to incorporate diverse kernel types at the same level (sizes of 1×1, 3×3, and pooling layers), effectively expanding the network's receptive field. These Inception modules enable the concurrent execution of numerous kernels, fostering greater feature extraction diversity. This core concept was introduced in the initial Inception-v1 model. Building upon its predecessor, Inception-v3 addresses the representational bottleneck issue through enhanced strategies. Notably, it incorporates kernel factorization and batch normalization within its auxiliary classifiers, leading to improved performance.
- VGG-16 and VGG-19 [72]: developed by the Visual Geometry Group (VGG) at the University of Oxford, VGG models represent a family of CNNs known for their simplicity and performance. Notably, VGG-16 and VGG-19, with 16 and 19 convolutional layers respectively, gained recognition at the ILSVRC 2014 competition as runners-up. These architectures feature relatively large numbers of parameters, with VGG-16 reaching approximately 138 million parameters. Additionally, both models incorporate fully connected layers containing 4096 hidden units each.
- SqueezeNet v1.1 [65]: network commences with a convolutional layer (conv1), followed by a sequence of eight blocks, each containing 2–9 fire modules. Each fire module employs a squeeze convolution layer with a filter size of 1×1, followed by two expand layers. One of these expand layers utilizes a filter size of 1×1, while the other utilizes a filter size of 3×3. The resulting feature maps from both expand layers are subsequently concatenated to form the input for the subsequent squeeze layer, which then feeds into the next fire module within the block.
- DenseNet-201 [68]: architecture leverages the concept of residual learning, introduced in ResNet, for network optimization. While ResNet employs element-wise addition of previous feature maps to the output, DenseNet utilizes depth concatenation of both the current and preceding outputs. This architecture comprises 32 dense blocks, each containing two distinct convolutional layers with kernel sizes of 1×1 and 3×3, respectively. Notably, these convolutional layers are preceded by batch normalization for improved convergence and training stability.
- ResNet-18 [67]: architecture leverages a series of eight basic building blocks, each containing a sequence of two convolutional layers. These convolutional layers utilize a fixed filter size of 3×3, ensuring consistent spatial feature extraction. Critically, each convolutional layer is followed by batch normalization, a technique that facilitates faster convergence and improved training stability. Notably, a key mechanism of ResNet-18 lies in the residual connection. This involves the element-wise addition of the current block's output to the output of the preceding block, allowing the information flow to propagate efficiently through the network.
- ResNet-50 and ResNet-101 [67]: architectures comprise variations of the ResNet-18 model, differentiating themselves through their respective depths of 50 and 101 layers. Both architectures leverage the bottleneck residual module, which processes the input signal through two distinct branches: (1) Convolutional Processing Branch: This branch applies a series of convolutions with varying kernel sizes (1×1 and 3×3) interspersed with batch normalization and ReLU activation functions; (2) Skip Connection Branch: This branch directly transmits the input signal unaltered, preserving crucial low-level feature information.
- Xception [75]: model stands out for its exclusive reliance on depthwise separable convolution layers. This architectural decision fosters computational efficiency while maintaining representational power. The network encompasses 36 convolutional layers, organized into 14 individual blocks. Only the first and final blocks deviate from the standard structure by lacking residual connections. In contrast, all remaining blocks incorporate linear residual connections. This strategic use of residual con-

nections facilitates gradient flow throughout the network, enhancing training and promoting optimal performance.
- MobileNet-v2 [66]: architecture incorporates two primary types of building blocks: (1) Linear Bottleneck Operations: These modules aim to achieve feature compression while maintaining representational power. (2) Skip Connections: These direct connections facilitate the flow of gradients and information across the network, mitigating the vanishing gradient problem that can occur in deep architectures. Both block types share fundamental operations, including convolution, batch normalization, and modified rectified linear unit (i.e., min (max (x, 0), 6)). The network comprises a total of 16 of these blocks, strategically arranged to achieve efficient feature extraction and classification performance.

CNN as Feature Extractor

The pre-trained CNN architectures, as detailed in Section 3.2, are utilized as unsupervised feature extractors. In this context, the deepest layer's output (directly preceding the classification layer) of each pre-trained model is flattened, generating a feature vector for each image. Fine-tuning of the pre-trained models is not conducted. Subsequently, a standard SVM classifier is trained and evaluated to assess the efficacy of these extracted deep features in classifying COVID-19 images. Table 2 presents a comprehensive overview of the feature vector lengths derived from each CNN model, thereby summarizing the salient features captured for each COVID-19 image.

Table 2. Details about the feature vector length of utilized CNN models as feature extractors.

Model's Name	Layer	Length
GoogLeNet	pool5-7x7_s1	1024
Inception-ResNet-v2	avg_pool	1536
Inception-v3	avg_pool	2048
VGG-16	fc6	4096
VGG-19	fc6	4096
ResNet-50	avg_pool	2048
ResNet-101	pool5	2048
SqueezeNet v1.1	pool10	1000
DenseNet-201	avg_pool	1920
ResNet-18	pool5	512
Xception	avg_pool	2048
MobileNet-v2	global_average_pooling2d_1	1280

4. Experiments and Results

This section provides a summary of the experiments conducted to evaluate and compare a wide range of hand-crafted and deep features for detecting COVID-19 infection from CT scan images. First, a description of the dataset utilized for the experimental studies is given at the beginning of this section. Then, a description of the nested cross-validation used to test my classification models and a list of utilized evaluation measures to assess the efficiency of the studied techniques. Followed by a brief highlight of the well-performing methods for the classification of COVID-19 CT images. Finally, I present a comprehensive review of the state-of-the-art methods with their corresponding performance utilizing the same benchmark dataset.

4.1. Dataset

For the purposes of this research investigation, I leverage the pre-processed COVID-19 dataset, SARS-CoV-2, as originally proposed by Angelov and Soares [32] and made publicly available on Kaggle (www.kaggle.com/plameneduardo/sarscov2-ctscan-dataset, accessed on 1 January 204). This dataset contains a total of 2482 chest CT-scans images that belong either to COVID-19 (i.e., 1252 images) or non-COVID-19 (i.e., 1230 images). The CT scans exhibit heterogeneity in their spatial dimensions, ranging from 104×119 to 416×512 pixels. Notably, all scans are grayscale and stored in the Portable Network Graphics (PNG) format. Angelov and Soares have classified the images based on the outcome of the RT-PCR tests. As such, patients with confirmed positive or negative RT-PCR tests for COVID-19 infection are included in the datasets.

4.2. Nested Cross-Validation (CV)

This work employs a five-fold nested CV strategy for hyperparameter optimization. The dataset is partitioned into five equally sized subsets. Within each outer fold, an inner loop utilizes four subsets for training and hyperparameter tuning via a classifier. The remaining subset in the inner loop serves as the validation set for hyperparameter selection. The geometric mean of the classifier performance serves as the objective function for hyperparameter optimization within the inner loop. Upon convergence, the optimized classifier is evaluated on the test set of the outer fold, utilizing a range of performance metrics including accuracy, sensitivity, specificity, F-measure, area under the receiver operating characteristic curve, positive predictive value, negative predictive value, and the geometric mean. This process is repeated for each of the five outer folds, ensuring a robust estimation of the classifier's generalizability.

Following the five-fold nested CV protocol outlined previously, I reiterate the entire procedure five times. Subsequently, across these five iterations, the average of the following classification metrics [16] is reported:

- Accuracy $ACC = (TP + TN)/(TP + TN + FP + FN)$, where TP, TN, FP, and FN indicate the number of true positives, true negatives, false positives, and false negatives, respectively.
- Sensitivity $SN = TP/(TP + FN)$.
- Specificity $SP = TN/(TN + FP)$.
- F-Measure $FM = (2 \times TP)/(2 \times TP + FP + FN)$
- The area under the curve (AUC) encapsulates the relationship between the true positive rate (sensitivity) and the false positive rate ($1 - $ specificity)
- Positive Predictive Value (PPV) = TP/(TP + FP).
- Negative Predictive Value (NPV) = TN/(TN + FN).
- Geometric mean (GM) is the square root of the product of sensitivity and specificity, or GM = $\sqrt{SN \times SP}$.

4.3. Results and Analysis

This section explores the classification results for COVID-19 detection. Initially, I present the performance of the SVM classifier utilizing hand-crafted descriptors. Subsequently, I report the SVM's performance when trained on diverse unsupervised deep features extracted from various CNN architectures. All built classification models in this study undergo training and validation within a nested CV framework.

4.3.1. Classification Results Using Handcrafted Descriptors

This section investigates the classification performance of hand-crafted features extracted from the SARS-CoV-2 dataset for COVID-19 detection. Each sample undergoes summarization via 27 distinct techniques, with their corresponding feature vector lengths outlined in Table 1. Subsequently, the resulting feature matrix, where each row represents an image's feature vector and its associated label, is fed into an SVM classifier. As detailed in the nested CV section, a 5-fold nested CV scheme is employed to evaluate the classifier's performance, with the optimal model selected based on the geometric mean score (See Figure 1). Table 3 presents the classification results achieved by the SVM models utilizing various hand-engineered descriptors. Notably, the Pyramid of Histogram of Oriented Gradients (PHOG) exhibits the highest performance. Both Gradient Local Ternary Pattern (GLTeP) and Local Ternary Pattern (LTeP) achieve comparable results; however, PHOG presents a distinct advantage due to its significantly lower feature vector dimensionality (168 compared to 512 for GLTeP and LTeP).

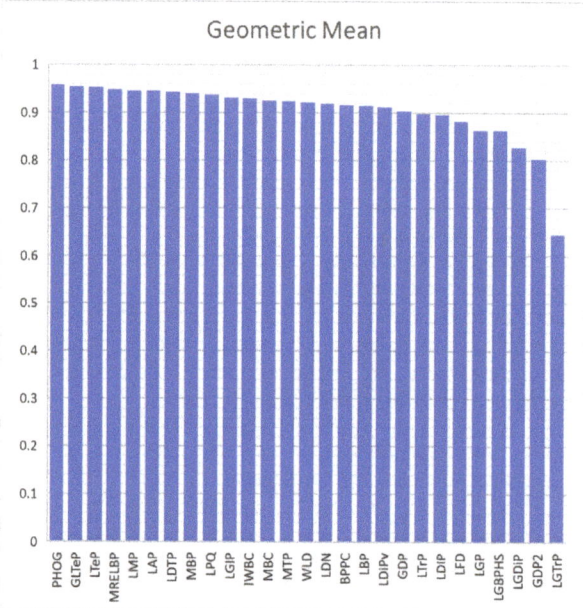

Figure 1. Achieved geometric mean by each hand-crafted method when used for training an SVM model in a 5-fold nested cross-validation fashion. Sorted from left to right.

Table 3. Average SVM performance trained using hand-engineered features for COVID-19 image classification. This table presents the mean values ± standard deviation.

ID	Method	ACC	SN	SP	FM	AUC	PPV	NPV	GM
1	BPPC	0.9175 ± 0.0104	0.9180 ± 0.0130	0.9171 ± 0.0202	0.9169 ± 0.0101	0.9175 ± 0.0103	0.9160 ± 0.0185	0.9194 ± 0.0113	0.9174 ± 0.0103
2	GDP	0.9045 ± 0.0245	0.8975 ± 0.0264	0.9114 ± 0.0251	0.9030 ± 0.0250	0.9045 ± 0.0245	0.9087 ± 0.0257	0.9007 ± 0.0251	0.9044 ± 0.0245
3	GDP2	0.8033 ± 0.0137	0.7990 ± 0.0173	0.8076 ± 0.0176	0.8010 ± 0.0137	0.8033 ± 0.0137	0.8031 ± 0.0152	0.8038 ± 0.0153	0.8032 ± 0.0137
4	GLTeP	0.9549 ± 0.0174	0.9508 ± 0.0123	0.9589 ± 0.0271	0.9544 ± 0.0171	0.9549 ± 0.0174	0.9583 ± 0.0263	0.9520 ± 0.0120	**0.9548 ± 0.0174**
5	IWBC	0.9301 ± 0.0235	0.9319 ± 0.0240	0.9284 ± 0.0231	0.9296 ± 0.0237	0.9301 ± 0.0235	0.9274 ± 0.0235	0.9329 ± 0.0236	0.9301 ± 0.0235
6	LAP	0.9456 ± 0.0192	0.9311 ± 0.0247	0.9598 ± 0.0180	0.9442 ± 0.0199	0.9454 ± 0.0192	0.9578 ± 0.0190	0.9344 ± 0.0228	0.9453 ± 0.0193
7	LBP	0.9155 ± 0.0228	0.9114 ± 0.0275	0.9195 ± 0.0188	0.9143 ± 0.0236	0.9155 ± 0.0229	0.9173 ± 0.0201	0.9138 ± 0.0258	0.9154 ± 0.0229
8	LDiP	0.8968 ± 0.0274	0.8868 ± 0.0229	0.9066 ± 0.0355	0.8950 ± 0.0270	0.8967 ± 0.0274	0.9036 ± 0.0344	0.8907 ± 0.0230	0.8966 ± 0.0273
9	LDiPv	0.9122 ± 0.0216	0.9212 ± 0.0202	0.9034 ± 0.0279	0.9123 ± 0.0211	0.9123 ± 0.0215	0.9038 ± 0.0259	0.9212 ± 0.0201	0.9122 ± 0.0216
10	LDN	0.9195 ± 0.0101	0.9033 ± 0.0181	0.9355 ± 0.0137	0.9175 ± 0.0096	0.9194 ± 0.0095	0.9323 ± 0.0136	0.9077 ± 0.0202	0.9192 ± 0.0095
11	LDTP	0.9439 ± 0.0143	0.9280 ± 0.0112	0.9594 ± 0.0192	0.9427 ± 0.0129	0.9437 ± 0.0141	0.9579 ± 0.0174	0.9308 ± 0.0157	0.9436 ± 0.0140
12	LFD	0.8826 ± 0.0180	0.8782 ± 0.0188	0.8872 ± 0.0247	0.8805 ± 0.0226	0.8827 ± 0.0181	0.8831 ± 0.0331	0.8816 ± 0.0145	0.8827 ± 0.0180
13	LGBPHS	0.8639 ± 0.0320	0.8496 ± 0.0395	0.8780 ± 0.0285	0.8602 ± 0.0359	0.8638 ± 0.0322	0.8713 ± 0.0367	0.8565 ± 0.0359	0.8636 ± 0.0322
14	LGDiP	0.8281 ± 0.0136	0.8323 ± 0.0123	0.8232 ± 0.0288	0.8273 ± 0.0148	0.8277 ± 0.0143	0.8225 ± 0.0229	0.8332 ± 0.0138	0.8276 ± 0.0145
15	LGIP	0.9313 ± 0.0136	0.9243 ± 0.0228	0.9389 ± 0.0115	0.9303 ± 0.0130	0.9316 ± 0.0135	0.9366 ± 0.0137	0.9258 ± 0.0259	0.9315 ± 0.0136
16	LGP	0.8639 ± 0.0151	0.8587 ± 0.0152	0.8690 ± 0.0181	0.8623 ± 0.0091	0.8638 ± 0.0150	0.8660 ± 0.0098	0.8611 ± 0.0270	0.8638 ± 0.0150
17	LGTrP	0.6477 ± 0.0303	0.5957 ± 0.0422	0.6987 ± 0.0205	0.6256 ± 0.0372	0.6472 ± 0.0302	0.6592 ± 0.0363	0.6379 ± 0.0408	0.6450 ± 0.0311
18	LMP	0.9455 ± 0.0151	0.9397 ± 0.0166	0.9515 ± 0.0162	0.9442 ± 0.0180	0.9456 ± 0.0154	0.9488 ± 0.0217	0.9420 ± 0.0123	0.9455 ± 0.0154
19	LPQ	0.9378 ± 0.0121	0.9249 ± 0.0173	0.9504 ± 0.0122	0.9359 ± 0.0154	0.9377 ± 0.0127	0.9474 ± 0.0172	0.9286 ± 0.0117	0.9376 ± 0.0128
20	LTeP	0.9529 ± 0.0160	0.9450 ± 0.0155	0.9606 ± 0.0176	0.9521 ± 0.0162	0.9528 ± 0.0160	0.9593 ± 0.0181	0.9468 ± 0.0150	**0.9528 ± 0.0160**
21	LTrP	0.8984 ± 0.0101	0.9026 ± 0.0154	0.8951 ± 0.0147	0.8976 ± 0.0131	0.8988 ± 0.0098	0.8931 ± 0.0227	0.9032 ± 0.0187	0.8988 ± 0.0098
22	MBC	0.9260 ± 0.0087	0.9102 ± 0.0151	0.9416 ± 0.0127	0.9237 ± 0.0129	0.9259 ± 0.0096	0.9378 ± 0.0179	0.9148 ± 0.0097	0.9257 ± 0.0096
23	MBP	0.9403 ± 0.0062	0.9374 ± 0.0142	0.9428 ± 0.0106	0.9394 ± 0.0077	0.9401 ± 0.0062	0.9415 ± 0.0111	0.9391 ± 0.0124	0.9400 ± 0.0063
24	MRELBP	0.9496 ± 0.0092	0.9124 ± 0.0229	0.9854 ± 0.0081	0.9468 ± 0.0121	0.9489 ± 0.0104	0.9842 ± 0.0084	0.9207 ± 0.0153	0.9481 ± 0.0108
25	MTP	0.9252 ± 0.0049	0.9077 ± 0.0115	0.9418 ± 0.0113	0.9230 ± 0.0075	0.9247 ± 0.0053	0.9389 ± 0.0097	0.9126 ± 0.0049	0.9245 ± 0.0053
26	PHOG	0.9581 ± 0.0049	0.9490 ± 0.0129	0.9670 ± 0.0098	0.9572 ± 0.0063	0.9580 ± 0.0053	0.9658 ± 0.0094	0.9510 ± 0.0111	**0.9579 ± 0.0054**
27	WLD	0.9228 ± 0.0106	0.9139 ± 0.0064	0.9317 ± 0.0160	0.9213 ± 0.0116	0.9228 ± 0.0105	0.9289 ± 0.0193	0.9165 ± 0.0115	0.9228 ± 0.0104

Top-3 best-performing hand-crafted methods are highlighted.

An unpaired t-test was conducted to investigate the statistical distinction between PHOG and GLTeP in their performance on the SARS-CoV-2 classification task. The test revealed that there is no statistically significant difference ($p < 0.05$), with a two-tailed p-value of 0.7135 and a t-value of 0.3805.

4.3.2. Classification Results Using Deep Features

This section leverages the SARS-CoV-2 dataset for my investigation. Each image within the dataset is processed through various pre-trained CNN models. Table 2 summarizes the feature vector lengths extracted from each of these CNN architectures. Subsequently, the resulting feature matrix, where each row represents an image's feature vector and its corresponding label, is fed into an SVM classifier. The performance of my classifier is evaluated using a 5-fold nested CV scheme, following the hyperparameter optimization strategy outlined in the nested CV section. The optimal model is chosen based on the geometric mean score (See Figure 2). As presented in Table 4, the deep features extracted from DenseNet-201 yield the highest classification performance. Notably, VGG-16 achieves comparable results to DenseNet-201.

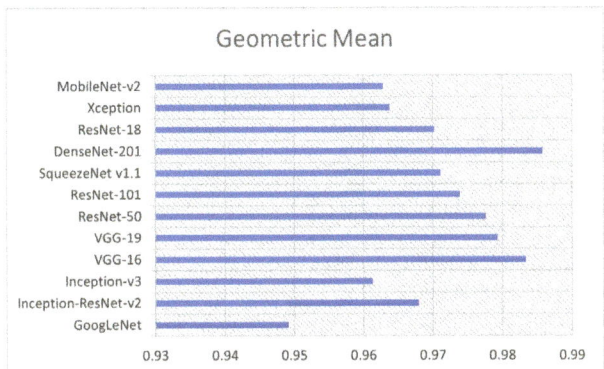

Figure 2. Achieved geometric mean by each CNN model's deep features when used for training an SVM model in a 5-fold nested cross-validation fashion.

Table 4. Average SVM performance trained using deep features for COVID-19 image classification. This table presents the mean values ± standard deviation.

Model's Name	ACC	SN	SP	FM	AUC	PPV	NPV	GM
GoogLeNet	0.9533 ± 0.0069	0.9475 ± 0.0090	0.9589 ± 0.0128	0.9526 ± 0.0069	0.9532 ± 0.0069	0.9579 ± 0.0126	0.9491 ± 0.0080	0.9491 ± 0.0080
Inception-ResNet-v2	0.9679 ± 0.0062	0.9672 ± 0.0127	0.9686 ± 0.0105	0.9676 ± 0.0063	0.9679 ± 0.0062	0.9681 ± 0.0100	0.9680 ± 0.0118	0.9680 ± 0.0118
Inception-v3	0.9614 ± 0.0092	0.9622 ± 0.0163	0.9606 ± 0.0164	0.9611 ± 0.0094	0.9614 ± 0.0092	0.9602 ± 0.0157	0.9632 ± 0.0152	0.9613 ± 0.0092
VGG-16	0.9833 ± 0.0048	0.9787 ± 0.0089	0.9879 ± 0.0090	0.9831 ± 0.0050	0.9833 ± 0.0049	0.9877 ± 0.0090	0.9793 ± 0.0084	**0.9833 ± 0.0049**
VGG-19	0.9821 ± 0.0049	0.9787 ± 0.0111	0.9855 ± 0.0105	0.9819 ± 0.0050	0.9821 ± 0.0049	0.9853 ± 0.0105	0.9793 ± 0.0103	0.9793 ± 0.0103
ResNet-50	0.9809 ± 0.0045	0.9770 ± 0.0074	0.9847 ± 0.0112	0.9807 ± 0.0044	0.9809 ± 0.0044	0.9844 ± 0.0111	0.9777 ± 0.0069	0.9777 ± 0.0069
ResNet-101	0.9768 ± 0.0018	0.9729 ± 0.0147	0.9807 ± 0.0138	0.9765 ± 0.0020	0.9768 ± 0.0018	0.9805 ± 0.0134	0.9739 ± 0.0137	0.9739 ± 0.0137
SqueezeNet v1.1	0.9712 ± 0.0098	0.9655 ± 0.0085	0.9767 ± 0.0149	0.9707 ± 0.0099	0.9711 ± 0.0097	0.9761 ± 0.0151	0.9666 ± 0.0081	0.9711 ± 0.0097
DenseNet-201	0.9858 ± 0.0029	0.9819 ± 0.0085	0.9895 ± 0.0073	0.9856 ± 0.0029	0.9857 ± 0.0029	0.9893 ± 0.0073	0.9825 ± 0.0080	**0.9857 ± 0.0029**
ResNet-18	0.9703 ± 0.0118	0.9655 ± 0.0160	0.9750 ± 0.0164	0.9699 ± 0.0119	0.9703 ± 0.0118	0.9745 ± 0.0164	0.9667 ± 0.0151	0.9702 ± 0.0118
Xception	0.9638 ± 0.0062	0.9664 ± 0.0034	0.9614 ± 0.0147	0.9636 ± 0.0059	0.9639 ± 0.0061	0.9611 ± 0.0143	0.9668 ± 0.0029	0.9638 ± 0.0061
MobileNet-v2	0.9630 ± 0.0075	0.9590 ± 0.0157	0.9670 ± 0.0112	0.9625 ± 0.0077	0.9630 ± 0.0075	0.9663 ± 0.0108	0.9603 ± 0.0144	0.9629 ± 0.0076

Top-2 best-performing set of deep features are highlighted.

To assess the statistical difference between DenseNet-201 and VGG-16 on the SARS-CoV-2 classification task, I performed an unpaired t-test. The results yielded no statistically significant difference ($p < 0.05$), with a two-tailed p-value equal to 0.3735 and a t-statistic of 0.9425.

4.4. Discussion

The landscape of COVID-19 classification using publicly available datasets, such as the SARS-CoV-2 dataset [32], is rapidly evolving. Researchers have proposed diverse conventional and DL techniques for classifying COVID-19 from CT-scan images. However, objective comparison across studies remains challenging due to several key factors impacting framework performance. These factors include: (1) Heterogeneity in CT-scan selection: Variations in acquisition protocols, scanners, and patient populations across datasets can significantly impact feature extraction and model generalization. (2) Varied image pre-processing techniques: Different pre-processing approaches, such as noise reduction, normalization, and segmentation, can significantly influence the extracted features and subsequent classification performance. (3) Divergence in training/testing protocols: Variations in data splitting (e.g., k-fold cross-validation, train/test ratio), evaluation metrics, and hyperparameter tuning strategies can hinder direct performance comparisons. Acknowledging these influencing factors is crucial for interpreting and comparing the results of COVID-19 classification studies.

A case study conducted by Maguolo and Nanni [76] examined various testing protocols while using COVID-19 X-ray 2D images. The authors showed that similar classification performance can be achieved while training a neural network using X-ray images that do not contain most of the lungs. Maguolo and Nanni removed the lungs from the images by inserting a black box into the center of the X-ray image. Then, these new images were used for training their classifiers only on the outer part of the images. The authors concluded that many of the testing protocols of published studies in the literature are not fair and the classifiers of neural networks were not learning patterns related to COVID-19. Hence, rigorous testing protocols should be established while training a DL model. As a result, one can conclude that assessing and comparing the performance of a method objectively is difficult because it is not clear which part of the technique (e.g., feature extraction/selection, pre-processing, or classification models) led to a tangible enhancement. Thus, my aim in my study is to learn from available published studies and to avoid potential mistakes (e.g., learning from the recommendation of Maguolo and Nanni to test with unbiased testing protocols).

Subsequently, I objectively planned to minimize the bias of the dataset by selecting SARS-CoV-2 dataset and this dataset is divided in the context of 5-fold nested cross-validation to rigorously evaluate wide range of hand-crafted descriptors and deep features from different number of CNN architectures, as shown in Tables 3 and 4. There are a large number of studies that utilized the SARS-CoV-2 dataset from which ten studies are summarized indicating their methodology essence, their training/testing protocol, and reported classification performance. Noteworthy, there is a vast number of studies that used the same dataset [30,77–98], but of similar nature, and thus, these studies are not summarized. Here is the summary of the ten studies:

- Halder and Datta [99] investigated the efficacy of transfer learning employing pre-trained CNN models, namely DenseNet201, VGG16, ResNet50V2, and MobileNet. Each model was independently trained and tested with a ratio of 8:2 on both unmodified and augmented datasets. Notably, DenseNet201 exhibited exceptional performance, achieving an AUC of 1.00 and 0.99 for the unaugmented and augmented datasets, respectively. Moreover, training DenseNet201 with the augmented data yielded a test set accuracy of 97%, surpassing ResNet50V2 (96%), MobileNet (95%), and VGG16 (94%).
- Alshazly et al. [100] investigated the application of transfer learning to various pre-trained CNN architectures, including SqueezeNet, Inception, ResNet, ResNeXt, Xception, ShuffleNet, and DenseNet. Five-fold cross-validation was utilized to evaluate the efficacy of their approach. Their ResNet101 model demonstrated remarkable performance, achieving average accuracy, precision, sensitivity, specificity, and F1-score values of 99.4%, 99.6%, 99.1%, 99.6%, and 99.4%, respectively.
- Ragab et al. [101] proposed a multi-modal fusion architecture for COVID-19 image classification. Their system leverages the pre-trained CNNs, namely AlexNet, GoogleNet, ShuffleNet, and ResNet-18, alongside hand-crafted features derived from statistical analysis, discrete wavelet transform, and grey-level co-occurrence matrix. They employed five-fold cross-validation to evaluate the efficacy of their approach. This

hybrid methodology achieved performance, attaining an average accuracy, sensitivity, specificity, and precision of approximately 99% across all evaluation metrics.
- Shaik and Cherukuri [102] presented an ensemble learning approach for COVID-19 image classification that leverages the combined prediction of diverse pre-trained CNN architectures. They employ a collection of eight models, including VGG16, VGG19, InceptionV3, ResNet50, ResNet50V2, InceptionResNetV2, Xception, and MobileNet. Each model is fine-tuned using an 80/20 data split for training and validation, respectively. This ensemble approach achieved an accuracy of 98.99%, precision of 98.98%, recall of 99.00%, and F-measure of 98.99%.
- Gaur et al. [103] presented a method that leverages the spectral information within each image channel (red, green, and blue) by applying a 2D-empirical wavelet transform. This decomposition generates five frequency sub-bands, which are subsequently augmented to enhance data variability. These augmented sub-bands then serve as the input for training a DenseNet121 classification model. To ensure a statistically robust evaluation, the dataset was randomly split into 1000 training, 100 validation, and 152 testing images prior to data augmentation. This strategy yielded a performance of accuracy of 85.50%, F-measure of 85.28%, and AUC of 96.6%.
- Canayaz et al. [104] explores the efficacy of Bayesian optimization in enhancing the performance of various machine learning algorithms for COVID-19 image classification. The authors propose and evaluate the application of this optimization technique to MobilNetv2, ResNet-50, SVM, and k-nearest neighbor (kNN) models. The proposed method consists of three steps: (1) train and optimize the deep learning models, (2) utilize trained models as feature extractors, and (3) train a machine learning algorithm. Notably, the ResNet-50 architecture, when optimized via Bayesian optimization and employed as a feature extractor for kNN (trained on 1968 COVID-19 images and tested on 492), yielded an accuracy of 99.37%, accompanied by a precision of 99.38%, recall of 99.36%, and F-score of 99.37%.
- Attallah and Samir [105] presented a two-stage framework for COVID-19 image classification that leverages spectral-temporal and spatial information. In the first stage, their method employs discrete wavelet decomposition (DWT) to extract frequency-domain features from the images, represented as heatmaps. These features are subsequently used to train a ResNet CNN model. Simultaneously, the original images are utilized to train a separate ResNet CNN model, capturing spatial information. Subsequently, both pipelines converge in a feature fusion stage, where spectral-temporal features are integrated with spatial features extracted from the second ResNet. To address dimensionality, the combined feature set is subjected to dimensionality reduction before being fed into support vector machine (SVM) classifiers. This strategy achieved a classification accuracy of 99.7% under a 5-fold cross-validation scheme.
- Kundu et al. [106] explored an ensemble learning approach by leveraging transfer learning. Their method, employing bootstrap aggregating (bagging) of three pre-trained architectures Inception v3, ResNet34, and DenseNet201 were examined under a 5-fold cross-validation scheme. The ensemble model achieved an accuracy of 97.81%, precision of 97.77%, sensitivity (recall) of 97.81%, and specificity of 97.77%.
- Islam and Nahiduzzaman [107] proposed employing a custom CNN architecture for extracting deep features. These features are subsequently fed into traditional machine learning algorithms, encompassing Gaussian Naive Bayes, Support Vector Machine, Decision Tree, Logistic Regression, and Random Forest. The output of these five learning algorithms is ensembled to find the final prediction. The proposed model undergoes training on 2109 COVID-19 images and evaluation on a separate set of 373 images. The model achieved an accuracy of 99.73%, an F1-score of 99.73%, a recall of 100%, and a precision of 99.46%.
- Choudhary et al. [108] introduced an approach for COVID-19 detection on resource-constrained devices, focusing on "important weights-only" transfer learning. This method optimizes pre-trained deep learning models for deployment on point-of-care

devices by selectively pruning less essential weight parameters. Their experiments were conducted on VGG16 and ResNet34 architectures. The proposed method was evaluated while using 1,687 samples for training, 420 samples for validation, and 375 samples for testing. The pruned ResNet34 model achieved an accuracy of 95.47%, a sensitivity of 0.9216, an F1-score of 0.9567, and a specificity of 0.9942 while exhibiting reductions in computational requirements: 41.96% fewer floating-point operations and 20.64% fewer weight parameters compared to the unpruned model.

It is noteworthy that the application of DL techniques has become predominant in SARS-CoV-2 research. Researchers often leverage various CNN architectures, opting for fine-tuning with or without data augmentation to address the inherent scarcity of medical datasets. While impressive results have been reported, inconsistencies in training and validation strategies across studies pose challenges for objective comparison. In contrast, my approach utilizes unsupervised deep features, minimizing computational demands. Furthermore, I employ a rigorous 5-fold nested CV scheme to evaluate the performance of my SVM classification models.

5. Conclusions and Future Studies

The main goal of this work is to compare and evaluate a wide range of conventional and DL-based techniques to identify effective and efficient approaches for classifying COVID-19 disease from CT scans. To achieve this goal, twenty-seven conventional techniques and 12 CNN architectures are examined. Thereafter each set of descriptors is fed as input to an SVM model, which is tested in the context of a 5-fold cross-validation scheme. The performance of the proposed methodologies is evaluated on the SARS-CoV-2 dataset. The empirical findings gleaned from this investigation posit that the proposed method holds promise for adoption as a pre-screening tool for COVID-19 cases, exhibiting competitive performance in comparison to established state-of-the-art methodologies. Additionally, the establishment of my framework requires minimal computational resources for conventional techniques, and particularly, for DL-based techniques as I avoid fine-tuning and data augmentation.

In the future, I plan to test my approach using other datasets with a similar nature of complexity, for example, the COVID-CT (COVID-CT benchmark can be accessed via https://www.kaggle.com/datasets/hgunraj/covidxct, accessed on 1 January 2024) dataset [109] and COVID multiclass dataset (The COVID-19 multicalss dataset can be accessed via https://www.kaggle.com/datasets/plameneduardo/a-covid-multiclass-dataset-of-ct-scans, accessed on 1 January 2024). Furthermore, I plan to combine both hand-crafted and deep features [110] in an attempt to deliver more robust classification models.

Funding: This research received no external funding.

Institutional Review Board Statement: Not applicable.

Data Availability Statement: This research leverages a publicly available dataset. COVID-19 dataset, SARS-CoV-2, can be accessed on Kaggle www.kaggle.com/plameneduardo/sarscov2-ctscan-dataset, accessed on 1 January 2024.

Conflicts of Interest: The author declares no conflict of interest.

References

1. Liu, J.; Liao, X.; Qian, S.; Yuan, J.; Wang, F.; Liu, Y.; Wang, Z.; Wang, F.S.; Liu, L.; Zhang, Z. Community transmission of severe acute respiratory syndrome coronavirus 2, Shenzhen, China, 2020. *Emerg. Infect. Dis.* **2020**, *26*, 1320. [CrossRef] [PubMed]
2. Ghinai, I.; McPherson, T.D.; Hunter, J.C.; Kirking, H.L.; Christiansen, D.; Joshi, K.; Rubin, R.; Morales-Estrada, S.; Black, S.R.; Pacilli, M.; et al. First known person-to-person transmission of severe acute respiratory syndrome coronavirus 2 (SARS-CoV-2) in the USA. *Lancet* **2020**, *395*, 1137–1144. [CrossRef] [PubMed]
3. Morens, D.M.; Fauci, A.S. Emerging pandemic diseases: How we got to COVID-19. *Cell* **2020**, *182*, 1077–1092. [CrossRef] [PubMed]
4. Poudel, A.N.; Zhu, S.; Cooper, N.; Roderick, P.; Alwan, N.; Tarrant, C.; Ziauddeen, N.; Yao, G.L. Impact of COVID-19 on health-related quality of life of patients: A structured review. *PLoS ONE* **2021**, *16*, e0259164. [CrossRef]

5. Arevalo-Rodriguez, I.; Buitrago-Garcia, D.; Simancas-Racines, D.; Zambrano-Achig, P.; Del Campo, R.; Ciapponi, A.; Sued, O.; Martinez-Garcia, L.; Rutjes, A.W.; Low, N.; et al. False-negative results of initial RT-PCR assays for COVID-19: A systematic review. *PLoS ONE* **2020**, *15*, e0242958. [CrossRef] [PubMed]
6. Axiaq, A.; Almohtadi, A.; Massias, S.A.; Ngemoh, D.; Harky, A. The role of computed tomography scan in the diagnosis of COVID-19 pneumonia. *Curr. Opin. Pulm. Med.* **2021**, *27*, 163–168. [CrossRef] [PubMed]
7. Fang, Y.; Zhang, H.; Xie, J.; Lin, M.; Ying, L.; Pang, P.; Ji, W. Sensitivity of chest CT for COVID-19: Comparison to RT-PCR. *Radiology* **2020**, *296*, E115–E117. [CrossRef] [PubMed]
8. Ai, T.; Yang, Z.; Hou, H.; Zhan, C.; Chen, C.; Lv, W.; Tao, Q.; Sun, Z.; Xia, L. Correlation of chest CT and RT-PCR testing for coronavirus disease 2019 (COVID-19) in China: A report of 1014 cases. *Radiology* **2020**, *296*, E32–E40. [CrossRef] [PubMed]
9. Hassan, H.; Ren, Z.; Zhao, H.; Huang, S.; Li, D.; Xiang, S.; Kang, Y.; Chen, S.; Huang, B. Review and classification of AI-enabled COVID-19 CT imaging models based on computer vision tasks. *Comput. Biol. Med.* **2022**, *141*, 105123. [CrossRef]
10. Saygılı, A. A new approach for computer-aided detection of coronavirus (COVID-19) from CT and X-ray images using machine learning methods. *Appl. Soft Comput.* **2021**, *105*, 107323. [CrossRef] [PubMed]
11. Shi, F.; Wang, J.; Shi, J.; Wu, Z.; Wang, Q.; Tang, Z.; He, K.; Shi, Y.; Shen, D. Review of artificial intelligence techniques in imaging data acquisition, segmentation, and diagnosis for COVID-19. *IEEE Rev. Biomed. Eng.* **2020**, *14*, 4–15. [CrossRef] [PubMed]
12. Elaziz, M.A.; Hosny, K.M.; Salah, A.; Darwish, M.M.; Lu, S.; Sahlol, A.T. New machine learning method for image-based diagnosis of COVID-19. *PLoS ONE* **2020**, *15*, e0235187. [CrossRef] [PubMed]
13. Ismael, A.M.; Şengür, A. The investigation of multiresolution approaches for chest X-ray image based COVID-19 detection. *Health Inf. Sci. Syst.* **2020**, *8*, 29. [CrossRef]
14. Alam, M.; Akram, M.U.; Fareed, W. Deep Learning-based Analysis and Classification of COVID Patients Through CT Images. In Proceedings of the IEEE 2023 3rd International Conference on Artificial Intelligence (ICAI), Wuhan, China, 17–19 November 2023; pp. 136–141.
15. Reis, H.C.; Turk, V. COVID-DSNet: A novel deep convolutional neural network for detection of coronavirus (SARS-CoV-2) cases from CT and Chest X-Ray images. *Artif. Intell. Med.* **2022**, *134*, 102427. [CrossRef] [PubMed]
16. Alinsaif, S. Unraveling Arrhythmias with Graph-Based Analysis: A Survey of the MIT-BIH Database. *Computation* **2024**, *12*, 21. [CrossRef]
17. Özkaya, U.; Öztürk, Ş.; Barstugan, M. Coronavirus (COVID-19) classification using deep features fusion and ranking technique. In *Big Data Analytics and Artificial Intelligence Against COVID-19: Innovation Vision and Approach*; Springer: Cham, Switzerland, 2020; pp. 281–295.
18. Rahmani, A.M.; Azhir, E.; Naserbakht, M.; Mohammadi, M.; Aldalwie, A.H.M.; Majeed, M.K.; Taher Karim, S.H.; Hosseinzadeh, M. Automatic COVID-19 detection mechanisms and approaches from medical images: A systematic review. *Multimed. Tools Appl.* **2022**, *81*, 28779–28798. [CrossRef] [PubMed]
19. Benameur, N.; Mahmoudi, R.; Zaid, S.; Arous, Y.; Hmida, B.; Bedoui, M.H. SARS-CoV-2 diagnosis using medical imaging techniques and artificial intelligence: A review. *Clin. Imaging* **2021**, *76*, 6–14. [CrossRef] [PubMed]
20. Serena Low, W.C.; Chuah, J.H.; Tee, C.A.T.; Anis, S.; Shoaib, M.A.; Faisal, A.; Khalil, A.; Lai, K.W. An overview of deep learning techniques on chest X-ray and CT scan identification of COVID-19. *Comput. Math. Methods Med.* **2021**, *2021*, 5528144. [CrossRef] [PubMed]
21. Hussain, L.; Nguyen, T.; Li, H.; Abbasi, A.A.; Lone, K.J.; Zhao, Z.; Zaib, M.; Chen, A.; Duong, T.Q. Machine-learning classification of texture features of portable chest X-ray accurately classifies COVID-19 lung infection. *BioMed. Eng. Online* **2020**, *19*, 88. [CrossRef] [PubMed]
22. Chen, Y. COVID-19 classification based on gray-level co-occurrence matrix and support vector machine. In *COVID-19: Prediction, Decision-Making, and Its Impacts*; Springer: Singapore, 2021; pp. 47–55.
23. Yosinski, J.; Clune, J.; Bengio, Y.; Lipson, H. How transferable are features in deep neural networks? In Proceedings of the Advances in Neural Information Processing Systems, Montreal, QC, Canada, 8–13 December 2014; Volume 27.
24. Ismael, A.M.; Şengür, A. Deep learning approaches for COVID-19 detection based on chest X-ray images. *Expert Syst. Appl.* **2021**, *164*, 114054. [CrossRef]
25. Haque, K.F.; Abdelgawad, A. A deep learning approach to detect COVID-19 patients from chest X-ray images. *AI* **2020**, *1*, 27. [CrossRef]
26. Jain, G.; Mittal, D.; Thakur, D.; Mittal, M.K. A deep learning approach to detect COVID-19 coronavirus with X-Ray images. *Biocybern. Biomed. Eng.* **2020**, *40*, 1391–1405. [CrossRef] [PubMed]
27. Saiz, F.; Barandiaran, I. COVID-19 detection in chest X-ray images using a deep learning approach. *Int. J. Interact. Multimed. Artif. Intell.* **2020**. . [CrossRef]
28. Sahin, M.E.; Ulutas, H.; Yuce, E.; Erkoc, M.F. Detection and classification of COVID-19 by using faster R-CNN and mask R-CNN on CT images. *Neural Comput. Appl.* **2023**, *35*, 13597–13611. [CrossRef] [PubMed]
29. Avola, D.; Bacciu, A.; Cinque, L.; Fagioli, A.; Marini, M.R.; Taiello, R. Study on transfer learning capabilities for pneumonia classification in chest-x-rays images. *Comput. Methods Programs Biomed.* **2022**, *221*, 106833. [CrossRef] [PubMed]
30. Kathamuthu, N.D.; Subramaniam, S.; Le, Q.H.; Muthusamy, S.; Panchal, H.; Sundararajan, S.C.M.; Alrubaie, A.J.; Zahra, M.M.A. A deep transfer learning-based convolution neural network model for COVID-19 detection using computed tomography scan images for medical applications. *Adv. Eng. Softw.* **2023**, *175*, 103317. [CrossRef] [PubMed]

31. Nanni, L.; Ghidoni, S.; Brahnam, S. Deep features for training support vector machines. *J. Imaging* **2021**, *7*, 177. [CrossRef] [PubMed]
32. Angelov, P.; Almeida Soares, E. SARS-CoV-2 CT-scan dataset: A large dataset of real patients CT scans for SARS-CoV-2 identification. *MedRxiv* **2020**. . [CrossRef]
33. Britain, R.S.G. *Machine Learning: The Power and Promise of Computers That Learn by Example: An Introduction*; Royal Society: London, UK, 2017.
34. Shawe-Taylor, J.; Sun, S. A review of optimization methodologies in support vector machines. *Neurocomputing* **2011**, *74*, 3609–3618. [CrossRef]
35. Kutyniok, G.; Labate, D. Introduction to shearlets. In *Shearlets: Multiscale Analysis for Multivariate Data*; Springer: New York, NY, USA, 2012; pp. 1–38.
36. Do, M.N.; Vetterli, M. The contourlet transform: An efficient directional multiresolution image representation. *IEEE Trans. Image Process.* **2005**, *14*, 2091–2106. [CrossRef] [PubMed]
37. Deng, J.; Dong, W.; Socher, R.; Li, L.J.; Li, K.; Fei-Fei, L. Imagenet: A large-scale hierarchical image database. In Proceedings of the 2009 IEEE Conference on Computer Vision and Pattern Recognition, Miami, FL, USA, 20–25 June 2009; pp. 248–255.
38. Turan, C.; Lam, K.M. Histogram-based local descriptors for facial expression recognition (FER): A comprehensive study. *J. Vis. Commun. Image Represent.* **2018**, *55*, 331–341. [CrossRef]
39. Sebe, N.; Tian, Q.; Loupias, E.; Lew, M.S.; Huang, T.S. Evaluation of salient point techniques. *Image Vis. Comput.* **2003**, *21*, 1087–1095. [CrossRef]
40. Shojaeilangari, S.; Yau, W.Y.; Li, J.; Teoh, E.K. Feature extraction through binary pattern of phase congruency for facial expression recognition. In Proceedings of the 2012 12th International Conference on Control Automation Robotics & Vision (ICARCV), Guangzhou, China, 5–7 December 2012; pp. 166–170.
41. Ahmed, F. Gradient directional pattern: A robust feature descriptor for facial expression recognition. *Electron. Lett.* **2012**, *48*, 1203–1204. [CrossRef]
42. Islam, M.S. Gender classification using gradient direction pattern. *Sci. Int.* **2013**, *25*, 797–799.
43. Valstar, M.; Pantic, M. Fully automatic facial action unit detection and temporal analysis. In Proceedings of the IEEE 2006 Conference on Computer Vision and Pattern Recognition Workshop (CVPRW'06), New York, NY, USA, 17–22 June 2006; p. 149.
44. Yang, B.Q.; Zhang, T.; Gu, C.C.; Wu, K.J.; Guan, X.P. A novel face recognition method based on IWLD and IWBC. *Multimed. Tools Appl.* **2016**, *75*, 6979–7002. [CrossRef]
45. Islam, M.S.; Auwatanamo, S. Facial expression recognition using local arc pattern. *Trends Appl. Sci. Res.* **2014**, *9*, 113. [CrossRef]
46. Ojala, T.; Pietikainen, M.; Maenpaa, T. Multiresolution gray-scale and rotation invariant texture classification with local binary patterns. *IEEE Trans. Pattern Anal. Mach. Intell.* **2002**, *24*, 971–987. [CrossRef]
47. Jabid, T.; Kabir, M.H.; Chae, O. Local directional pattern (LDP)–A robust image descriptor for object recognition. In Proceedings of the 2010 7th IEEE International Conference on Advanced Video and Signal Based Surveillance, Boston, MA, USA, 29 August–1 September 2010; pp. 482–487.
48. Kabir, M.H.; Jabid, T.; Chae, O. A local directional pattern variance (LDPv) based face descriptor for human facial expression recognition. In Proceedings of the 2010 7th IEEE International Conference on Advanced Video and Signal Based Surveillance, Boston, MA, USA, 29 August–1 September 2010; pp. 526–532.
49. Rivera, A.R.; Castillo, J.R.; Chae, O.O. Local directional number pattern for face analysis: Face and expression recognition. *IEEE Trans. Image Process.* **2012**, *22*, 1740–1752. [CrossRef] [PubMed]
50. Rivera, A.R.; Castillo, J.R.; Chae, O. Local directional texture pattern image descriptor. *Pattern Recognit. Lett.* **2015**, *51*, 94–100. [CrossRef]
51. Lei, Z.; Ahonen, T.; Pietikäinen, M.; Li, S.Z. Local frequency descriptor for low-resolution face recognition. In Proceedings of the 2011 IEEE International Conference on Automatic Face & Gesture Recognition (FG), Santa Barbara, CA, USA, 21–23 March 2011; pp. 161–166.
52. Zhang, W.; Shan, S.; Gao, W.; Chen, X.; Zhang, H. Local gabor binary pattern histogram sequence (lgbphs): A novel non-statistical model for face representation and recognition. In Proceedings of the 10th IEEE International Conference on Computer Vision (ICCV'05), Washington, DC, USA, 17–20 October 2005; Volume 1, pp. 786–791.
53. Ishraque, S.Z.; Banna, A.H.; Chae, O. Local Gabor directional pattern for facial expression recognition. In Proceedings of the 2012 IEEE 15th International Conference on Computer and Information Technology (ICCIT), Chittagong, Bangladesh, 22–24 December 2012; pp. 164–167.
54. Zhou, L.; Wang, H. Local gradient increasing pattern for facial expression recognition. In Proceedings of the 2012 19th IEEE International Conference on Image Processing, Orlando, FL, USA, 30 September–3 October 2012; pp. 2601–2604.
55. Islam, M.S. Local gradient pattern—A novel feature representation for facial expression recognition. *J. AI Data Min.* **2014**, *2*, 33–38.
56. Ahsan, T.; Jabid, T.; Chong, U.P. Facial expression recognition using local transitional pattern on Gabor filtered facial images. *IETE Tech. Rev.* **2013**, *30*, 47–52. [CrossRef]
57. Mohammad, T.; Ali, M.L. Robust facial expression recognition based on local monotonic pattern (LMP). In Proceedings of the 14th International Conference on Computer and Information Technology (ICCIT 2011), Dhaka, Bangladesh, 22–24 December 2011; pp. 572–576.

58. Ojansivu, V.; Heikkilä, J. Blur insensitive texture classification using local phase quantization. In Proceedings of the Image and Signal Processing: 3rd International Conference (ICISP 2008), Cherbourg-Octeville, France, 1–3 July 2008; Springer: Berlin/Heidelberg, Germany, 2008; pp. 236–243.
59. Bashar, F.; Khan, A.; Ahmed, F.; Kabir, M.H. Robust facial expression recognition based on median ternary pattern (MTP). In Proceedings of the IEEE 2013 International Conference on Electrical Information and Communication Technology (EICT), Khulna, Bangladesh, 13–15 February 2014; pp. 1–5.
60. Jabid, T.; Chae, O. Local transitional pattern: A robust facial image descriptor for automatic facial expression recognition. In Proceedings of the International Conference on Computer Convergence Technology, Seoul, Republic of Korea, 28–30 September 2011; pp. 333–344.
61. Lu, J.; Liong, V.E.; Zhou, J. Cost-sensitive local binary feature learning for facial age estimation. *IEEE Trans. Image Process.* **2015**, *24*, 5356–5368. [CrossRef] [PubMed]
62. Liu, L.; Lao, S.; Fieguth, P.W.; Guo, Y.; Wang, X.; Pietikäinen, M. Median robust extended local binary pattern for texture classification. *IEEE Trans. Image Process.* **2016**, *25*, 1368–1381. [CrossRef] [PubMed]
63. Bosch, A.; Zisserman, A.; Munoz, X. Representing shape with a spatial pyramid kernel. In Proceedings of the 6th ACM International Conference on Image and Video Retrieval, Amsterdam, The Netherlands, 9–11 July 2007; pp. 401–408.
64. Li, S.; Gong, D.; Yuan, Y. Face recognition using Weber local descriptors. *Neurocomputing* **2013**, *122*, 272–283. [CrossRef]
65. Iandola, F.N.; Han, S.; Moskewicz, M.W.; Ashraf, K.; Dally, W.J.; Keutzer, K. SqueezeNet: AlexNet-level accuracy with 50x fewer parameters and < 0.5 MB model size. *arXiv* **2016**, arXiv:1602.07360.
66. Sandler, M.; Howard, A.; Zhu, M.; Zhmoginov, A.; Chen, L.C. Mobilenetv2: Inverted residuals and linear bottlenecks. In Proceedings of the IEEE Conference on Computer Vision and Pattern Recognition, Salt Lake City, UT, USA, 18–22 June 2018; pp. 4510–4520.
67. He, K.; Zhang, X.; Ren, S.; Sun, J. Deep residual learning for image recognition. In Proceedings of the IEEE Conference on Computer Vision and Pattern Recognition, Las Vegas, NV, USA, 27–30 June 2016; pp. 770–778.
68. Huang, G.; Liu, Z.; Van Der Maaten, L.; Weinberger, K.Q. Densely connected convolutional networks. In Proceedings of the IEEE Conference on Computer Vision and Pattern Recognition, Honolulu, HI, USA, 21–26 July 2017; pp. 4700–4708.
69. Zhang, X.; Zhou, X.; Lin, M.; Sun, J. Shufflenet: An extremely efficient convolutional neural network for mobile devices. In Proceedings of the IEEE Conference on Computer Vision and Pattern Recognition, Salt Lake City, UT, USA, 18–22 June 2018; pp. 6848–6856.
70. Zoph, B.; Vasudevan, V.; Shlens, J.; Le, Q.V. Learning transferable architectures for scalable image recognition. In Proceedings of the IEEE Conference on Computer Vision and Pattern Recognition, Salt Lake City, UT, USA, 18–22 June 2018; pp. 8697–8710.
71. Szegedy, C.; Vanhoucke, V.; Ioffe, S.; Shlens, J.; Wojna, Z. Rethinking the inception architecture for computer vision. In Proceedings of the IEEE Conference on Computer Vision and Pattern Recognition, Las Vegas, NV, USA, 27–30 June 2016; pp. 2818–2826.
72. Simonyan, K.; Zisserman, A. Very deep convolutional networks for large-scale image recognition. *arXiv* **2014**, arXiv:1409.1556.
73. Szegedy, C.; Liu, W.; Jia, Y.; Sermanet, P.; Reed, S.; Anguelov, D.; Erhan, D.; Vanhoucke, V.; Rabinovich, A. Going deeper with convolutions. In Proceedings of the IEEE Conference on Computer Vision and Pattern Recognition, Boston, MA, USA, 7–12 June 2015; pp. 1–9.
74. Szegedy, C.; Ioffe, S.; Vanhoucke, V.; Alemi, A. Inception-v4, inception-resnet and the impact of residual connections on learning. In Proceedings of the AAAI Conference on Artificial Intelligence, San Francisco, CA, USA, 4–9 February 2017; Volume 31.
75. Chollet, F. Xception: Deep learning with depthwise separable convolutions. In Proceedings of the IEEE Conference on Computer Vision and Pattern Recognition, Honolulu, HI, USA, 21–26 July 2017; pp. 1251–1258.
76. Maguolo, G.; Nanni, L. A critic evaluation of methods for covid-19 automatic detection from x-ray images. *Inf. Fusion* **2021**, *76*, 1–7. [CrossRef] [PubMed]
77. Aswathy, A.; Hareendran, A.; SS, V.C. COVID-19 diagnosis and severity detection from CT-images using transfer learning and back propagation neural network. *J. Infect. Public Health* **2021**, *14*, 1435–1445.
78. Li, C.; Yang, Y.; Liang, H.; Wu, B. Transfer learning for establishment of recognition of COVID-19 on CT imaging using small-sized training datasets. *Knowl.-Based Syst.* **2021**, *218*, 106849. [CrossRef] [PubMed]
79. Lahsaini, I.; Daho, M.E.H.; Chikh, M.A. Deep transfer learning based classification model for COVID-19 using chest CT-scans. *Pattern Recognit. Lett.* **2021**, *152*, 122–128. [CrossRef]
80. Pinki, F.T.; Masud, M.A.; Ferdousi, J.; Eva, F.Y.; Rana, M.M.R. SVM Based COVID-19 Detection from CT Scan Image using Local Feature. In Proceedings of the IEEE 2021 International Conference on Electronics, Communications and Information Technology (ICECIT), Online, 14–16 September 2021; pp. 1–4.
81. Kaur, T.; Gandhi, T.K. Classifier fusion for detection of COVID-19 from CT scans. *Circuits Syst. Signal Process.* **2022**, *41*, 3397–3414. [CrossRef]
82. Islam, M.K.; Habiba, S.U.; Khan, T.A.; Tasnim, F. COV-RadNet: A Deep Convolutional Neural Network for Automatic Detection of COVID-19 from Chest X-rays and CT Scans. *Comput. Methods Programs Biomed. Update* **2022**, *2*, 100064. [CrossRef] [PubMed]
83. Peng, L.; Wang, C.; Tian, G.; Liu, G.; Li, G.; Lu, Y.; Yang, J.; Chen, M.; Li, Z. Analysis of CT scan images for COVID-19 pneumonia based on a deep ensemble framework with DenseNet, Swin transformer, and RegNet. *Front. Microbiol.* **2022**, *13*, 995323. [CrossRef] [PubMed]

84. Singh, V.K.; Kolekar, M.H. Deep learning empowered COVID-19 diagnosis using chest CT scan images for collaborative edge-cloud computing platform. *Multimed. Tools Appl.* **2022**, *81*, 3–30. [CrossRef] [PubMed]
85. DOLMA, Ö. COVID-19 and Non-COVID-19 Classification from Lung CT-Scan Images Using Deep Convolutional Neural Networks. *Int. J. Multidiscip. Stud. Innov. Technol.* **2023**, *7*, 53–60. [CrossRef]
86. Tiwari, S.; Jain, A.; Chawla, S.K. Diagnosing COVID-19 From Chest CT Scan Images Using Deep Learning Models. *Int. J. Reliab. Qual. e-Healthc. (IJRQEH)* **2022**, *11*, 1–15. [CrossRef]
87. Premamayudu, B.; Bhuvaneswari, C. COVID-19 Automatic Detection from CT Images through Transfer Learning. *Int. J. Image Graph. Signal Process.* **2022**, *14*, 48–95. [CrossRef]
88. Zahir, M.J.B.; Azim, M.A.; Chy, A.N.; Islam, M.K. A Fast and Reliable Approach for COVID-19 Detection from CT-Scan Images. *J. Inf. Syst. Eng. Bus. Intell.* **2023**, *9*, 288–304. [CrossRef]
89. Ibrahim, W.R.; Mahmood, M.R. Classified COVID-19 by densenet121-based deep transfer learning from CT-scan images. *Sci. J. Univ. Zakho* **2023**, *11*, 571–580. [CrossRef]
90. Ibrahim, M.R.; Youssef, S.M.; Fathalla, K.M. Abnormality detection and intelligent severity assessment of human chest computed tomography scans using deep learning: A case study on SARS-COV-2 assessment. *J. Ambient. Intell. Humaniz. Comput.* **2023**, *14*, 5665–5688. [CrossRef] [PubMed]
91. Gupta, K.; Bajaj, V. Deep learning models-based CT-scan image classification for automated screening of COVID-19. *Biomed. Signal Process. Control* **2023**, *80*, 104268. [CrossRef] [PubMed]
92. Duong, L.T.; Nguyen, P.T.; Iovino, L.; Flammini, M. Automatic detection of COVID-19 from chest X-ray and lung computed tomography images using deep neural networks and transfer learning. *Appl. Soft Comput.* **2023**, *132*, 109851. [CrossRef] [PubMed]
93. Ali, N.G.; El Sheref, F.K. A Hybrid Model for COVID-19 Detection using CT-Scans. *Int. J. Adv. Comput. Sci. Appl.* **2023**, *14*, 627–633. [CrossRef]
94. Farjana, A.; Liza, F.T.; Al Mamun, M.; Das, M.C.; Hasan, M.M. SARS CovidAID: Automatic detection of SARS CoV-19 cases from CT scan images with pretrained transfer learning model (VGG19, RESNet50 and DenseNet169) architecture. In Proceedings of the IEEE 2023 International Conference on Smart Applications, Communications and Networking (SmartNets), Istanbul, Turkey, 25–27 July 2023; pp. 1–6.
95. Lim, Y.J.; Lim, K.M.; Lee, C.P.; Chang, R.K.Y.; Lim, J.Y. COVID-19 identification and analysis with CT scan images using densenet and support vector machine. In Proceedings of the IEEE 2023 11th International Conference on Information and Communication Technology (ICoICT), Melaka, Malaysia, 23–24 August 2023; pp. 254–259.
96. Motwani, A.; Shukla, P.K.; Pawar, M.; Kumar, M.; Ghosh, U.; Alnumay, W.; Nayak, S.R. Enhanced framework for COVID-19 prediction with computed tomography scan images using dense convolutional neural network and novel loss function. *Comput. Electr. Eng.* **2023**, *105*, 108479. [CrossRef] [PubMed]
97. Perumal, M.; Srinivas, M. DenSplitnet: Classifier-invariant neural network method to detect COVID-19 in chest CT data. *J. Vis. Commun. Image Represent.* **2023**, *97*, 103949. [CrossRef]
98. Krishnan, A.; Rajesh, S.; Gollapinni, K.; Mohan, M.; Srinivasa, G. A Comparative Analysis of Chest X-rays and CT Scans Towards COVID-19 Detection. In Proceedings of the IEEE 2023 4th International Conference for Emerging Technology (INCET), Belgaum, India, 26–28 May 2023; pp. 1–7.
99. Halder, A.; Datta, B. COVID-19 detection from lung CT-scan images using transfer learning approach. *Mach. Learn. Sci. Technol.* **2021**, *2*, 045013. [CrossRef]
100. Alshazly, H.; Linse, C.; Barth, E.; Martinetz, T. Explainable COVID-19 detection using chest CT scans and deep learning. *Sensors* **2021**, *21*, 455. [CrossRef]
101. Ragab, D.A.; Attallah, O. FUSI-CAD: Coronavirus (COVID-19) diagnosis based on the fusion of CNNs and handcrafted features. *PeerJ Comput. Sci.* **2020**, *6*, e306. [CrossRef]
102. Shaik, N.S.; Cherukuri, T.K. Transfer learning based novel ensemble classifier for COVID-19 detection from chest CT-scans. *Comput. Biol. Med.* **2022**, *141*, 105127. [CrossRef] [PubMed]
103. Gaur, P.; Malaviya, V.; Gupta, A.; Bhatia, G.; Pachori, R.B.; Sharma, D. COVID-19 disease identification from chest CT images using empirical wavelet transformation and transfer learning. *Biomed. Signal Process. Control* **2022**, *71*, 103076. [CrossRef]
104. Canayaz, M.; Şehribanoğlu, S.; Özdağ, R.; Demir, M. COVID-19 diagnosis on CT images with Bayes optimization-based deep neural networks and machine learning algorithms. *Neural Comput. Appl.* **2022**, *34*, 5349–5365. [CrossRef] [PubMed]
105. Attallah, O.; Samir, A. A wavelet-based deep learning pipeline for efficient COVID-19 diagnosis via CT slices. *Appl. Soft Comput.* **2022**, *128*, 109401. [CrossRef]
106. Kundu, R.; Singh, P.K.; Ferrara, M.; Ahmadian, A.; Sarkar, R. ET-NET: An ensemble of transfer learning models for prediction of COVID-19 infection through chest CT-scan images. *Multimed. Tools Appl.* **2022**, *81*, 31–50. [CrossRef] [PubMed]
107. Islam, M.R.; Nahiduzzaman, M. Complex features extraction with deep learning model for the detection of COVID19 from CT scan images using ensemble based machine learning approach. *Expert Syst. Appl.* **2022**, *195*, 116554. [CrossRef] [PubMed]
108. Choudhary, T.; Gujar, S.; Goswami, A.; Mishra, V.; Badal, T. Deep learning-based important weights-only transfer learning approach for COVID-19 CT-scan classification. *Appl. Intell.* **2023**, *53*, 7201–7215. [CrossRef]

109. Gunraj, H.; Wang, L.; Wong, A. Covidnet-ct: A tailored deep convolutional neural network design for detection of covid-19 cases from chest ct images. *Front. Med.* **2020**, *7*, 608525. [CrossRef] [PubMed]
110. Alinsaif, S.; Lang, J. 3D shearlet-based descriptors combined with deep features for the classification of Alzheimer's disease based on MRI data. *Comput. Biol. Med.* **2021**, *138*, 104879. [CrossRef] [PubMed]

Disclaimer/Publisher's Note: The statements, opinions and data contained in all publications are solely those of the individual author(s) and contributor(s) and not of MDPI and/or the editor(s). MDPI and/or the editor(s) disclaim responsibility for any injury to people or property resulting from any ideas, methods, instructions or products referred to in the content.

Review

Survey of Recent Deep Neural Networks with Strong Annotated Supervision in Histopathology

Dominika Petríková [1,2,*] and Ivan Cimrák [1,2]

1. Cell-in-Fluid Biomedical Modelling & Computations Group, Faculty of Management Science and Informatics, University of Žilina, Univerzitná 8215/1, 010 26 Žilina, Slovakia
2. Research Centre, University of Žilina, Univerzitná 8215/1, 010 26 Žilina, Slovakia
* Correspondence: dominika.petrikova@fri.uniza.sk

Abstract: Deep learning (DL) and convolutional neural networks (CNNs) have achieved state-of-the-art performance in many medical image analysis tasks. Histopathological images contain valuable information that can be used to diagnose diseases and create treatment plans. Therefore, the application of DL for the classification of histological images is a rapidly expanding field of research. The popularity of CNNs has led to a rapid growth in the number of works related to CNNs in histopathology. This paper aims to provide a clear overview for better navigation. In this paper, recent DL-based classification studies in histopathology using strongly annotated data have been reviewed. All the works have been categorized from two points of view. First, the studies have been categorized into three groups according to the training approach and model construction: 1. fine-tuning of pre-trained networks for one-stage classification, 2. training networks from scratch for one-stage classification, and 3. multi-stage classification. Second, the papers summarized in this study cover a wide range of applications (e.g., breast, lung, colon, brain, kidney). To help navigate through the studies, the classification of reviewed works into tissue classification, tissue grading, and biomarker identification was used.

Keywords: classification; convolutional neural networks; deep learning; digital pathology; histology image analysis

1. Introduction

Traditionally, pathology diagnosis has been performed by a human pathologist observing stained specimens from tumors on glass slides using a microscope to diagnose cancer. In recent years, deep learning has rapidly developed, and more and more entire tissue slides are being captured digitally by scanners and saved as whole slide images (WSIs) [1]. Since a large amount of WSIs are being digitized, it is only natural that many attempts have been made to explore the potential of deep learning on histopathological image analysis. Histological images and tasks have unique characteristics, and specific processing techniques are often required [2]. The authors in [3] carried out an extensive and comprehensive overview of deep neural network models developed in the context of computational histopathology image analysis. Their survey covers the period up to December 2019. Since the volume of research in this domain is rapidly growing, the aim of this review is to complement their overview with papers published since 2020. In contrast to their survey, the focus of this review is on a specific area of supervised learning only, namely classification using strongly annotated data.

The rest of this paper is organized as follows. In Section 2, a basic overview of neural networks used in the context of computational histopathology is presented. Section 3 discusses in detail supervised deep learning models and approaches used in digital pathology for classification tasks. These approaches have been grouped into three main categories: one-stage classification using fine-tuning, one-stage classification training models from

scratch, and the multi-stage classification approach. In Section 4, we discuss the histopathological point of view by classifying the methods according to their area of application. In Section 5, we conclude the paper.

2. Materials and Methods—Convolutional Neural Network

For this survey, only papers that performed classification of histological images with common convolutional neural network models and used strongly annotated datasets were selected. Other articles that used more complex deep learning models or weak annotations were not included in this review. The review was carried out by searching mostly through PubMed and also arXiv for articles containing deep learning (DL) keywords such as "convolutional neural networks", "classification", "deep learning", and histology keywords such as "hematoxylin and eosin", "H&E", and "histopathology" in the title or abstract. To narrow down the selection, combinations of deep learning keywords with histology keywords were used, for example, "CNN hematoxylin and eosin". The combination "deep learning histopathology" was omitted since both words are too general. Moreover, only articles published since 2020 have been searched. The subsequent filtering process can be described in four steps. The first two steps were designed to quickly filter out articles that were obviously irrelevant to the topic of this review and thus reduce as much as possible the number of articles that needed to be analyzed in more detail in the remaining two steps. In the first step, articles were filtered based on the title. Papers that were obviously not related to CNN's application for histological image data classification were excluded. This resulted in approximately 700 papers. Articles that could not be unambiguously excluded based on the title were filtered in a second step based on reading the abstract. In the third step, the introduction was analyzed. The main purpose was to exclude studies that did not meet the criteria of this review, such as papers using more complex deep learning approaches than convolutional neural networks or datasets not only consisting of histological images. In the last step, approximately 100 articles were fully read. This part was mainly focused on filtering out studies that only worked with strongly annotated datasets. We also included some papers that were missing from the initial search but were cross-referenced in selected articles.

The purpose of this chapter is to explain the concepts and models of deep neural networks (DNNs) used for classification tasks in digital pathology. Machine learning is a type of artificial intelligence that allows computers to learn and modify their behavior based on training data [4]. Supervised learning methods are the most commonly used, where the dataset consists of input features and corresponding labels. In the case of classification, the label represents one of a fixed number of classes. The algorithm learns patterns and connections in the data to find a suitable function that maps inputs to outputs, creating a model that captures hidden properties in the data and can be used to predict outputs for new inputs. Training a model involves finding the best model parameters that predict the data based on a defined loss function [5,6].

Neural networks are the foundation of most DNN algorithms, consisting of interconnected units called neurons organized into layers, including input, hidden, and output layers. DNNs have multiple hidden layers. A neuron's output, or activation, is a linear combination of its inputs and parameters (weights and bias) transformed by an activation function. Common activation functions in neural networks include sigmoid, hyperbolic tangent, and ReLU functions. At the final output layer, activations are mapped to a distribution over classes using the softmax function [6,7].

One of the most popular and commonly used supervised deep learning networks is CNNs, which are often employed for visual data processing of images and video sequences [8–10]. CNNs consist of three types of layers: convolutional layers, pooling layers, and fully connected layers, as shown in Figure 1. The convolutional layer is the most significant component of the CNN architecture. It consists of several filters, also called kernels, which are represented as a grid of discrete values. These values are referred to as kernel weights and are tuned during the training phase. The convolution operation consists of the kernel sliding over the whole image horizontally and vertically. Additionally, the dot product is calculated between the image and kernel by multiplying corresponding values and summing up to create a scalar value at each position. In particular, each kernel is convolved over the input matrix to obtain a feature map. Subsequently, the feature maps generated by the convolutional operation are sub-sampled in the pooling layer. The convolution and pooling layers together form a pipeline called feature extraction. Above all, the fully connected layers combine the features extracted by the previous layers to perform the final classification task [8,11,12].

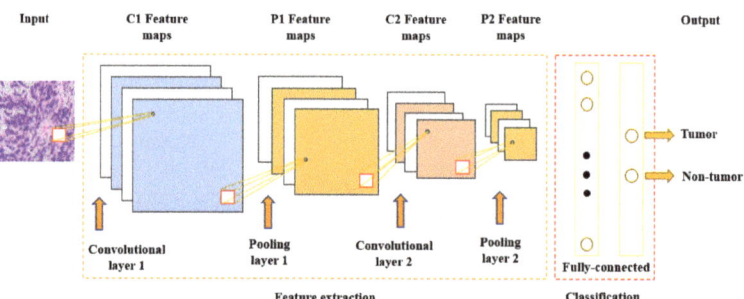

Figure 1. Convolutional neural network architecture.

3. Classification of Histopathology Images

This section provides a general overview of recent publications using deep learning and convolutional neural networks (CNNs) in digital pathology. The focus of this work is solely on supervised learning tasks applied for the classification of histological images. This category includes models that perform image-level classification, such as tumor subtype classification and grading, or use a sliding window approach to identify tissue types. Most deep learning approaches do not use the whole-slide image (WSI) as input because it would be computationally expensive (high dimensionality). Instead, they extract small square patches and assign a label to them. Existing methods can be grouped according to the level of annotations they employ. Based on the type of annotations used for training, two subcategories may be identified: the strong-annotations approach (patch-level annotations) and the weak-annotations approach (slide-level annotations) [13]. The first approach relies on the identification of regions of interest and the detailed localization of tumors by certified pathologists, while for the latter approach, it is sufficient to assign a specific class to a whole-slide image. In this work, a survey of the strong-annotations approach is conducted.

3.1. Strong-Annotations Approach (Patch-Level Annotation)

Referring to patch-level annotations as strong means that all extracted patches have their own label class. Typically, patch labels are derived from pixel-level annotations. Manually annotating pixels is very time-consuming and laborious work requiring an expert approach. For instance, pathologists have to localize and annotate all pixels or cells in WSI by contouring the whole tumor. This approach is shown in Figure 2. Therefore, there are currently very few strongly annotated histological images. Besides whole-slide image classification, pixel-wise/patch-wise predictions with the sliding window method enable spatial predictions such as localization and detection of cancerous cells/tissue. In addition, stacking patch predictions next to each other builds a WSI heatmap, so the model can be considered interpretable. Multiple examples of using CNNs in the problem of patch classification employ a single-stage approach when the patch is classified using one CNN architecture. In contrast, several approaches use a multi-stage workflow, where typically the output of one CNN architecture is fed into another CNN that delivers the final decision. Of course, even more CNN models can be included in such a workflow that can be labeled as multi-stage classification. For the one-stage approach, one can differentiate between models that have been trained from scratch with artificially initiated weights and models that use pre-trained CNN architectures on data often not related to the original problem. For multi-stage problems, such differentiation becomes difficult due to many possibilities, since some CNNs from the multi-stage workflow may be trained from scratch, while others may be pre-trained. In Figure 3, the top graphic shows the categorization of CNN methods used in this section.

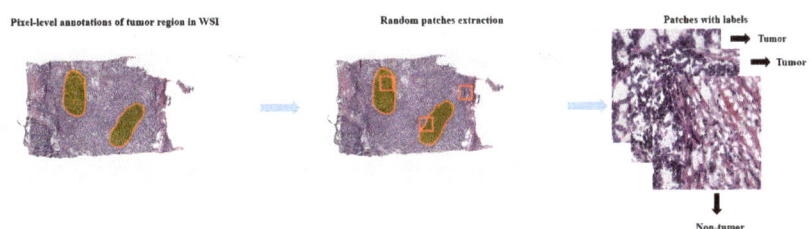

Figure 2. Construction of patches from pixel-level annotations of WSI.

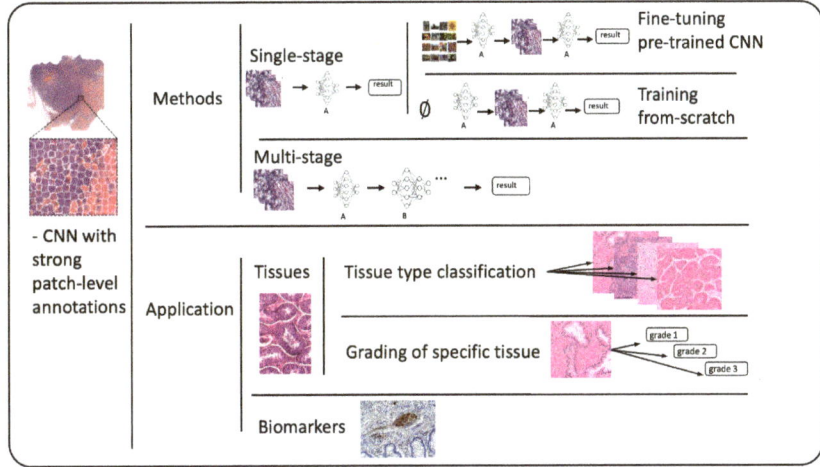

Figure 3. Methods: Categorization of CNN methods used in Section 3. Application: Categorization of application areas used in Section 4.

3.2. Fine-Tuning

The easiest way of training CNNs with a limited amount of data is using one of the well-known pre-trained architectures. Typically, models are initialized using weights pre-trained on ImageNet and fine-tuned on histopathological images. Papers using this approach are summarized in Table 1. In [14], the authors fine-tuned VGGNet [15], ResNet [16], and InceptionV4 [17] models to obtain the probabilities of small patches (100 × 100 pixels), being tumor-infiltrating lymphocyte (TIL)-positive or TIL-negative extracted from WSIs of 23 cancer types. For the region classification performance, they extracted bigger super-patches (800 × 800 pixels) and annotated them with three categories (Low TIL, Medium TIL, or High TIL) based on the ratio of TIL-positive area. To obtain a prediction of the category, super-patches were divided into an 8x8 grid and each square (100 × 100 pixel patch) was classified as TIL-positive or TIL-negative. Subsequently, the correlation between the score of CNN (number of positive patches in super-patch) and pathologists' annotations was observed. In [18], they developed a deep learning-based six-type classifier for the identification of a wider spectrum of lung lesions including lung cancer. Furthermore, they also included pulmonary tuberculosis and organizing pneumonia, which often needs to be surgically inspected to be differentiated from cancer. EfficientNet [19] and ResNet were employed to carry out patch-level classification. To aggregate patch predictions into slide-level classification, two methods were compared: majority voting and mean pooling. Moreover, two-stage aggregation was implemented to prioritize cancer tissues in slides.

In [20], scholars proposed three steps to develop an AI-based screening method for lymph node metastases. First, they trained a segmentation model to obtain lymph node tissue from WSI and broke it into patches. Next, they used a fine-tuned Xception model to classify patches into metastasis-positive/negative. Finally, the absence or presence of two connected patches classified as positive determined the final result of WSI. In [21], the authors compared the accuracies of stand-alone VGG-16 and VGG-19 models with ensemble models consisting of both architectures in classifying breast cancer histopathological images as carcinoma and non-carcinoma. In [22], the authors compared the performance of the VGG19 architecture with methods used in supervised learning with weakly labeled data to classify ovarian carcinoma histotype. The problem of binary classification into benign and malignant lesions, with subsequent division into eight subtypes with modified EfficientNetV2 architecture on images from the BreakHis dataset, was addressed by the authors in [23]. Similarly, Xception was employed in [24] for subtyping breast cancer into four categories. The binary subtype classification of eyelid carcinoma was performed in [25]. They used DenseNet-161 to make predictions for every patch in WSI and then used a patch voting strategy to decide the WSI subtype. In [26], the authors used AlexNet [27], GoogLeNet [28], and VGG-16 to detect histopathology images with cancer cells and to classify ovarian cancer grade. Since neural networks behave like black-box models, the authors employed the Grad-CAM method to demonstrate that CNN models attended to the cancer cell organization patterns when differentiating histopathology tumor images of different grades. Grad-CAM was also employed in [29], where the authors used this method to provide interpretability and approximate visual diagnosis for the presentation of the model's results to pathologists. The model consisted of three neural networks fine-tuned on a custom dataset to classify H&E stained tissue patches into five types of liver lesions, cirrhosis, and nearly normal tissue. A decision algorithm consisting of three networks was also proposed in [30] to detect odontogenic cyst recurrence using binary classifiers. The procedure consisted of letting the first two models make predictions. If the predictions did not match, a third model was loaded to obtain the final decision. Another example of using Grad-CAM is [31] to visualize classification results of the VGG16 network in grading bladder non-invasive carcinoma.

Hematoxylin-eosin (H&E) is considered as the gold standard for evaluating many cancer types. However, it contains only basic morphological information. In clinical practice, to obtain molecular information, immunohistochemical (IHC) staining is often employed. Such staining can visualize the expressions of different proteins (e.g., Ki67) on the cell membrane or nucleus. This approach is referred to as double staining. Many recent studies have shown that there is a correlation between H&E and IHC staining [32–34].

In [35], the authors addressed the problem of double staining in determining the number of Ki67-positive cells for cancer treatment. They employed matching pairs of IHC- and H&E-stained images and fine-tuned ResNet-18 at the cell-level from H&E images. Subsequently, to create a heat map, they transformed the CNN into a fully convolutional network without fully connected layers. As a result, the fine-tuned ResNet-18 was able to handle WSI as input and produce a heat map as output.

In [36], the authors proposed a modified Xception network called HE-HER2Net by adding global average pooling, batch normalization layers, dropout layers, and dense layers with a Swish activation function. The network was designed to classify H&E images into four categories based on Human epidermal growth factor receptor 2 (HER2) positivity from 0 to 3+. In addition to routine model evaluation, the authors compared their modified network to other existing architectures and claimed that HE-HER2Net surpassed all existing models in terms of accuracy, precision, recall, and AUC score.

To produce accurate models capable of generalization, it is essential to obtain large amounts of diversified data. Typically, this problem is addressed by pooling all necessary data to a centralized location. However, due to the nature of medical data, this approach has many obstacles regarding privacy and data ownership, as well as various regulatory policies (e.g., the General Data Protection Regulation GDPR of the European Union [37]). The authors of [38] simulated a Federated Learning (FL) environment to train a deep learning model that classifies cells and nuclei to identify TILs in WSI. They generated a dataset from WSIs of cancer from 12 anatomical sites and partitioned it into eight different nodes. To evaluate the performance of FL, they also trained a CNN using a centralized approach and compared the results. The study shows that the FL approach achieves similar performance to the model trained with data pooled at a centralized location.

Table 1. Summary of fine-tuning papers.

Reference	Cancer Types	Staining	Dataset	Neural Networks in Models	Method
Abousamra et al. (2022) [14]	23 cancer types	H&E	The Cancer Genome Atlas (TCGA)	Vgg-16, ResNet-34, InceptionV4	Patch-level classification of Tumor infiltrating lymphocytes (TIL)
Yang et al. (2021) [18]	Lung cancer	H&E	Custom dataset of 1271 WSIs and 422 WSIs from TCGA	ResNet-50, EfficientNet-B5	Six-type classification of lung lesions including pulmonary tuberculosis and Organizing pneumonia
Hameed et al. (2020) [21]	Breast cancer	H&E	Custom dataset of 544 WSIs	VGG-16, VGG-19	Ensemble of neural networks to classify carcinoma and non-carcinoma images
Yu et.al (2020) [26]	Ovarian cancer	H&E	TCGA	AlexNet, GoogLeNet, VGG-16	Cancerous regions identification and grades classification
Liu et al. (2020) [35]	Different types of cancer	H&E, IHC (Ki67)	Custom dataset from 300 Regions of interest	ResNet-18	Classification of Ki67 positive and negative cells
Baid et al. (2022) [38]	12 types	H&E	TCGA	VGG-16	Federated learning for classification of tumor infiltrating lymphocytes
Cheng et al. (2022) [29]	Liver cancer	H&E	Custom dataset	ResNet50, InceptionV3, Xception	Ensemble of 3 networks pretrained on ImageNet used to differentiate Hepatocellular nodular lesions (5 types) with nodular cirrhosis and nearly normal liver tissue
Shovon et al. (2022) [36]	Breast cancer	H&E	BCI dataset	Modified Xception	Four class classification of HER2 with modified Xception model pretrained on ImageNet
Rao et al. (2022) [30]	Odontogenic cysts	H&E	Custom dataset	Inception-V3, DenseNet-121, Inception-Resnet-V2	Binary classification of cyst recurrence based on decision algorithm consisting of 3 models
Farahani et al. (2022) [22]	Ovarian cancer	H&E	Custom dataset	VGG19	Comparison of classification of ovarian carcinoma histotype by four models
Sarker et al. (2023) [23]	Breast cancer	H&E	BreakHis dataset	Modified EfficientNetV2	Binary classification of malignant and benign tissue and multi-class subtyping using fused mobile inverted bottleneck convolutions and mobile inverted bottleneck convolutions with dual squeeze and excitation network and EfficientNetV2 as backbone
Luo et al. (2022) [25]	Eyelid carcinoma	H&E	Custom dataset	DenseNet161	The differential diagnosis of eyelid basal cell carcinoma and sebaceous carcinoma based on patch prediction by the DenseNet161 architecture and WSI differentiation by an average-probability strategy-based integration module
Mundhada et al. (2023) [31]	Bladder cancer	H&E	Custom dataset	VGG16	Grading of non-invasive carcinoma
Khan et al. (2023) [20]	Breast and colon cancer	H&E	PatchCamelyon	Xception	Segmentation of lymph node tissue with subsequent classification to detect metastases
Hameed et al. (2022) [24]	Breast cancer	H&E	Colsanitas dataset	Xception	Using Xception networks as feature extractor to classify breast cancer into four categories: normal tissue, benign lesion, in situ carcinoma, and invasive carcinoma

3.3. Training from Scratch

As already stated, fine-tuning is a promising method for training deep neural networks. On the other hand, it can only be applied to well-known architectures that are already pre-trained. When designing a custom CNN architecture, it needs to be trained from scratch. Table 2 summarizes studies in which neural networks were trained from scratch. In [39], the authors proposed a method based on CNN with residual blocks (Res-Net) referred to as DeepLRHE to predict lung cancer recurrence and the risk of metastasis. Later in [40], scholars established the new DeepIMHL model consisting of CNN and Res-Net to predict mutated genes as biomarkers for targeted-drug therapy of lung cancer. In addition, the authors in [41] trained and optimized EfficientNet models on images of non-Hodgkin lymphoma and evaluated its potential to classify tumor-free reference lymph nodes, nodal small lymphocytic lymphoma/chronic lymphocytic leukemia, and nodal diffuse large B-cell lymphoma. In [42], the authors proposed three architectures of ResNet differing in the construction of residual blocks trained from scratch. Their suggested model achieved accuracy comparable to other state-of-the-art approaches in the classification of oral cancer histological images into three stages. To classify kidney cancer subtypes, in [43] the authors developed an ensemble-pyramidal model consisting of three CNNs that process images of different sizes. The authors in [44] demonstrated that CNN-based DL can predict the gBRCA mutation status from H&E-stained WSIs in breast cancer. According to researchers in [45], CNN can be employed to differentiate non-squamous Non-Small Cell Lung Cancer versus squamous cell carcinoma. To classify the tumor slide, they pooled information using the max-pooling strategy. Moreover, they added quality check with a threshold for predictions to select only tiles with a high prediction level. Additionally, to improve the prediction, they also used a virtual tissue microarray (circle from the centroid based on the pathologist's hand-drawn tumor annotations) instead of WSI.

To compare the performance of pre-trained networks with the custom ones trained from scratch, researchers in [46] used images of three cancer types: melanoma, breast cancer, and neuroblastoma. Unlike others using patches, the authors applied the simple linear iterative clustering (SLIC) to segment images into superpixels which group together similar neighboring pixels, as shown in Figure 4. Thus, these superpixels were classified into multiple subtype categories based on the type of cancer. To make WSI-level predictions, they used multiple specific quantification metrics such as stroma-to-tumor ratio. Although the custom NN achieved comparable results, pre-trained networks performed better on all three cancer types. A similar comparison was carried out in [47] for the classification of subtypes in lung cancer biopsy slides. Results showed that a CNN model built from scratch fitted to the specific pathological task could produce better performances than fine-tuning pre-trained CNNs.

A comparison of training from scratch versus transfer learning was performed in [48]. The authors compared three approaches for training the VGG16 network: training from scratch, transfer learning as a feature extractor, and fine-tuning on images of breast cancer to detect Invasive Ductal Carcinoma. According to the results, the model trained from scratch achieved better results in terms of accuracy (0.85). However, using transfer learning, they were able to train a comparable model (accuracy 0.81) ten times faster. Furthermore, among the transfer learning approaches, transfer learning via feature extraction (accuracy 0.81), which involved retraining some of the convolutional blocks, yielded better results in less time compared to transfer learning via fine-tuning (accuracy 0.51).

Table 2. Summary of papers training neural networks from scratch.

Reference	Cancer Types	Staining	Dataset	Neural Networks in Models	Method
Wu et al. (2020) [39]	Lung cancer	H&E	211 samples from TCGA	Custom CNN with residual blocks	Prediction of lung cancer recurrence
Huang et al. (2021) [40]	Lung cancer	H&E	TCGA	Custom CNN with residual blocks	Identification of the bio-markers of lung cancer
Steinbuss et al. (2021) [41]	Blood cancer	H&E	Custom dataset from 629 patients	EfficientNet	Classification of tumor-free lymph nodes, nodal small lymphocytic lymphoma/chronic lymphocytic leukemia, and nodal diffuse large B-cell lymphoma
Panigrahi et al. (2022) [42]	Oral cancer	H&E	Custom dataset	Three ResNet architectures	Classification of 3 grades
Wang et al. (2021) [44]	Breast cancer	H&E	Custom dataset of 222 images	ResNet-18	BRCA gene mutations prediction
Le Page et al. (2021) [45]	Lung cancer	H&E	Custom dataset of 197 images and 60 images from TCGA	InceptionV3	Classification of patches (tiles) into cancer subtypes. For final case classification they used majority-vote method or highest probability class
Zormpas-Petridis et al. (2021) [46]	Melanoma, breast cancer and childhood neuroblastoma	H&E	Custom dataset	Custom CNN	Classification of the: melanoma (tumor tissue, stroma, cluster of lymphocytes, normal epidermis, fat, and empty/white space) breast cancer (tumor, necrosis, stroma, cluster of lymphocytes, fat, and lumen/empty space) neuroblastoma (undifferentiated neuroblasts, tissue damage (necrosis/apoptosis), areas of differentiation, cluster of lymphocytes, hemorrhage, muscle, kidney, and empty/white space)
Abdolahi et al. (2020) [48]	Breast cancer	H&E	Kaggle	Custom CNN, VGG-16	Classification of invasive ductal carcinoma
Yang et al. (2022) [47]	Lung cancer	H&E	Custom dataset	Custom CNN	Comparison of classification lung cancer by fine-tuned models and models trained from scratch
Abdeltawab et al. (2022) [43]	Kidney cancer	H&E	Custom dataset	Custom CNN	An ensemble-pyramidal deep learning model consisting of three CNNs processing different image sizes to differentiate 4 tissue subtypes

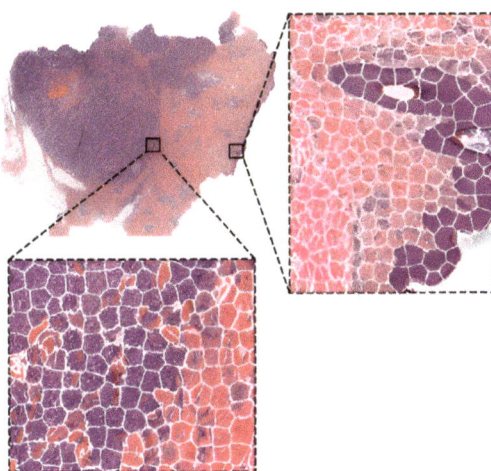

Figure 4. WSI image segmentation using the SLIC superpixels algorithm. Reprinted from [46], with permission according to Creative Commons Attribution License.

3.4. Multi-Stage Classification

In [49], scholars tackled the complex problem of computer-aided disease diagnosis by designing a two-stage system to determine the Tumor Mutation Burden (TMB) status, which is an important biomarker for predicting the response to immunotherapy in lung cancer. For the first stage, they developed a CNN based on InceptionV3 [50] to classify known histologic features for individual patches across H&E-stained WSIs. In the second stage, the patch-level CNN predictions were aggregated over the entire slide and combined with clinical features such as smoking status, age, stage, and sex to classify the TMB status. The final model was obtained by ensembling 10 independently trained networks.

In [51], the authors proposed a diagnostic framework for generating a whole-case report consisting of the detection of renal cancer regions, classification of cancer subtypes, and cancer grades. From every stain-normalized WSI, patches were selected from tumor and non-tumor regions to form a dataset. For tumor region classification, they fine-tuned several different architectures and identified InceptionV3 as the most suitable one. Thus, they also used this architecture for the remaining tasks. Patches classified as containing a tumor were further classified into three tumor subtypes and four grade classes.

It should be noted that CNNs have proven to be successful classifiers in the field of histology. Nevertheless, they can also be employed in conjunction with other machine learning (ML) classifiers. The authors in [52] developed a CNN model for the automated classification of pathology glioma (brain tumor) images into six subtypes. The images pass through the CNN to obtain patch-level output categories. At this point, those patch labels go through a hierarchical decision tree for patient-level diagnosis based on the amounts and proportions of tumor types. The outcome thus includes results for both the image patch-label and the patient-level label. In [53], researchers developed a three-step approach to HER2 status tissue classification in breast cancer. Firstly, they used a pre-trained UNet-based nucleus detector [54] to create patches. Secondly, they trained a CNN to identify tumor nuclei and further classified them as HER2-positive or HER2-negative. In [55], the authors proposed a classification method for subtype differentiation of liver cancer based on a stacking classifier with deep neural networks as feature extractors. They used four pre-trained deep convolutional neural networks, ResNet50, VGG16, DenseNet201 [56], and InceptionResNetV2 [17], to extract deep features from histopathological images. After fusing extracted deep features from different architectures, they applied multiple ML classifiers (Support-vector machines (SVMs), k-Nearest Neighbor (k-NN), Random Forest (RF)) on the feature vector to obtain final classification.

To predict 5-year overall survival in renal cell carcinoma, scholars in [57] fine-tuned a ResNet18 pre-trained on the ImageNet dataset. The CNN assigned a probability score for every patch, and to determine the class for an entire WSI, the scores of all associated patches were averaged and classified. In addition to single-stage classification, they also used CNN prediction with other clinicopathological variables for multivariable logistic regression analysis. The authors in [58] presented an approach that combines a deep convolutional neural network as a patch-level classifier and XGBoost [59] as a WSI-level classifier to automatically classify H&E-stained breast digital pathology images into four classes: normal tissue, benign lesion, ductal carcinoma in situ, and invasive carcinoma. InceptionV3 was trained as the Patch-Level Classifier to generate four predicted probability values combined into a heatmap. By comparing the classification accuracy of different classifiers, they chose XGBoost as the WSI-level classifier.

In [60], researchers trained a deep learning classifier and applied it to classify lung tumor samples into nine tissue classes. From the extracted features, they computed spatial features that describe the composition of the tumor microenvironment and used them in combination with clinical data to predict patient survival, as well as to predict tumor mutation. The authors of [61] claim that they were the first to propose a method for detecting Pancreatic ductal adenocarcinoma in WSIs based on CNNs. They employed InceptionV3 as a patch-level classifier and predicted patches combined with a malignancy probability heatmap. At this point, statistical features were extracted from WSI heatmaps and applied to train a Light Gradient Boosting Machine [62] for slide-level classification. Similar approaches were taken by researchers in [63]. On histological images of gastric cancer (GC), they made both binary and multi-class classifications. Firstly, InceptionV3 was used for both malignant and benign patch classification as well as discriminating normal mucosa, gastritis, and gastric cancer. Secondly, they separated all WSIs into categories, "complete normal WSIs" and "mixture WSIs" with gastritis or GC, and used 44 features extracted from the malignancy probability heatmap generated by CNN to train and fine-tune the RF classifier.

The addition of attention mechanisms to CNNs for increased performance has become increasingly popular nowadays. In [64], the Divide-and-Attention Network (DANet) was proposed for breast cancer classification and grading of both breast and colorectal cancers. This network has three inputs: the original pathological image, the nuclei image, and the non-nuclei image. The nuclei and non-nuclei images are obtained as a result of a nuclei segmentation model. A similar approach was used in [65], where the authors developed the Nuclei-Guided Network (NGNet) for grading of breast invasive ductal carcinoma. Compared to DANet, NGNet has only two input images: the original image and the nuclei image obtained from segmentation.

Medulloblastoma (MB) is a dangerous malignant pediatric brain tumor that can lead to death [66]. In [67], the authors proposed a mixture of deep learning and machine learning methods called MB-AI-His for the automatic diagnosis and classification of four subtypes of pediatric MB. The diagnosis is performed in two levels. The first level classifies the images into normal and abnormal (binary classification level), while the second level classifies the abnormal images containing MB tumor into the four subtypes of childhood MB tumor (multi-classification level). Three pre-trained deep CNNs are utilized with transfer learning (ResNet-50, DenseNet-201, and MobileNet [68]) to extract spatial features. These features are combined with time-frequency features extracted using the discrete wavelet transform (DWT) method. Finally, a combination of spatial features and five popular classifiers is used to perform multi-class classification, including SVM, k-NN, Linear Discriminant Analysis, and Ensemble Subspace Discriminant. A similar approach is introduced by the authors in [69]. Multi-class classification of the four classes of childhood MB is much more complicated than binary classification. Few research articles have investigated this multi-class classification problem. Their pipeline consists of spatial DL feature extraction from 10 fine-tuned CNN architectures, feature fusion and reduction using the DWT method, and subsequent selection of features. Classification is accomplished using a bidirectional Long-Short-Term Memory classifier. All papers using multistage classification are listed in Table 3.

Table 3. Summary of studies using multi-stage classification.

Reference	Cancer Types	Staining	Dataset	Neural Networks in Models	Method
Sadhwani et al. (2021) [49]	Lung cancer	H&E	TCGA and custom dataset of 50 WSIs	Custom CNN	Multiclassification into subtypes and binary classification of Tumor Mutation Burden
Wu et al. (2021) [51]	Renal cell cancer (RCC)	H&E	667 WSIs from TCGA + new RCC dataset of 632 WSIs	InceptionV3	Identification of tumor regions and classification into tumor subtypes and different grades
Jin et.al (2021) [52]	Brain cancer	H&E	slides of 323 patients from the Central Nervous System Disease Biobank	custom CNN based on DenseNet	Classification into 5 subtypes of glioma
Anand et al. (2020) [53]	Breast cancer	H&E, IHC	dataset from University of Warwick and TCGA	Custom neural network	Identification of tumor patches and classification of HER2 into positive or negative
Dong et al. (2022) [55]	Liver cancer	H&E	Custom dataset of 73 images	ResNet-50, VGG-16, DenseNet-201, InceptionResNetV2	Classification of three differentiation states
Mi et al. (2021) [58]	Breast cancer	H&E	Custom dataset of 540 WSIs	InceptionV3	Multi-class classification of normal tissue, benign lesion, ductal carcinoma in situ, and invasive carcinoma
Fu et al. (2021) [61]	Pancreas	H&E	Custom dataset of 231 WSIs	InceptionV3	Classification of patches into cancerous or normal
Ma et al. (2020) [63]	Gastric cancer	H&E	Custom dataset of 763 WSIs	InceptionV3	Classification of normal mucosa, chronic gastritis, and intestinal-type

Table 3. Cont.

Attallah (2021) [67]	Brain cancer	H&E	Custom dataset of 204 images	ResNet-50, DenseNet-201, MobileNet	Classification of normal and abnormal Medulloblastoma
Attallah (2021) [69]	Brain cancer	H&E	Custom dataset of 204 images	10 CNN architectures	Multi-class classification of 4 medulloblastoma subtypes
Yan et al. (2022) [64]	Breast and colorectal cancer	H&E	BACH dataset and datasets avaiable from different articles	Xception	Classification of breast cancer, colorectal and breast cancer grading based on Divide-and-Attention Network using Xception CNN as backbone
Yan et al. (2022) [65]	Breast cancer	H&E	Custom dataset	NGNet	Grading of breast cancer using attention modules and segmentation. Classification is done with two images: original image and corresponding nuclei image)
Raczkowski et al. (2022) [60]	Lung cancer	H&E	Custom dataset	ARA-CNN	Classification of mutation based on tissue prevalence and tumor microenvironment composition computed from ARA-CNN output. CNN was used to classify patches into 9 tissue subtypes
Wessels et al. (2022) [57]	Kidney cancer	H&E	TCGA	ResNet18	Pretrained ResNet18 CNN was used to predict 5-year overal survival in renal cell carcinoma. Furthermore, the CNN-based classification was an independent predictor in a multivariable clinicopathological model

4. Discussion

Based on the studies described in the previous chapter, it is clear that there are many approaches to successfully use neural networks for many classification tasks in histology and a variety of cancer types. Most commonly, DL has been applied to lung and breast cancer. Breast cancer is a leading cause of cancer-related deaths in women worldwide, and lung cancer was the second most commonly diagnosed cancer worldwide in 2020, behind female breast cancer [24,47]. From a histological point of view, the tasks in which neural networks were successfully applied have been divided into the following three groups: tissue types, grading, and biomarker classification. The articles mentioned in this review are arranged according to this categorization in Table 4.

Table 4. Overview of all studies classified according to the application area.

Tissue	Tissue type	Yang et al. (2021) [18] Hameed et al. (2020) [21] Farahani et al. (2022) [22] Luo et al. (2022) [25] Hameed et al. (2022) [24] Le Page et al. (2021) [45] Abdolahi et al. (2020) [48] Abdeltawab et al. (2022) [43] Wu et al. (2021) [51] Anand et al. (2020) [53] Mi et al. (2021) [58] Ma et al. (2020) [63] Yan et al. (2022) [64]	Yu et.al (2020) [26] Cheng et al. (2022) [29] Sarker et al. (2023) [23] Khan et al. (2023) [20] Steinbuss et al. (2021) [41] Zormpas-Petridis et al. (2021) [46] Yang et al. (2022) [47] Sadhwani et al. (2021) [49] Jin et.al (2021) [52] Dong et al. (2022) [55] Fu et al. (2021) [61] Attallah (2021) [69] Attallah (2021) [67]	
	Tissue grading	Yu et.al (2020) [26] Wu et al. (2021) [51] Panigrahi et al. (2022) [42]	Mundhada et al. (2023) [31] Yan et al. (2022) [65]	
Biomarkers		Abousamra et al. (2022) [14] Shovon et al. (2022) [36] Anand et al. (2020) [53]	Liu et al. (2020) [35] Huang et al. (2021) [40] Raczkowski et al. (2022) [60]	Baid et al. (2022) [38] Wang et al. (2021) [44]

4.1. Tissue Types

One of the most fundamental tasks in histology is the classification of tissue types. It is possible to look at this task in two ways. The first aspect and the complete basis is to identify tumor tissue and other tissue types. This may involve a binary division into tumor and non-tumor tissue (this approach was used in [21]) as well as multi-class detection of tumor, stroma, lymphocytes, fat, necrosis, and other. In [46], the authors demonstrated that their proposed SuperHistopath framework succeeded in tissue multi-classification of three different cancer types and was able to achieve high accuracy (98.8% in melanomas, 93.1% in breast cancer, and 98.3% in childhood neuroblastoma).

The second aspect is classifying tumor tissue into cancer subtypes. This could be the classification of malign vs. benign carcinoma, invasive vs. non-invasive carcinoma, or various subtypes of a certain cancer type. This subtyping is an important part of determining a treatment plan; however, it often needs special IHC staining to be done. Therefore, the ability to perform subtype classification directly from H&E images could be of great benefit in terms of clinical application. Authors in [23] proposed a method for subdividing breast cancer into eight subtypes, four for benign (adenosis, fibroadenoma, phyllodes tumour, and tubular adenoma) and four for malignant (carcinoma, lobular carcinoma, mucinous carcinoma, and papillary carcinoma). They showed that their model achieved significant results compared to other state-of-the-art models mentioned in the study.

It should be noted that the two approaches are not always clearly separable, and the classification of tissue type is often associated with the classification of tumor subtypes. This approach was demonstrated in [24], where the breast tissue was categorized as normal tissue, benign lesion, in situ carcinoma, or invasive carcinoma. Another example is [29], where researchers managed to obtain models with accuracy over 0.95% in classifying five types of liver lesions, cirrhosis, and nearly normal tissue.

4.2. Tissue Grading

Cancer grading has its origins in 1914 when pathologist Albert Broders began collecting data showing that cancers of the same histologic type behaved differently. By the late 1930s, tumor grading was considered a state-of-the-art prognostic technique for scientific cancer care. Today, there are hundreds of grading schemes for various types of cancer [70]. However, in comparison with subtype classification, pathological image grading is considered a fine-grained task [64,65].

Researchers in [42] used residual networks to grade images of squamous cell carcinoma, since it accounts for about 90% of oral disorders. To demonstrate the deep learning capability of grading different cancer types, the authors of [64] developed a model with an average classification accuracy of 95% and 91% for colorectal and breast cancer grading, respectively. The breast cancer grading task was also addressed in [65].

4.3. Bio-Marker Classification

A bio-marker is a biological molecule found in tissues that is a sign of a normal or abnormal process or of a condition or disease, such as cancer. Typically, bio-markers differentiate a person without disease from an affected patient. There is a tremendous variety of bio-markers, including proteins, antibodies, nucleic acids, gene expression, and others. They can be used in clinical treatment for multiple tasks, such as estimating the risk of disease, differential diagnosis, predicting response to therapy, determining the prognosis of the disease, and so on [71].

In [36], the authors presented the architecture HE-HER2Net, which surpassed the accuracy of other common architectures in the multiclassification of HER2 into four categories. Following this, researchers in [40] developed a CNN to predict the mutated genes, which are potential candidates for targeted-drug therapy for lung cancer. The average probability of the bio-markers of lung cancer was received through the model, with the highest accuracy of 86.3%.

Ki67 is a protein that is found in the cancer cell nucleus and can be found only in cells that are actively growing and dividing, which is typical for cells mutated into cancer. Therefore, Ki67 is sometimes considered a good marker of proliferation (rapid increase in the number of cells) [72]. In [35], scholars fine-tuned an NN to classify cell images into Ki67-positive, Ki67-negative, and as a background image with an accuracy of 93%.

5. Conclusions

The article presents a detailed survey of recent DL models based on neural networks in the context of classification tasks for the analysis of histological images. The analysis of approximately 70 articles published in the last three years shows that automated processing and classification of histopathological images by deep learning methods have been applied to a wide range of histological tasks, such as tumor tissue classification or biomarker evaluation to determine treatment plans. The survey reveals several conclusions:

Application Areas: Deep learning has been applied to several types of cancer (e.g., breast, lung, colon, brain, kidney) and has proven to be capable of assisting pathologists with visual tasks in the treatment of various diseases. The reviewed works have identified the following three groups of specific tasks: classification of tissue type, grading of specific tissue, and identification of the presence of biomarkers.

Single- and Multi-Stage Approaches: Convolutional neural networks can be applied either as a stand-alone classifier or can be used as a feature extractor whose outputs will proceed into another machine learning model to carry out the final classification.

Pre-Training: Training networks from scratch requires a large dataset and a lot of computing time. Therefore, it is recommended to experiment with well-known architectures pre-trained on ImageNet. If the results are not sufficient, then one can design their own custom network and train it from scratch.

Author Contributions: Conceptualization, D.P. and I.C.; resources, D.P.; writing—original draft preparation, D.P.; writing—review and editing, D.P. and I.C.; visualization, D.P.; supervision, I.C.; funding acquisition, I.C. All authors have read and agreed to the published version of the manuscript.

Funding: This research was supported by the Operational Program "Integrated Infrastructure" of the project "Integrated strategy in the development of personalized medicine of selected malignant tumor diseases and its impact on life quality", ITMS code: 313011V446, co-financed by resources of European Regional Development Fund.

Data Availability Statement: Not applicable

Conflicts of Interest: The authors declare no conflict of interest.

Abbreviations

The following abbreviations are used in this manuscript:

AI	Artificial Intelligence
AUC	Area Under Curve
BCI	Breast Cancer Immunohistochemical
DL	Deep Learning
CNNs	Convolutional Neural Networks
WSIs	Whole Slide Images
NNs	Neural Networks
TIL	Tumor Infiltrating Lymphocytes
H&E	Hematoxylin and Eosin
IHC	Immunohistochemical
FL	Federated Learning
GDPR	General Data Protection Regulation
SLIC	Simple Linear Iterative Clustering
TMB	Tumor Mutation Burden
ML	Machine Learning
HER2	Human Epidermal Growth Factor Receptor 2
SVM	Support-vector machines
k-NN	k-Nearest Neighbor
RF	Random Forest
GC	Gastric Cancer

MB	Medulloblastoma
DWT	Discrete Wavelet Transform
RCC	Renal cell cancer

References

1. Pantanowitz, L. Digital images and the future of digital pathology: From the 1st Digital Pathology Summit, New Frontiers in Digital Pathology, University of Nebraska Medical Center, Omaha, Nebraska 14–15 May 2010. *J. Pathol. Inform.* **2010**, *1*, 15. [CrossRef]
2. Komura, D.; Ishikawa, S. Machine Learning Methods for Histopathological Image Analysis. *Comput. Struct. Biotechnol. J.* **2018**, *16*, 34–42. [CrossRef] [PubMed]
3. Srinidhi, C.L.; Ciga, O.; Martel, A.L. Deep neural network models for computational histopathology: A survey. *Med. Image Anal.* **2021**, *67*, 101813. [CrossRef] [PubMed]
4. Alzubi, J.; Nayyar, A.; Kumar, A. Machine Learning from Theory to Algorithms: An Overview. *J. Phys. Conf. Ser.* **2018**, *1142*, 012012. [CrossRef]
5. Jordan, M.I.; Mitchell, T.M. Machine learning: Trends, perspectives, and prospects. *Science* **2015**, *349*, 255–260. [CrossRef]
6. Litjens, G.; Kooi, T.; Bejnordi, B.E.; Setio, A.A.A.; Ciompi, F.; Ghafoorian, M.; van der Laak, J.A.; van Ginneken, B.; Sánchez, C.I. A survey on deep learning in medical image analysis. *Med. Image Anal.* **2017**, *42*, 60–88. [CrossRef]
7. Wang, M.; Lu, S.; Zhu, D.; Lin, J.; Wang, Z. A High-Speed and Low-Complexity Architecture for Softmax Function in Deep Learning. In Proceedings of the 2018 IEEE Asia Pacific Conference on Circuits and Systems (APCCAS), Chengdu, China, 26–30 October 2018; pp. 223–226. [CrossRef]
8. Ahmad, J.; Farman, H.; Jan, Z., Deep Learning Methods and Applications. In *Deep Learning: Convergence to Big Data Analytics*; Springer: Singapore, 2019; pp. 31–42. [CrossRef]
9. Yao, G.; Lei, T.; Zhong, J. A review of Convolutional-Neural-Network-based action recognition. *Pattern Recognit. Lett.* **2019**, *118*, 14–22. [CrossRef]
10. Dhillon, A.; Verma, G.K. Convolutional neural network: A review of models, methodologies and applications to object detection. *Prog. Artif. Intell.* **2019**, *9*, 85–112. [CrossRef]
11. Alzubaidi, L.; Zhang, J.; Humaidi, A.J.; Al-Dujaili, A.; Duan, Y.; Al-Shamma, O.; Santamaría, J.; Fadhel, M.A.; Al-Amidie, M.; Farhan, L. Review of deep learning: Concepts, CNN architectures, challenges, applications, future directions. *J. Big Data* **2021**, *8*, 53. [CrossRef]
12. O'Shea, K.; Nash, R. An Introduction to Convolutional Neural Networks. *arXiv* **2015**, arXiv:1511.08458. [CrossRef]
13. Dimitriou, N.; Arandjelović, O.; Caie, P.D. Deep Learning for Whole Slide Image Analysis: An Overview. *Front. Med.* **2019**, *6*, 00264. [CrossRef] [PubMed]
14. Abousamra, S.; Gupta, R.; Hou, L.; Batiste, R.; Zhao, T.; Shankar, A.; Rao, A.; Chen, C.; Samaras, D.; Kurc, T.; et al. Deep Learning-Based Mapping of Tumor Infiltrating Lymphocytes in Whole Slide Images of 23 Types of Cancer. *Front. Oncol.* **2022**, *11*, 806603. [CrossRef]
15. Simonyan, K.; Zisserman, A. Very deep convolutional networks for large-scale image recognition. Computational and Biological Learning Society. *arXiv* **2015**, arXiv:1409.1556.
16. He, K.; Zhang, X.; Ren, S.; Sun, J. Deep Residual Learning for Image Recognition. In Proceedings of the 2016 IEEE Conference on Computer Vision and Pattern Recognition (CVPR), Los Alamitos, CA, USA, 27–30 June 2016; pp. 770–778. [CrossRef]
17. Szegedy, C.; Ioffe, S.; Vanhoucke, V.; Alemi, A. Inception-v4, Inception-ResNet and the Impact of Residual Connections on Learning. In Proceedings of the AAAI Conference on Artificial Intelligence, San Francisco, CA, USA, 4–9 February 2017; Volume 31. [CrossRef]
18. Yang, H.; Chen, L.; Cheng, Z.; Yang, M.; Wang, J.; Lin, C.; Wang, Y.; Huang, L.; Chen, Y.; Peng, S.; et al. Deep learning-based six-type classifier for lung cancer and mimics from histopathological whole slide images: A retrospective study. *BMC Med.* **2021**, *19*, 80. [CrossRef]
19. Tan, M.; Le, Q. EfficientNet: Rethinking Model Scaling for Convolutional Neural Networks. In Proceedings of the 36th International Conference on Machine Learning, PMLR, Long Beach, CA, USA, 9–15 June 2019; Chaudhuri, K., Salakhutdinov, R., Eds.; JMLR: Cambridge, MA, USA, 2019; Volume 97, pp. 6105–6114.
20. Khan, A.; Brouwer, N.; Blank, A.; Müller, F.; Soldini, D.; Noske, A.; Gaus, E.; Brandt, S.; Nagtegaal, I.; Dawson, H.; et al. Computer-assisted diagnosis of lymph node metastases in colorectal cancers using transfer learning with an ensemble model. *Mod. Pathol.* **2023**, *36*, 100118. [CrossRef]
21. Hameed, Z.; Zahia, S.; Garcia-Zapirain, B.; Javier Aguirre, J.; María Vanegas, A. Breast Cancer Histopathology Image Classification Using an Ensemble of Deep Learning Models. *Sensors* **2020**, *20*, 4373. [CrossRef] [PubMed]
22. Farahani, H.; Boschman, J.; Farnell, D.; Darbandsari, A.; Zhang, A.; Ahmadvand, P.; Jones, S.J.M.; Huntsman, D.; Köbel, M.; Gilks, C.B.; et al. Deep learning-based histotype diagnosis of ovarian carcinoma whole-slide pathology images. *Mod. Pathol.* **2022**, *35*, 1983–1990. [CrossRef] [PubMed]
23. Sarker, M.M.K.; Akram, F.; Alsharid, M.; Singh, V.K.; Yasrab, R.; Elyan, E. Efficient Breast Cancer Classification Network with Dual Squeeze and Excitation in Histopathological Images. *Diagnostics* **2023**, *13*, 103. [CrossRef]

24. Hameed, Z.; Garcia-Zapirain, B.; Aguirre, J.J.; Isaza-Ruget, M.A. Multiclass classification of breast cancer histopathology images using multilevel features of deep convolutional neural network. *Sci. Rep.* **2022**, *12*, 15800. [CrossRef]
25. Luo, Y.; Zhang, J.; Yang, Y.; Rao, Y.; Chen, X.; Shi, T.; Xu, S.; Jia, R.; Gao, X. Deep learning-based fully automated differential diagnosis of eyelid basal cell and sebaceous carcinoma using whole slide images. *Quant. Imaging Med. Surg.* **2022**, , 4166–4175. [CrossRef]
26. Yu, K.H.; Hu, V.; Wang, F.; Matulonis, U.A.; Mutter, G.L.; Golden, J.A.; Kohane, I.S. Deciphering serous ovarian carcinoma histopathology and platinum response by convolutional neural networks. *BMC Med.* **2020**, *18*, 236. [CrossRef] [PubMed]
27. Krizhevsky, A.; Sutskever, I.; Hinton, G.E. ImageNet Classification with Deep Convolutional Neural Networks. In Proceedings of the Advances in Neural Information Processing Systems, Virtual, 6–12 December 2020; Pereira, F., Burges, C., Bottou, L., Weinberger, K., Eds.; Curran Associates, Inc.: New York, NY, USA, 2012; Volume 25.
28. Szegedy, C.; Liu, W.; Jia, Y.; Sermanet, P.; Reed, S.; Anguelov, D.; Erhan, D.; Vanhoucke, V.; Rabinovich, A. Going deeper with convolutions. In Proceedings of the 2015 IEEE Conference on Computer Vision and Pattern Recognition (CVPR), Boston, MA, USA, 7–12 June 2015; pp. 1–9. [CrossRef]
29. Cheng, N.; Ren, Y.; Zhou, J.; Zhang, Y.; Wang, D.; Zhang, X.; Chen, B.; Liu, F.; Lv, J.; Cao, Q.; et al. Deep learning-based classification of hepatocellular nodular lesions on whole-slide histopathologic images. *Gastroenterology* **2022**, *162*, 1948–1961.e7. [CrossRef] [PubMed]
30. Rao, R.S.; Shivanna, D.B.; Lakshminarayana, S.; Mahadevpur, K.S.; Alhazmi, Y.A.; Bakri, M.M.H.; Alharbi, H.S.; Alzahrani, K.J.; Alsharif, K.F.; Banjer, H.J.; et al. Ensemble Deep-Learning-Based Prognostic and Prediction for Recurrence of Sporadic Odontogenic Keratocysts on Hematoxylin and Eosin Stained Pathological Images of Incisional Biopsies. *J. Pers. Med.* **2022**, *12*, 1220. [CrossRef] [PubMed]
31. Mundhada, A.; Sundaram, S.; Swaminathan, R.; D' Cruze, L.; Govindarajan, S.; Makaram, N. Differentiation of urothelial carcinoma in histopathology images using deep learning and visualization. *J. Pathol. Inform.* **2023**, *14*, 100155. [CrossRef]
32. Naik, N.; Madani, A.; Esteva, A.; Keskar, N.S.; Press, M.F.; Ruderman, D.; Agus, D.B.; Socher, R. Deep learning-enabled breast cancer hormonal receptor status determination from base-level H&E stains. *Nat. Commun.* **2020**, *11*, 5727. [CrossRef]
33. Seegerer, P.; Binder, A.; Saitenmacher, R.; Bockmayr, M.; Alber, M.; Jurmeister, P.; Klauschen, F.; Müller, K.R., Interpretable Deep Neural Network to Predict Estrogen Receptor Status from Haematoxylin-Eosin Images. In *Artificial Intelligence and Machine Learning for Digital Pathology: State-of-the-Art and Future Challenges*; Holzinger, A.; Goebel, R.; Mengel, M.; Müller, H.; Eds.; Springer International Publishing: Cham, Switzerland, 2020; pp. 16–37. [CrossRef]
34. Rawat, R.R.; Ortega, I.; Roy, P.; Sha, F.; Shibata, D.; Ruderman, D.; Agus, D.B. Deep learned tissue "fingerprints" classify breast cancers by ER/PR/Her2 status from H&E images. *Sci. Rep.* **2020**, *10*, 7275. [CrossRef]
35. Liu, Y.; Li, X.; Zheng, A.; Zhu, X.; Liu, S.; Hu, W.; Luo, Q.; Liao, H.; Liu, M.; He, Y.; et al. Predict Ki-67 Positive Cells in H&E-Stained Images Using Deep Learning Independently From IHC-Stained Images. *Front. Mol. Biosci.* **2020**, *7*, 00183. [CrossRef]
36. Shovon, M.S.H.; Islam, M.J.; Nabil, M.N.A.K.; Molla, M.M.; Jony, A.I.; Mridha, M.F. Strategies for Enhancing the Multi-Stage Classification Performances of HER2 Breast Cancer from Hematoxylin and Eosin Images. *Diagnostics* **2022**, *12*, 2825. [CrossRef]
37. Voigt, P.; von dem Bussche, A. *The EU General Data Protection Regulation (GDPR)*, 1st ed.; Springer International Publishing: Cham, Switzerland, 2017.
38. Baid, U.; Pati, S.; Kurc, T.M.; Gupta, R.; Bremer, E.; Abousamra, S.; Thakur, S.P.; Saltz, J.H.; Bakas, S. Federated Learning for the Classification of Tumor Infiltrating Lymphocytes. *arXiv* **2022**, arXiv:2203.16622. [CrossRef]
39. Wu, Z.; Wang, L.; Li, C.; Cai, Y.; Liang, Y.; Mo, X.; Lu, Q.; Dong, L.; Liu, Y. DeepLRHE: A deep convolutional neural network framework to evaluate the risk of lung cancer recurrence and metastasis from histopathology images. *Front. Genet.* **2020**, *11*, 768. [CrossRef]
40. Huang, K.; Mo, Z.; Zhu, W.; Liao, B.; Yang, Y.; Wu, F.X. Prediction of Target-Drug Therapy by Identifying Gene Mutations in Lung Cancer With Histopathological Stained Image and Deep Learning Techniques. *Front. Oncol.* **2021**, *11*, 642945. [CrossRef]
41. Steinbuss, G.; Kriegsmann, M.; Zgorzelski, C.; Brobeil, A.; Goeppert, B.; Dietrich, S.; Mechtersheimer, G.; Kriegsmann, K. Deep Learning for the Classification of Non-Hodgkin Lymphoma on Histopathological Images. *Cancers* **2021**, *13*, 2419. [CrossRef]
42. Panigrahi, S.; Bhuyan, R.; Kumar, K.; Nayak, J.; Swarnkar, T. Multistage classification of oral histopathological images using improved residual network. *Math. Biosci. Eng.* **2022**, *19*, 1909–1925. [CrossRef] [PubMed]
43. Abdeltawab, H.A.; Khalifa, F.A.; Ghazal, M.A.; Cheng, L.; El-Baz, A.S.; Gondim, D.D. A deep learning framework for automated classification of histopathological kidney whole-slide images. *J. Pathol. Inform.* **2022**, *13*, 100093. [CrossRef]
44. Wang, X.; Zou, C.; Zhang, Y.; Li, X.; Wang, C.; Ke, F.; Chen, J.; Wang, W.; Wang, D.; Xu, X.; et al. Prediction of BRCA Gene Mutation in Breast Cancer Based on Deep Learning and Histopathology Images. *Front. Genet.* **2021**, *12*, 661109. [CrossRef]
45. Le Page, A.L.; Ballot, E.; Truntzer, C.; Derangère, V.; Ilie, A.; Rageot, D.; Bibeau, F.; Ghiringhelli, F. Using a convolutional neural network for classification of squamous and non-squamous non-small cell lung cancer based on diagnostic histopathology HES images. *Sci. Rep.* **2021**, *11*, 23912. [CrossRef] [PubMed]
46. Zormpas-Petridis, K.; Noguera, R.; Ivankovic, D.K.; Roxanis, I.; Jamin, Y.; Yuan, Y. SuperHistopath: A Deep Learning Pipeline for Mapping Tumor Heterogeneity on Low-Resolution Whole-Slide Digital Histopathology Images. *Front. Oncol.* **2021**, *10*, 586292. [CrossRef] [PubMed]

47. Yang, J.W.; Song, D.H.; An, H.J.; Seo, S.B. Classification of subtypes including LCNEC in lung cancer biopsy slides using convolutional neural network from scratch. *Sci. Rep.* **2022**, *12*, 1830. [CrossRef]
48. Abdolahi, M.; Salehi, M.; Shokatian, I.; Reiazi, R. Artificial intelligence in automatic classification of invasive ductal carcinoma breast cancer in digital pathology images. *Med. J. Islam. Repub. Iran* **2020**, *34*, 140. [CrossRef]
49. Sadhwani, A.; Chang, H.W.; Behrooz, A.; Brown, T.; Auvigne-Flament, I.; Patel, H.; Findlater, R.; Velez, V.; Tan, F.; Tekiela, K.; et al. Comparative analysis of machine learning approaches to classify tumor mutation burden in lung adenocarcinoma using histopathology images. *Sci. Rep.* **2021**, *11*, 16605. [CrossRef] [PubMed]
50. Szegedy, C.; Vanhoucke, V.; Ioffe, S.; Shlens, J.; Wojna, Z. Rethinking the Inception Architecture for Computer Vision. In Proceedings of the 2016 IEEE Conference on Computer Vision and Pattern Recognition (CVPR), Las Vegas, NV, USA, 26 June–1 July 2016; pp. 2818–2826. [CrossRef]
51. Wu, J.; Zhang, R.; Gong, T.; Bao, X.; Gao, Z.; Zhang, H.; Wang, C.; Li, C. A Precision Diagnostic Framework of Renal Cell Carcinoma on Whole-Slide Images using Deep Learning. In Proceedings of the 2021 IEEE International Conference on Bioinformatics and Biomedicine (BIBM), Houston, TX, USA, 9–12 December 2021; pp. 2104–2111. [CrossRef]
52. Jin, L.; Shi, F.; Chun, Q.; Chen, H.; Ma, Y.; Wu, S.; Hameed, N.U.F.; Mei, C.; Lu, J.; Zhang, J.; et al. Artificial intelligence neuropathologist for glioma classification using deep learning on hematoxylin and eosin stained slide images and molecular markers. *Neuro. Oncol.* **2021**, *23*, 44–52. [CrossRef] [PubMed]
53. Anand, D.; Kurian, N.C.; Dhage, S.; Kumar, N.; Rane, S.; Gann, P.H.; Sethi, A. Deep Learning to Estimate Human Epidermal Growth Factor Receptor 2 Status from Hematoxylin and Eosin-Stained Breast Tissue Images. *J. Pathol. Inform.* **2020**, *11*, 19. [CrossRef] [PubMed]
54. Ronneberger, O.; Fischer, P.; Brox, T. U-Net: Convolutional Networks for Biomedical Image Segmentation. *arXiv* **2015**, arXiv:1505.04597. [CrossRef]
55. Dong, X.; Li, M.; Zhou, P.; Deng, X.; Li, S.; Zhao, X.; Wu, Y.; Qin, J.; Guo, W. Fusing pre-trained convolutional neural networks features for multi-differentiated subtypes of liver cancer on histopathological images. *BMC Med. Inform. Decis. Mak.* **2022**, *22*, 122. [CrossRef]
56. Huang, G.; Liu, Z.; van der Maaten, L.; Weinberger, K.Q. Densely Connected Convolutional Networks. *arXiv* **2016**, arXiv:1608.06993. [CrossRef]
57. Wessels, F.; Schmitt, M.; Krieghoff-Henning, E.; Kather, J.N.; Nientiedt, M.; Kriegmair, M.C.; Worst, T.S.; Neuberger, M.; Steeg, M.; Popovic, Z.V.; et al. Deep learning can predict survival directly from histology in clear cell renal cell carcinoma. *PLoS ONE* **2022**, *17*, e0272656. [CrossRef] [PubMed]
58. Mi, W.; Li, J.; Guo, Y.; Ren, X.; Liang, Z.; Zhang, T.; Zou, H. Deep learning-based multi-class classification of breast digital pathology images. *Cancer Manag. Res.* **2021**, *13*, 4605–4617. [CrossRef]
59. Chen, T.; Guestrin, C. XGBoost. In Proceedings of the 22nd ACM SIGKDD International Conference on Knowledge Discovery and Data Mining, ACM, San Francisco, CA, USA, 13–17 August 2016. [CrossRef]
60. Rączkowski, Ł.; Paśnik, I.; Kukiełka, M.; Nicoś, M.; Budzinska, M.A.; Kucharczyk, T.; Szumiło, J.; Krawczyk, P.; Crosetto, N.; Szczurek, E. Deep learning-based tumor microenvironment segmentation is predictive of tumor mutations and patient survival in non-small-cell lung cancer. *BMC Cancer* **2022**, *22*, 1001. [CrossRef] [PubMed]
61. Fu, H.; Mi, W.; Pan, B.; Guo, Y.; Li, J.; Xu, R.; Zheng, J.; Zou, C.; Zhang, T.; Liang, Z.; et al. Automatic Pancreatic Ductal Adenocarcinoma Detection in Whole Slide Images Using Deep Convolutional Neural Networks. *Front. Oncol.* **2021**, *11*, 665929. [CrossRef] [PubMed]
62. Ke, G.; Meng, Q.; Finley, T.; Wang, T.; Chen, W.; Ma, W.; Ye, Q.; Liu, T.Y. LightGBM: A Highly Efficient Gradient Boosting Decision Tree. In Proceedings of the Advances in Neural Information Processing Systems; Long Beach, CA, USA, 4–9 December 2017; Guyon, I., Luxburg, U.V., Bengio, S., Wallach, H., Fergus, R., Vishwanathan, S., Garnett, R., Eds.; Curran Associates, Inc.: New York, NY, USA, 2017; Volume 30.
63. Ma, B.; Guo, Y.; Hu, W.; Yuan, F.; Zhu, Z.; Yu, Y.; Zou, H. Artificial Intelligence-Based Multiclass Classification of Benign or Malignant Mucosal Lesions of the Stomach. *Front. Pharmacol.* **2020**, *11*, 572372. [CrossRef]
64. Yan, R.; Yang, Z.; Li, J.; Zheng, C.; Zhang, F. Divide-and-Attention Network for HE-Stained Pathological Image Classification. *Biology* **2022**, *11*, 982. [CrossRef]
65. Yan, R.; Ren, F.; Li, J.; Rao, X.; Lv, Z.; Zheng, C.; Zhang, F. Nuclei-Guided Network for Breast Cancer Grading in HE-Stained Pathological Images. *Sensors* **2022**, *22*, 4061. [CrossRef]
66. Grist, J.T.; Withey, S.; MacPherson, L.; Oates, A.; Powell, S.; Novak, J.; Abernethy, L.; Pizer, B.; Grundy, R.; Bailey, S.; et al. Distinguishing between paediatric brain tumour types using multi-parametric magnetic resonance imaging and machine learning: A multi-site study. *arXiv* **2019**, arXiv:1910.09247. [CrossRef]
67. Attallah, O. MB-AI-His: Histopathological Diagnosis of Pediatric Medulloblastoma and its Subtypes via AI. *Diagnostics* **2021**, *11*, 359. [CrossRef] [PubMed]
68. Howard, A.G.; Zhu, M.; Chen, B.; Kalenichenko, D.; Wang, W.; Weyand, T.; Andreetto, M.; Adam, H. MobileNets: Efficient Convolutional Neural Networks for Mobile Vision Applications. *arXiv* **2017**, arXiv:1704.04861. [CrossRef]
69. Attallah, O. CoMB-Deep: Composite Deep Learning-Based Pipeline for Classifying Childhood Medulloblastoma and Its Classes. *Front. Neuroinform.* **2021**, *15*, 663592. [CrossRef]

70. Wright, J.R., Jr. Albert C. Broders, tumor grading, and the origin of the long road to personalized cancer care. *Cancer Med.* **2020**, *9*, 4490–4494. [CrossRef]
71. Henry, N.L.; Hayes, D.F. Cancer biomarkers. *Mol. Oncol.* **2012**, *6*, 140–146. Personalized cancer medicine. [CrossRef]
72. Kos, Z.; Dabbs, D.J. Biomarker assessment and molecular testing for prognostication in breast cancer. *Histopathology* **2016**, *68*, 70–85. [CrossRef] [PubMed]

Disclaimer/Publisher's Note: The statements, opinions and data contained in all publications are solely those of the individual author(s) and contributor(s) and not of MDPI and/or the editor(s). MDPI and/or the editor(s) disclaim responsibility for any injury to people or property resulting from any ideas, methods, instructions or products referred to in the content.

Review

Machine Learning in X-ray Diagnosis for Oral Health: A Review of Recent Progress

Mónica Vieira Martins [1,*], Luís Baptista [1], Henrique Luís [1,2,3,4], Victor Assunção [1,2,3,4], Mário-Rui Araújo [1] and Valentim Realinho [1,5]

1. Polytechnic Institute of Portalegre, 7300-110 Portalegre, Portugal; lmtb@ipportalegre.pt (L.B.); henrique.luis@ipportalegre.pt (H.L.); victorassuncao@ipportalegre.pt (V.A.); mra@ipportalegre.pt (M.-R.A.); vrealinho@ipportalegre.pt (V.R.)
2. Faculdade de Medicina Dentária, Universidade de Lisboa, Unidade de Investigação em Ciências Orais e Biomédicas (UICOB), Rua Professora Teresa Ambrósio, 1600-277 Lisboa, Portugal
3. Faculdade de Medicina Dentária, Universidade de Lisboa, Rede de Higienistas Orais para o Desenvolvimento da Ciência (RHODes), Rua Professora Teresa Ambrósio, 1600-277 Lisboa, Portugal
4. Center for Innovative Care and Health Technology (ciTechcare), Polytechnic of Leiria, 2410-541 Leiria, Portugal
5. VALORIZA—Research Center for Endogenous Resource Valorization, 7300-555 Portalegre, Portugal
* Correspondence: mvmartins@ipportalegre.pt

Abstract: The past few decades have witnessed remarkable progress in the application of artificial intelligence (AI) and machine learning (ML) in medicine, notably in medical imaging. The application of ML to dental and oral imaging has also been developed, powered by the availability of clinical dental images. The present work aims to investigate recent progress concerning the application of ML in the diagnosis of oral diseases using oral X-ray imaging, namely the quality and outcome of such methods. The specific research question was developed using the PICOT methodology. The review was conducted in the Web of Science, Science Direct, and IEEE Xplore databases, for articles reporting the use of ML and AI for diagnostic purposes in X-ray-based oral imaging. Imaging types included panoramic, periapical, bitewing X-ray images, and oral cone beam computed tomography (CBCT). The search was limited to papers published in the English language from 2018 to 2022. The initial search included 104 papers that were assessed for eligibility. Of these, 22 were included for a final appraisal. The full text of the articles was carefully analyzed and the relevant data such as the clinical application, the ML models, the metrics used to assess their performance, and the characteristics of the datasets, were registered for further analysis. The paper discusses the opportunities, challenges, and limitations found.

Keywords: machine learning; artificial intelligence; oral health; X-ray imaging; diagnosis; convolutional neural networks; deep learning

1. Introduction

Dental caries and periodontal disease are two of the most common dental conditions that affect people worldwide. Dental caries, also known as tooth decay, is a multifactorial disease mainly caused by the interaction of the bacteria present in dental plaque and sugars from the diet, which produces acids that erode the tooth structure [1]. Periodontitis, on the other hand, is a chronic inflammatory condition that affects the supporting structures of the teeth, including the gums, periodontal ligament, dental root cement, and alveolar bone. It is also multifactorial and is caused by the accumulation of bacterial plaque and dental calculus around the teeth, which triggers an immune response that leads to tissue destruction [2].

X-ray exams are essential diagnostic tools in dentistry. They allow oral health professionals to visualize the internal structures of the teeth and jaws, which are not visible during a clinical examination. There are several types of dental X-ray exams, including bitewing, periapical, panoramic, and cone beam computed tomography (CBCT). Bitewing X-rays are used to detect dental caries. Periapical X-rays are used to detect dental caries and bone loss due to periodontitis and periapical lesions, while panoramic and CBCT X-rays are used to evaluate the overall condition of the teeth and the upper and lower jaws, including the presence of periodontal disease and other abnormalities [3].

Dental X-rays have revolutionized the practice of dentistry by providing detailed information about oral structures. They allow dental professionals to detect dental caries, periodontal diseases, and other conditions at an early stage, which can prevent further complications and improve treatment outcomes. Dental X-rays can also reveal other conditions, such as impacted teeth, tumors, and cysts, which may not be visible during a clinical examination. Additionally, they are useful in treatment planning and monitoring the progress of ongoing treatments [4,5].

Fast-emerging artificial intelligence (AI) technology is changing many scenarios in our society. The oral health field is not an exception, mainly because of its regular use of digitized imaging and electronic health records which facilitate AI algorithms [6,7]. The science is recent and caution should be used. Human supervision is needed, but the door is open and it is important to understand the real benefits of this technology in health activities [8].

The availability of clinical dental images and the development of deep learning algorithms in recent years has led to significant improvements in the accuracy and robustness of these algorithms in supporting the diagnosis of various dental conditions.

Convolutional neural networks (CNN) [9] are a type of deep learning neural network that are considered the most prominent algorithm used, due to their high accuracy and ability to learn and extract features from images. A CNN consists of multiple layers, including convolutional, pooling, and fully connected layers. CNNs have shown remarkable performance in image classification tasks and have been widely used in a variety of fields, including medical image analysis, object detection, and natural language processing.

Transfer learning is a machine learning technique that involves the use of a pretrained model (e.g., a CNN model), which has already learned relevant features from a large image dataset, such as ImageNet [10], COCO [11], MNIST [12], CIFAR-10/100 [13], or VOC [14]. It is then fine-tuned on a smaller dataset for a specific task. Pretrained image models are used as a starting point for training the new model and the most popular pretrained image architectures include GoogLeNet Inception [15], ResNet [16], VGG [17], and Xception [18]. Among these, GoogLeNEt Inception and ResNet hold special significance in oral health applications. GoogLeNet Inception–v3 architecture was introduced in 2014 and demonstrated excellent performance in the ImageNet Large Scale Visual Recognition Challenge. It was trained with more than a million images of 1000 object categories from the ImageNet dataset. The original architecture has 22 deep layers, allowing different scale features to be obtained by applying convolutional filters of different sizes in the same layers.

ResNet was introduced in 2015, and it has since become a foundational architecture in the field of deep learning, serving as a basis for many subsequent advancements. It addresses the problem of vanishing gradients that can occur when training very deep neural networks by using residual connections, where shortcut connections are added to bypass one or more layers.

Other works use a mixed approach that applies traditional machine learning methods, such as support vector machine (SVM) [19], k-nearest neighbors (kNN) [20], random forest [21], or extreme gradient boosting (XGBOOST) [22] for classification, using the image features previously extracted employing a CNN.

This scoping review aims to explore the current state of the art of AI-assisted diagnosis in oral health using X-ray-based images, focusing on the last five years. The specific objectives are to summarize several aspects of the current state of the art in the field and to identify limitations and research gaps that must be addressed to advance the field.

By providing a comprehensive overview of the quality and advancements of predictive models developed using artificial intelligence-based methods for oral X-ray diagnosis, this scoping review identifies trends, challenges, and gaps in the development and evaluation of these models. The review's findings offer valuable insights into the feasibility and effectiveness of AI-based approaches in dental imaging, potentially improving diagnostic accuracy and patient outcomes in oral healthcare.

The rest of the paper is organized as follows: Section 2 describes the Methods used for information search and analysis; Section 3 summarizes the results obtained; Section 4 provides a discussion of the findings and Section 5 presents the Conclusions.

2. Methods

This review aims to obtain important insights into scientific production to identify the status of machine learning in diagnosis using X-ray-based images in oral health.

Our research questions were built using the PICOT [23] framework. The PICOT framework is widely used in healthcare research to generate specific research questions and concisely guide study design. It is an acronym that stands for population, intervention, control, outcome, and time. The PICOT elements for this review are presented in Table 1.

Table 1. Description of the PICOT elements.

	Study Question
Population	Oral X-ray diagnostic images of patients (radiography, CBCT)
Intervention	Artificial intelligence-based forms of diagnosis
Control	Oral health
Outcome	Quality of the predictive models
Time	Last five years

Therefore, the research question was formulated as follows:

"What is the quality of the predictive models being used for diagnosis in oral health using X-ray-based images?"

According to the formulated question, the systematic literature search was performed with the following inclusion criteria:

1. Studies between 1 January 2018 to 31 December 2022, since the goal was to access the most recent progress in a rapidly evolving field;
2. Studies with a focus on dental/oral imaging techniques based on X-rays, including cone beam computed tomography (CBCT);
3. Studies with a focus on diagnostic applications. To our knowledge, this is the first paper that exclusively reviews the application of ML methods in oral health diagnosis.

The three different databases shown in Table 2 were used for information retrieval.

Table 2. Databases used to conduct the search.

Name	Acronym	URL
IEEE Xplore	IEEEXplore	https://ieeexplore.ieee.org/Xplore/home.jsp (accessed on 6 March 2023)
Science Direct	SciDir	https://www.sciencedirect.com/ (accessed on 6 March 2023)
Web of Science	WoS	https://www.webofscience.com/wos/ (accessed on 6 March 2023)

The search strategy was built by logical operators used for query search in the databases. Since each database uses different syntaxes for queries, a specific query was built for each one. An example of a query used is as follows:

(Dental OR Dentistry) AND (Imaging OR Images) AND ("Machine Learning" OR "Artificial Intelligence")

The search was limited to journal articles written in the English language excluding conference papers, reviews, and editorials. The search was conducted by one reviewer (M.V.M.), who also evaluated the search results for relevance based on their title and abstract. After the remotion of duplicates, the abstracts of the papers selected for screening were evaluated by blinded pairs of researchers (M.V.M., L.B., H.L., V.A., M.R.A., V.R.) using the web app Rayyan [24]. After individual evaluation, discrepancies were solved by reaching a consensus. The full text of the selected studies was examined in detail for eligibility (M.V.M., L.B., H.L., V.A., M.R.A., V.R.). At this stage, a few papers were excluded for not meeting the inclusion criteria. Data extraction from the included publications was then performed (M.V.M., L.B., H.L., V.A., M.R.A., V.R.) and recorded in a spreadsheet. At all stages, there was complete consensus among the evaluators on the literature selection process and the classification of the publications.

The study characteristics recorded included the year of publication, country, the aim of the study, clinical application, type of X-ray images used, data source, size and partition (training, test, and validation sizes), if augmentation strategies were used, the type of task (classification, regression), machine learning models used, the metrics used to evaluate the models and their best reported values, and if human comparators were employed.

3. Results

3.1. Search and Study Selection

The Prisma [23] diagram presented in Figure 1 shows the flowchart for the study search and selection process. The initial search identified 104 papers. After the remotion of duplicates, a total of 92 papers were left for screening. During the screening phase, 52 papers were excluded. Reasons for exclusion included the study not dealing with diagnosis questions; the study did not use X-ray-based images; the metrics of the developed models were not reported. A total of 40 papers were then accessed for eligibility, and a further 18 papers were excluded for not dealing with diagnosis issues or not using X-ray images. A total of 22 papers were included in this review.

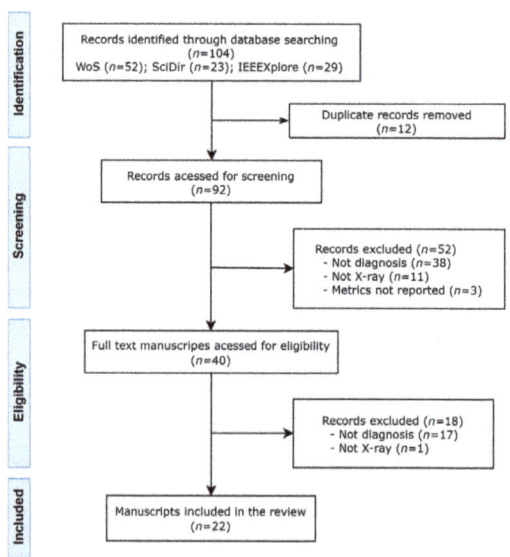

Figure 1. Flowchart of the search, where n represents the number of papers.

3.2. Included Studies

Some of the characteristics of the selected papers are presented in Table 3. The complete information can be found in Supplementary Table S1.

Table 3. Characteristics of the selected papers.

Study	Country, Year	Diagnosis of	ImageType	Data Source	Dataset Size	Machine Learning Task	Metrics	Models
[25]	South Korea, 2018	Dental caries	Periapical	Hospital	24,600	Classification	Acc, Sens, Spec, PPV, NPV, ROC-AUC	GoogLeNet
[26]	Germany, 2019	Apical lesions	Panoramic	University	2877	Classification	ROC-AUC, Sens, Spec, PPV, NPV	Proprietary CNN
[27]	Germany, 2019	Periodontal diseases	Panoramic	University	2538	Classification	Acc, ROC-AUC, F1, Sens, Spec, PPV, NPV	Proprietary CNN
[28]	India, 2020	Dental caries	Periapical	University	105	Classification	Acc, FPR, PRC, MCC	BPNN
[29]	Germany, 2020	Apical lesions	Panoramic	University	3099	Classification	PPV, Sens, F1, Prec, TPR	U-Net
[30]	South Korea, 2020	Oral lesions	CBCT, Panoramic	University	170,525	Classification	ROC-AUC, Sens, Specificity	GoogLeNet
[31]	Saudi Arabia, 2020	Apical lesions, dental caries, periodontal diseases	Periapical	Database	120	Classification	Acc, Spec, Prec, Rec, F1	Proprietary CNN
[32]	USA, 2021	Oral lesions	CBCT	University	100	Classification	Prec, Rec, Dice, Acc	Proprietary CNN
[33]	South Korea, 2020	Implant defects	Periapical, Panoramic	Hospital	1,292,360	Classification	ROC-AUC, Sens, Spec, YI	VGG, GoogLeNet, Proprietary CNN
[34]	Japan, 2021	Dental caries	Panoramic	Hospital	533	Classification	Acc, Sens, Spec, PPV, NPV, F1	Alexnet, GoogLeNet, VGG, ResNet, Xception, SVM, KNN, DT, NB, RF
[35]	South Korea, 2021	Periodontal diseases	Periapical	University	708	Classification	Prec, Rec, mOKS	Mask R-CNN, ResNet
[36]	USA, 2022	Periodontal diseases	Bitewing, Periapical	Private clinic	133,304	Generative; Regression	MAE, MBE	Proprietary CNN, DeepLabV3, DETR
[37]	China, 2022	Periodontal diseases, Dental caries	Periapical	Hospital	7924	Classification	Sens, Spec, PPV, NPV, F1, ROC-AUC	Modified ResNet-18
[38]	China, 2022	Ectopic eruption	Panoramic	Hospital	3160	Classification	Sens, Spec, PPV, NPV, ROC-AUC, F1	Proprietary CNN

Table 3. Cont.

Study	Country, Year	Diagnosis of	Image Type	Data Source	Dataset Size	Machine Learning Task	Metrics	Models
[39]	Saudi Arabia, 2022	Impacted tooth	Panoramic	University	416	Classification	Acc, Prec, Rec, Spec, F1	DenseNet, VGG, Inception V3, ResNet-50
[40]	China, 2022	Dental caries	Periapical	University	840	Classification	DICE, Prec, Sens, Spec	Proprietary CNN
[41]	Turkey, 2022	Dental caries	Periapical	Private clinic	340	Classification	Acc, ROC-AUC, CM	Proprietary CNN, VGG, SqueezeNet, GoogleNet, ResNet, ShuffleNet, Xception, MobileNet, DarkNet
[42]	Japan, 2022	Oral lesions	Panoramic	Hospital	7260	Classification	Acc, Sens, Spec, Prec, Rec, F1	YOLO v3
[43]	Germany, 2022	Oral lesions	Panoramic	University	1239	Classification	Prec, Rec, NPV, Spec, F1	ResNet, RF
[44]	Netherlands, 2022	Periodontal diseases	Periapical	University	1546	Regression	MSE	Proprietary CNN
[45]	China, 2022	Dental caries	Periapical	University	800	Classification	Prec, F1	Proprietary CNN
[46]	Turkey, 2022	Periodontal diseases	X-ray, type not defined	Database	1432	Classification	Acc, Sens, Spec, Prec, F1	AlexNet, SqueezeNet, EfficientNet, DT, KNN, NB, RUSBoost, SVM

Acc: accuracy; CM: confusion matrix; DT: decision tree; FPR: false positives ratio; KNN: K-nearest neighbor; NB: naïve Bayes; MAE: mean absolute error; MBE: mean bias error; MCC: Matthews correlation coefficient; mOKS: mean object keypoint similarity; MSE: mean squared error; NPV: negative predictive value; Prec: precision; PPV: positive predictive value; PRC: precision-recall curve; RF: random forest; Rec: recall; ROC-AUC: receiver operator characteristic–area under the curve; Sens: sensitivity; Spec: specificity; SVM: support vector machine; TPR: true positives ratio; USA: United States of America; YI: Youden index.

All studies included were published between 2018 and 2022, with a notable increase in the last year considered, which represented 50% of all studies (Figure 2).

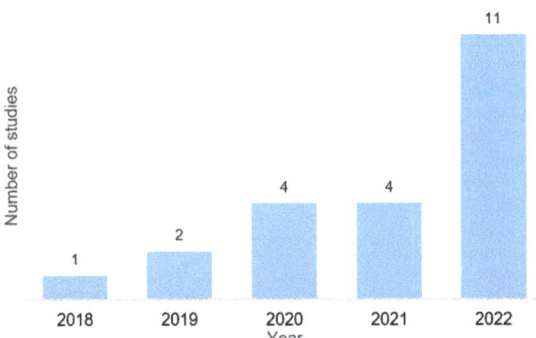

Figure 2. Number of included studies per year of publication.

The 22 included studies involved a total number of 153 researchers affiliated with 17 countries. Of these 153 researchers, 65% ($n = 100$) had their affiliation with institutions related to health (colleges or departments of oral health and similar, hospitals and clinics), and the rest (35%, $n = 53$) with institutions from areas related to computer science, physics, engineering and similar. The majority of the first authors were affiliated in China and South Korea with four papers, and the United States with three papers (Figure 3). These three countries represent 50% of the included studies.

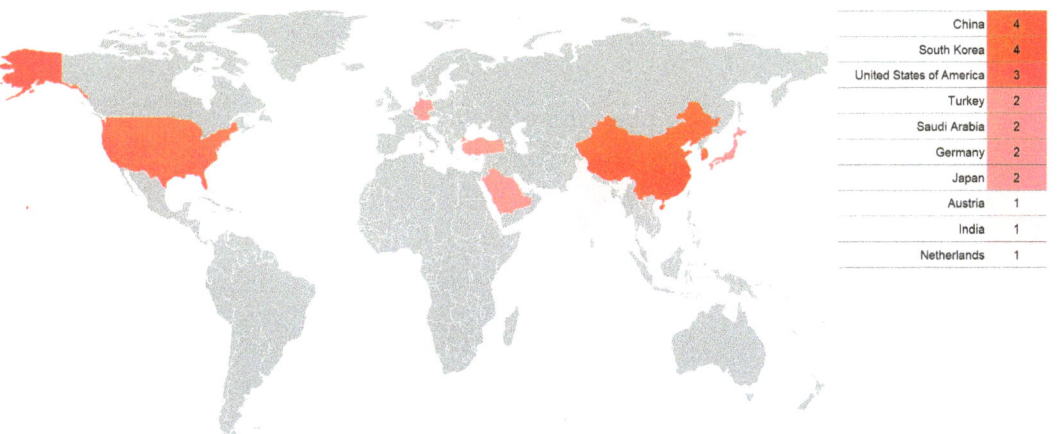

Figure 3. Geographic distribution of the country affiliation of the first author of the studies.

The studies were published in sixteen different journals, with the Journal of Dentistry being the one that published the most articles, with 23% of the total ($n = 5$), and Diagnostics being the second one with 14% ($n = 3$). The remaining papers were distributed by the fourteen other sources shown in Table 4.

Table 4. Journal sources of the included papers.

Journal	n	%
Journal of Dentistry	5	23%
Diagnostics	3	14%
Biomedical Signal Processing and Control	1	5%
Scientific Reports	1	5%
Journal of Oral and Maxillofacial Surgery, Medicine, and Pathology	1	5%
Informatics in Medicine Unlocked	1	5%
Cluster Computing	1	5%
International Dental Journal	1	5%
Journal of Clinical Medicine	1	5%
Displays	1	5%
Journal of Endodontics	1	5%
Health Information Science and Systems	1	5%
Oral Diseases	1	5%
IEEE Access	1	5%
Applied Sciences	1	5%
IEEE Transactions on Automation Science and Engineering	1	5%

The keywords used in the studies (Figure 4) totalized 120 terms, with the most used being "artificial intelligence" ($n = 13$), followed by "machine learning" and "deep learning" (each $n = 11$). Less used keywords were "convolutional neural network" ($n = 4$), "radiography", "supervised machine learning", "dental caries" (each $n = 3$), "classification", "digital image/radiology", "endodontics", "diagnosis", "panoramic radiograph" and "cysts" (each $n = 2$). The rest of the terms, a total of $n = 60$, each appeared only in one study.

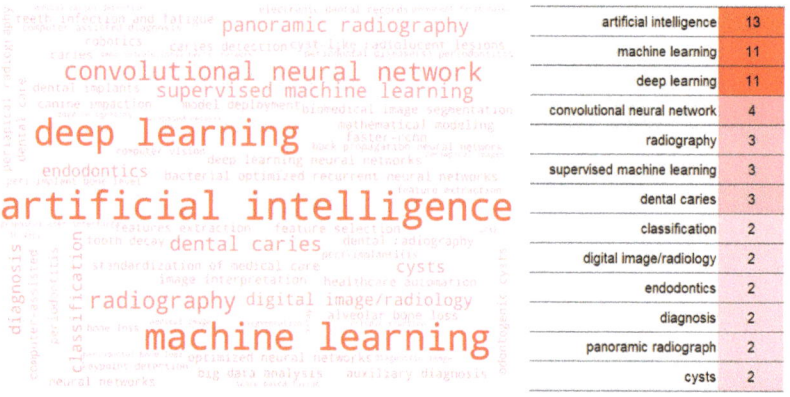

Figure 4. Word cloud of keywords.

3.3. Clinical Applications, Image Types, Data Sources and Labeling

Most of the studies analyzed ($n = 8$) applied the machine learning models to the diagnosis of dental caries, followed by the diagnosis of periodontal diseases ($n = 7$), diagnosis of oral lesions ($n = 4$), and diagnosis of apical lesions ($n = 3$). A small number of papers addressed the diagnosis of implant defects ($n = 1$), ectopic eruption ($n = 1$), and impacted teeth ($n = 1$).

The vast majority of the studies considered in this review used periapical ($n = 10$) or panoramic images ($n = 10$), while one paper used both periapical and bitewing images. Only two papers used CBCT images. One of these used both CBCT and panoramic images. One paper did not specify the type of X-ray image being used.

Universities were the most common source of data ($n = 12$), followed by hospitals ($n = 6$). There were also studies based on external datasets ($n = 2$) and a small number used data from private clinics (n = 2). The majority of datasets consisted of data from a single institution. However, there was one particular paper [43] that constructed its dataset by incorporating information from two different hospitals. The papers that used external data sources did not include a description of the labeling process. Among the remaining papers, two did not describe the labeling process. Only 15 papers provided information regarding the number of annotators. They ranged from one to six annotators, with varying degrees of experience. Only seven papers provided information regarding the seniority of the annotators, which ranged from 3 to 33 years of experience. In three of those papers, the annotators had a minimum of 3 years of experience, while in four they had at least 10 years of experience.

3.4. Datasets Size, Partitions, and Data Augmentation

The majority of the 22 papers (68%, 15), use data augmentation, namely zooming, rotation, shearing, flipping, and shifting. For a reliable comparison between papers, the dataset size must consider the data augmentation process. So, for each paper where data augmentation was used, we considered the actual number of examples that fed the machine learning algorithm, instead of the original dataset size. In practical terms, data augmentation corresponds to an increment in the dataset size.

Table 5 sums up the dataset size distribution. The sizes ranged from small datasets of one hundred examples to an enormous dataset of 1,292,360 examples. Half of the datasets were below 1500 instances; only three datasets were above 100,000 instances and all the other nineteen datasets were below 50,000 instances.

Table 5. Dataset size distribution.

Dataset Size	Number of Datasets
<500	5
500–1000	4
1000–1500	2
1500–2000	1
2000–5000	4
5000–10,000	2
10,000–50,000	1
50,000–100,000	0
10,000–500,000	2
500,000–1,000,000	0
>1,000,000	1

Dataset images typically have many teeth, but five (23%) datasets used images with only one tooth.

Datasets are split into three sets: training, validation, and test. There were two papers that did not have information regarding the division of the dataset. In these cases, we assumed that the training set was the dataset. Half of the training sets had sizes above 87% of the dataset size, and the training set size with the minimum percentage was 60% of the dataset size.

There were four papers that did not use or had no information regarding the test set. All the other 18 papers used a test set for evaluating the ML algorithms. Usually, this is a subset from the original dataset. However, there was a particular paper [38] that used an external dataset as a test set.

Regarding the validation set, 36% (n = 8) of the papers had no information, 32% (n = 7) of the papers used cross-validation and the other 32% (n = 7) used a validation set.

3.5. Machine Learning Tasks and Models

Most of the papers addressed the machine learning application to the diagnosis in dental health as a classification task (n = 20). One study addressed the problem as a regression task, and another study used a combination of regression and generative machine learning models.

The huge majority of studies used exclusively convolutional neural networks ($n = 19$), but three studies used a combination of CNN and traditional algorithms. In these three studies, the approach was to use CNN for feature extraction and then traditional algorithms, such as support vector machine ($n = 2$), k-nearest neighbors ($n = 2$), naïve Bayes ($n = 2$), and random forest ($n = 1$) for classification.

Among the papers that used CNN, 41% ($n = 9$) used exclusively proprietary architectures. One study used both a proprietary CNN and pretrained CNNs via transfer learning.

Transfer learning was used by a considerable number of studies ($n = 10$), usually by changing the last layers in the original architectures and fine tuning the model with the dataset used in the paper. The preferred pretrained models were GoogLeNet Inception ($n = 6$), ResNet ($n = 6$), different versions of VGG ($n = 4$), Xception ($n = 2$) and AlexNet ($n = 2$). Other architectures used were DeepLab [47], Mask R-CNN [48], DETR [47], DenseNet [49], Yolo [50], MobileNet [51], and DarkNet [51] (each $n = 1$).

The preferred pretrained model was the GoogLeNet Inception. In several studies, GoogLeNet Inception V3 was used as the main model for their respective classification tasks [25,30,34]. Hashem et al. used the Inception original architecture, and adjusted output layers to classify the images in one of three kinds of cysts [30]. The weights of the model were optimized by adjusting the hyperparameters including the learning rate, batch size, dropout rate, and by using batch normalization. Lee et al. also adapted the last layer for an adequate number of categories (presence or absence of dental caries), but provided less detail about the process of hyperparameter tuning [25].

Some papers used the results obtained with GoogLeNet Inception V3 for comparison with other models, such as a proprietary model developed in the paper for a specific diagnosis task ([33] for the detection and classification of dental implants), a specific model which was optimized ([41] for AlexNet), or other pretrained models (such as DenseNet, VGG, and ResNEt-50 in [39]). Very often, when the pretrained model was used for comparison, there was a lack of detail in the description of the adaptation of the original model to the specific task being handled.

ResNet was the other preferred pretrained model used in the selected papers. ResNet was used by Cha et al. for training a classification model created for sorting upper and lower periapical radiographs [35]. The weights of the pretrained model were used, with the last connected layer modified to meet the number of classes (upper and lower maxillary). The radiograph image was then fed into another model trained specially for the upper or lower maxillary. This second set of models used a version of the R-CNN architecture for localizing the implants and finding key points, thus allowing the calculation of the marginal bone loss ratio. Li et al. use a modification of the ResNet-18 to detect the crown categories (caries or normal) and root categories (periapical periodontitis/normal) of the tooth. For a single tooth, the model needed to be executed twice: the first time to obtain the dental root results and the second time to obtain the dental crown results. It was, however, not clear how the modified model was trained on the available dataset [37].

In the study conducted by Feher et al., the authors employed an approach that combined object detection and image segmentation of anatomical structures to predict two classes of cysts: odontogenic and non-odontogenic [43]. The object detection model consisted of a feature pyramid network using a pretrained ResNet as the backbone that outputs a bounding box with the location of the cysts. In parallel, a pretrained U-Net segmentation model was used to obtain relevant anatomical structures, such as the maxilla, mandible, mandibular canal, maxillary sinuses, dentition, and individual teeth. The overlap of the detection boxes and segmented anatomical structures was computed, and fed into a random forest classifier for cyst classification. Tsoromokos et al. used an architecture named faster R-CNN, an object detection network based on R-CNN and fast-RCNN [45]. The objective was to classify teeth in periapical images as caries or non-caries. The main architecture was composed of a feature extraction network, a regional proposal network, and a prediction and localization network. The feature extraction component used the pretrained weights of ResNet, and the global model was trained with a small dataset of 720 instances. The paper omitted the details of how validation was performed.

Bui et al. focused on extracting pertinent features to optimize the classification of tooth images as either caries or non-caries [34]. Several well-known pretrained models, such as AlexNet, Inception, VGG, ResNet, and Xception were used to extract deep-activated features. Experiments were performed to find out which deep layer (before the prediction layer) provided the highest performance features. At this stage, it is worth noting that no details were provided regarding the parameters used for feature extraction with each model. The extracted features were then fused with statistical and texture features computed at the pixel level, such as mean, contrast, entropy, or correlation. The fused set of features was fed into traditional machine learning algorithms, such as SVM, NB, KNN, DT, and RF to obtain a prediction of the two categories. Sunnetci et al. had a similar approach but with the aim of classifying the images as periodontal bone loss or non-periodontal bone loss [46]. The paper used pretrained AlexNet and SqueezeNet to extract features from a defined deep layer in each model. The deep image features were then fed to algorithms such as kNN, NB, SVM, and tree ensemble algorithms that performed the classification task. The paper also referred to the use of efficient net for comparison purposes, but no further details were provided. Geetha et al. had a similar but simpler approach, where a segmentation algorithm using an adaptive threshold and morphological processing was used for statistical feature extraction [28]. The extracted features were then fed into a neural network with one hidden layer used to classify the images as either caries or normal. The results were compared with the results from methods such as SVM, kNN, and XGBoost. It is worth mentioning that this was one of the papers where it was not possible to identify the test set used. It was also one of the papers with the smallest training dataset, which justified the simple neural network used.

In the study conducted by Endres et al., a 26-layer U-net-based architecture was employed for image segmentation [29]. This methodology was specifically designed to detect radiolucent alteration in panoramic images. Those alterations are common radiographic findings that have a differential diagnosis including infections, granuloma, cysts, and tumors. The model outputs an intensity map indicating regions of high or low confidence for containing a radiolucent periapical alteration.

The YOLO algorithm was used by Tajima et al. to detect cyst-like radiolucent lesions of the jaws [42]. The YOLO algorithm has gained significant attention in the field of computer vision and medical imaging, as it predicts the bounding boxes and class probabilities directly from the full image in one pass. The model described in the paper used 75 convolutional layers and the ResNet structure for feature extraction, followed by a deep learning network to generate the bounding boxes where the lesions were present. The metrics reported were all above 90%, but few details were provided regarding the deep network employed.

Ekert et al. developed a seven layer neural network to classify panoramic images into apical or non-apical lesions [26]. The network contained four convolutional layers and two dense layers. The architecture was optimized for the numbers of neuronal units, the number of filters for each particular convolutional layer, the kernel sizes, the configurations of the max pooling layers and the dropout layers. A relatively small dataset with fewer than 3000 images was used. The authors justified the preference for custom-made architecture by the fact that more complex, state-of-the-art pretrained models caused overfitting with their limited-size dataset. Similar work was performed by Kros et al., but for the task of detecting periodontal bone loss [27].

In the study conducted by Hashem et al., conventional procedures for image segmentation and feature extraction were employed [31]. Subsequently, these extracted features were then fed into neural networks to classify the images and determine the presence of infection. The authors referred to the use of four different models of deep neural networks.

However, the information provided does not allow us to understand the architectures, or how the models were trained with a small dataset of 80 images.

Liu et al. devised a deep neural architecture specifically designed for the identification of ectopic eruptions from panoramic images [38]. It consisted of one first and three last plain convolutional layers, with middle layers for feature extraction. These middle layers used specific kernels for position-wise and channel-wise feature extraction. The model was trained with defined parameters with a dataset of 2960 region images from children's panoramic images. No information was provided on how the validation was performed. Interestingly, this paper used an independent dataset collected from another hospital as an external testing set.

The authors of [32] used a simplified adaptation of DenseNet to develop a model for segmentation and lesion detection with CBCT images. The input for the model was both images and oral-anatomical knowledge, such as constraints regarding the spatial location of lesions, the connection of restorative material, or the location of the background. The rationale behind the incorporation of anatomical knowledge was to limit the search space for the deep learning algorithm to find the optimal parameters. The model was trained on a very small dataset of 100 slices of CBCT images. It is not clear what test set was used.

One paper employed generative adversarial networks (GAN) to facilitate the measurement of clinical attachment levels [36]. GANs are a class of machine learning models that consist of a generator and a discriminator, competing against each other to generate realistic data and distinguish it from real data, respectively. The authors developed a GAN to predict the out-of-view anatomy in bitewing images for the measurement of clinical attachment levels. The generative adversarial network with partial convolutions comprises two generators and three discriminator CNNs. An encoder-decoder generator focuses the network on the missing regions of the images and fills in missing anatomy, while an encoder-decoder generator encourages the overall realism of the image and helps refine the predictions. The intermediate prediction images resulting from the GAN are fed into a refined encoder-decoder generator, a pretrained VGG discriminator and a final discriminator. The resulting images are then fed into deep learning open-source prediction algorithms (DETR and DeepLab). The model was trained, validated, and tested in a large set of some thousand teeth images.

A deep neural network based on UNet and Trans-UNET was developed by Ying et al. for carie segmentation [40]. Trans-UNet is an extension of UNet introduced in 2001 [52] that incorporates transformer modules, inspired by the success of transformers in natural language processing tasks. Trans-UNet combines convolutional and self-attention mechanisms to improve the modeling capability of UNet. The proposed model was trained with a small dataset of 800 teeth images extracted from periapical images. Despite the high metric values obtained, the authors recognized that the training set might be too small to train the deep architecture. There was no information on how validation was performed.

3.6. Outcome Metrics and Model Performance

The studies based on classification tasks all used a combination of two or more metrics to evaluate the model's performance. The minimum number of metrics used was two, the maximum was seven, and the mean was 4.75. Recall, also referred to as sensitivity or true positive rate ($n = 17$), precision, also referred to as positive predictive value ($n = 16$), specificity, also referred to as true negative rate ($n = 14$), and F1 score, also referred to as the Dice coefficient ($n = 13$), were the most used metrics. Other metrics commonly used were accuracy ($n = 9$), receiver operating characteristic–area under curve ($n = 8$), and negative predictive value ($n = 7$). Confusion matrices, false positive rate, precision-recall curve, Youden's index, and Matthews correlation coefficient were also used in the classification studies.

The regression studies used a smaller group of metrics to access model performance, namely mean absolute error, mean bias error ($n = 1$), and mean squared error ($n = 1$).

The values reported for model performance vary widely. Table 6 presents the average, minimum, and maximum values of the most used metrics, computed over the best reported values in each study. Average values were above 0.81 and below 0.93; maximum values were very high and between 0.96 and 1.0; minimum values ranged between 0.51 and 0.85. The lowest average values were obtained for precision and F1 score and the highest was obtained for ROC-AUC.

Table 6. Average, minimum, and maximum values of the most used metrics, considering the best reported values in each manuscript.

Metric	n	Average	Minimum	Maximum
Recall	17	0.84	0.51	0.96
Precision	16	0.81	0.67	0.99
Specificity	14	0.85	0.51	1.00
F1 score	13	0.81	0.58	0.97
Accuracy	9	0.92	0.81	0.98
ROC-AUC *	8	0.93	0.85	0.98
NPV **	7	0.83	0.68	0.95

* ROC-AUC: receiver operating characteristic–area under curve; ** NPV: negative predictive value.

The lowest values for recall, precision, and F1 score were obtained in a study using panoramic images for the diagnosis of apical lesions and a dataset size of 3099. The lowest values of specificity and negative predictive value were reported in a study using panoramic images for the diagnosis of oral lesions and a dataset of 800 images. The lowest values of accuracy and ROC-AUC were also obtained with panoramic images, for the diagnosis of periodontal diseases (dataset size 2538), and of oral lesions (dataset size 120), respectively.

The highest value of recall was obtained in a study that used CBCT images for the diagnosis of oral lesions, and a dataset size of 170,525. The highest values of precision, specificity, F1 score, and accuracy were obtained in a study using panoramic images for the diagnosis of oral lesions, with a dataset size of 1546. The best value for ROC-AUC was reported in a study using periapical images for the diagnosis of implant defects, with a dataset size of 533. The highest value of NPV was reported in a study using panoramic images for the diagnosis of apical lesions, with a dataset size of 2877.

The study that used both CBCT and panoramic images obtained higher performance metrics for the models that used CBCT images. The study that used periapical and panoramic images obtained higher performance models using the periapical images.

3.7. Human Comparators

Only a small number of studies ($n = 5$) compared the machine learning model's performance with human performance. Those were all classification tasks, with dataset sizes ranging from 708 to 7924 instances, and using either proprietary CNN or pretrained models via transfer learning [10–14]. The number of dentists ranged from one junior dentist to twenty-four oral and maxillofacial surgeons (OMF). The reported experience ranged from 3 to 10 years. Most of the studies ($n = 4$) concluded that the machine learning models reached a similar diagnostic performance to experienced dentists. One of the studies that used a high number of experts [29] additionally concluded that the ML model outperformed 58% of OMF surgeons. Another of these studies [38] additionally found that the ML algorithm was much faster at reaching a similar to human performance and that the best detection performance was obtained by human experts assisted by the automatic model.

One study [37] found that the ML model achieved significantly higher performance than that of young dentists, and, with the assistance of the model, the experts not only reached a higher diagnostic accuracy but also increased interobserver agreement.

4. Discussion

The growth in the number of published studies that investigate the use of machine learning techniques in X-ray diagnostics for oral health demonstrates the growing interest that the field has aroused in the scientific community. Most of the researchers involved in these publications are affiliated with clinical institutions and the majority of the papers were published in clinical journals, as opposed to technical journals. Moreover, the majority of those clinical journals belong to the specific clinical field of oral health (Journal of Dentistry, Journal of Oral and Maxillofacial Surgery, Medicine, and Pathology, International Dental Journal, Journal of Endodontics, Oral Diseases). These facts are in contrast with previous literature reviews [53], and seem to indicate an evolution in the maturity of the field. The focus of the research is slowly being displaced from the technical development of the models to an initial stage in the evaluation of their use as a potential clinical tool.

The sizes of the datasets and the ML strategies used vary widely in the studies analyzed in this review. There seems to be a relation between the dataset sizes and the use of pretrained machine learning models. For instance, the average size of the datasets when pretrained models were used was above 21,000, instances; even if the largest dataset was not considered, while slightly below 2000 instances when proprietary architectures were used. Interestingly, there were two small datasets with less than 500 instances that used pretrained models with accuracy results above 0.95.

Several limitations regarding the data were identified in the reviewed studies. Some of these problems are common in the application of ML to other areas of medical imaging as well. One major limitation is that datasets are often constructed using data from a single institution, which limits their generality and heterogeneity. To minimize potential biases, datasets should be as diverse as possible. Additionally, a significant number of studies rely on small datasets with poorly described curation processes. There is often a lack of adequate description of dataset characteristics, such as category distribution. For large datasets collected over long periods of time, the diversity of data acquisition (clinical protocols and equipment) was not always clear. The issue of labeling is also relevant. Usually, multiple annotators are necessary to obtain a gold standard label for the data. In the revised studies, it was not always identified how the quality of the labeling process through multiple annotators was assured. For instance, in some cases the task was performed by a single annotator. In other cases, it was unclear how disagreements were resolved. Additionally, some studies lacked information on the annotation procedures employed.

The analysis of the performance of the models did not allow us to draw plain conclusions, either concerning the type of image being used, the ML approach, the clinical application or the dataset size. For instance, some of the best results were obtained for panoramic images, and some of the worst results were also obtained for panoramic images. Some of the highest performance models were obtained with big datasets, but some others with datasets with as low as 533 instances, data augmentation included. On the other hand, some of the worst performance models were obtained with datasets with several thousand instances. These results are in line with the findings of other reviews [53] and seem to indicate the need for the standardization of procedures.

Some of the studies analyzed displayed a few limitations in their described methodology. Frequently, there was a lack of information regarding the validation procedure or the nature of the test set used. These are two aspects that serve as reference in machine learning, ensuring the prevention of data leakage that can lead to falsely inflated metric values. The absence of such information raises concerns about the actual quality of the reported models. It was also observed that the information provided on model training was not always sufficiently comprehensive. In some cases, there was a lack of information on the hyperparameters used or the strategy employed to select specific parameters.

No single ML approach could be identified as "the best" approach in the analyzed papers. They encompass a wide range of ML methods, including vanilla methods using transfer learning from pretrained models, as well as custom state-of-the-art approaches using transformers or GANs. Due to the diverse characteristics of the datasets, tasks, and metrics employed, making meaningful comparisons becomes challenging. Unlike other areas where ML is used in medical imaging, the absence of large, curated datasets that can serve as benchmarks also hinders any comparison. The lack of reporting standards further complicates this task.

Indeed, the lack of standardized experimental design and reporting in machine learning research, including oral health applications, contrasts with the presence of reporting guidelines commonly used in the medical field. While existing standards, such as TRIPOD [54] and PROBAST [55], might not perfectly fit ML research in medical imaging, efforts should be made to adhere to reporting guidelines. The upcoming extension to TRIPOD and PROBAST for AI applications [56], which are also relevant for oral health applications, is a positive development. In the meantime, there are checklists available that can and should serve as guidance for researchers and reviewers [57,58]. One approach that might contribute to the progressive adoption and acceptance of ML technology in oral health is the application of formal methods [59]. Formal verification techniques can provide guarantees on the robustness and generalizability of the models, aiding in the detection of potential biases, and therefore contributing to enhancing the reliability, explainability and trustworthiness of the diagnostic systems. However, collaboration and further research are necessary to refine and expand the use of formal methods of ML in healthcare, namely in oral health diagnosis. Only a small number of studies compared the performance of the machine learning models with dentists. Notably, in all cases, the models matched or outperformed the dentists. The main conclusion to be drawn is that the assistance of AI seems to help experts improve their diagnosis performance, especially in interpreting difficult cases [38]. These are very interesting results, which need to be confirmed by future investigations, along with their implications in the clinical setting.

Indeed, the majority of studies focused primarily on the technical aspects of the automated diagnosis of oral conditions, with limited exploration of the broader healthcare implications. While the technical components of these systems are unquestionably important, it is crucial to also consider the impact of these innovations on patient care and clinical decision-making as the field progresses. Adopting a more comprehensive approach that takes into account both technological advancements and healthcare perspectives could be beneficial for future research endeavors.

Finally, it is crucial to address the complex ethical considerations surrounding privacy and algorithm biases. These issues require careful attention and consideration to ensure that patient privacy is protected and that the algorithms used do not perpetuate biases. Addressing these ethical concerns is essential for the responsible development and deployment of AI technologies in oral healthcare.

This paper acknowledges some limitations. First, our query, although capturing a considerable number of papers, was relatively simple, might not have captured some relevant articles on the subject while including many unrelated papers not pertaining to diagnosis in oral health. Second, the omission of more specific terms in the query may have resulted in overlooking potentially relevant literature that could have provided further insights into our research topic. Additionally, to enhance the comprehensiveness of the review, it would have been beneficial to supplement the systematic search with snowballing techniques. These techniques involve reviewing the reference lists of identified articles and conducting citation searches to identify additional relevant studies that may have been missed in the initial search. Moreover, by not including the PubMed database, we may have overlooked papers published in biomedical or clinical journals. Future work should consider incorporating both snowballing techniques and a more specific query, including a search in the PubMed database, to address these limitations and enhance the quality of the research.

5. Conclusions

The application of AI in the diagnosis of oral health issues using X-ray-based images is a rapidly developing field. There is still a clear need for further investigation of the role of AI in dental diagnosis in the clinical setting. The present review of the literature seems to indicate that the field should naturally evolve toward the use of predictive models as an effective, stable and sustainable beneficial tool for oral health professionals performing diagnosis.

Supplementary Materials: The following supporting information can be downloaded at: https://www.mdpi.com/article/10.3390/computation11060115/s1, Table S1: Papers included in the review and their characteristics.

Author Contributions: Conceptualization, methodology, investigation: M.V.M., L.B., H.L., V.A., M.-R.A. and V.R.; formal analysis: M.V.M., L.B. and V.R.; data curation: M.V.M.; writing—original draft preparation: M.V.M., L.B. and V.A.; writing—review and editing: H.L., V.A. and M.-R.A.; visualization: V.R.; project administration: M.V.M. All authors have read and agreed to the published version of the manuscript.

Funding: This research was funded by national funds through the Fundação para a Ciência e a Tecnologia. I.P. (Portuguese Foundation for Science and Technology) by the project UIDB/05064/2020 (VALORIZA—Research Centre for Endogenous Resource Valorization).

Data Availability Statement: All relevant data are available through the paper and Supplementary Material. Additional information is available from the authors upon reasonable request.

Conflicts of Interest: The authors declare no conflict of interest. The funders had no role in the design of the study; in the collection. analyses. or interpretation of data; in the writing of the manuscript; or in the decision to publish the results.

References

1. Pitts, N.B.; Zero, D.T.; Marsh, P.D.; Ekstrand, K.; Weintraub, J.A.; Ramos-Gomez, F.; Tagami, J.; Twetman, S.; Tsakos, G.; Ismail, A. Dental Caries. *Nat. Rev. Dis. Prim.* **2017**, *3*, 17030. [CrossRef]
2. Kinane, D.F.; Stathopoulou, P.G.; Papapanou, P.N. Periodontal Diseases. *Nat. Rev. Dis. Prim.* **2017**, *3*, 17038. [CrossRef]
3. The Use of Dental Radiographs: Update and Recommendations. *J. Am. Dent. Assoc.* **2006**, *137*, 1304–1312. [CrossRef]
4. Ludlow, J.B.; Ivanovic, M. Comparative Dosimetry of Dental CBCT Devices and 64-Slice CT for Oral and Maxillofacial Radiology. *Oral Surg. Oral Med. Oral Pathol. Oral Radiol.* **2008**, *106*, 106–114. [CrossRef]
5. Tadinada, A. Dental Radiography BT. In *Evidence-Based Oral Surgery: A Clinical Guide for the General Dental Practitioner*; Ferneini, E.M., Goupil, M.T., Eds.; Springer International Publishing: Cham, Switzerland, 2019; pp. 67–90, ISBN 978-3-319-91361-2.
6. Shan, T.; Tay, F.R.; Gu, L. Application of Artificial Intelligence in Dentistry. *J. Dent. Res.* **2021**, *100*, 232–244. [CrossRef]
7. Carrillo-Perez, F.; Pecho, O.E.; Morales, J.C.; Paravina, R.D.; Della Bona, A.; Ghinea, R.; Pulgar, R.; Pérez, M.D.M.; Herrera, L.J. Applications of Artificial Intelligence in Dentistry: A Comprehensive Review. *J. Esthet. Restor. Dent.* **2022**, *34*, 259–280. [CrossRef]
8. Mahdi, S.S.; Battineni, G.; Khawaja, M.; Allana, R.; Siddiqui, M.K.; Agha, D. How Does Artificial Intelligence Impact Digital Healthcare Initiatives? A Review of AI Applications in Dental Healthcare. *Int. J. Inf. Manag. Data Insights* **2023**, *3*, 100144. [CrossRef]
9. LeCun, Y.; Bottou, L.; Bengio, Y.; Haffner, P. Gradient-Based Learning Applied to Document Recognition. *Proc. IEEE* **1998**, *86*, 2278–2324. [CrossRef]
10. Deng, J.; Dong, W.; Socher, R.; Li, L.-J.; Li, K.; Li, F.-F. ImageNet: A Large-Scale Hierarchical Image Database. In Proceedings of the 2009 IEEE Conference on Computer Vision and Pattern Recognition, Miami, FL, USA, 20–25 June 2009.
11. Lin, T.-Y.; Maire, M.; Belongie, S.; Hays, J.; Perona, P.; Ramanan, D.; Dollár, P.; Zitnick, C.L. Microsoft COCO: Common Objects in Context BT. In *Computer Vision—ECCV 2014*; Fleet, D., Pajdla, T., Schiele, B., Tuytelaars, T., Eds.; Springer International Publishing: Cham, Switzerland, 2014; pp. 740–755.
12. Deng, L. Digit Images for Machine Learning Research. *IEEE Signal Process. Mag.* **2012**, *29*, 141–142. [CrossRef]
13. Krizhevsky, A. Learning Multiple Layers of Features from Tiny Images. 2009. Available online: https://www.cs.toronto.edu/~kriz/learning-features-2009-TR.pdf (accessed on 5 May 2023).
14. Everingham, M.; Eslami, S.M.A.; Van Gool, L.; Williams, C.K.I.; Winn, J.; Zisserman, A. The Pascal Visual Object Classes Challenge: A Retrospective. *Int. J. Comput. Vis.* **2015**, *111*, 98–136. [CrossRef]
15. Szegedy, C.; Liu, W.; Jia, Y.; Sermanet, P.; Reed, S.; Anguelov, D.; Erhan, D.; Vanhoucke, V.; Rabinovich, A. Going Deeper with Convolutions. In Proceedings of the 2015 IEEE Conference on Computer Vision and Pattern Recognition (CVPR), Boston, MA, USA, 7–12 June 2015.
16. He, K.; Zhang, X.; Ren, S.; Sun, J. Deep Residual Learning for Image Recognition. In Proceedings of the 2016 IEEE Conference on Computer Vision and Pattern Recognition (CVPR), Las Vegas, NV, USA, 27–30 June 2016.
17. Simonyan, K.; Zisserman, A. Very Deep Convolutional Networks for Large-Scale Image Recognition. In Proceedings of the 3rd International Conference on Learning Representations, ICLR 2015, San Diego, CA, USA, 7–9 May 2015.

18. Chollet, F. Xception: Deep Learning with Depthwise Separable Convolutions. In Proceedings of the 2017 IEEE Conference on Computer Vision and Pattern Recognition (CVPR), Honolulu, HI, USA, 21–26 July 2017.
19. Cortes, C.; Vapnik, V. Support-Vector Networks. *Mach. Learn.* **1995**, *20*, 273–297. [CrossRef]
20. Fix, E.; Hodges, J.L. Discriminatory Analysis. Nonparametric Discrimination: Consistency Properties. *Int. Stat. Rev.* **1989**, *57*, 238–247. [CrossRef]
21. Breiman, L. Random Forests. *Mach. Learn.* **2001**, *45*, 5–32. [CrossRef]
22. Chen, T.; Guestrin, C. XGBoost: A Scalable Tree Boosting System. In Proceedings of the ACM SIGKDD International Conference on Knowledge Discovery and Data Mining, Washington DC USA, 14–18 August 2016.
23. Huang, X.; Lin, J.; Demner-Fushman, D. Evaluation of PICO as a Knowledge Representation for Clinical Questions. *AMIA Annu. Symp. Proc. AMIA Symp.* **2006**, *2006*, 359–363.
24. Ouzzani, M.; Hammady, H.; Fedorowicz, Z.; Elmagarmid, A. Rayyan-a Web and Mobile App for Systematic Reviews. *Syst. Rev.* **2016**, *5*, 210. [CrossRef]
25. Lee, J.H.; Kim, D.H.; Jeong, S.N.; Choi, S.H. Detection and Diagnosis of Dental Caries Using a Deep Learning-Based Convolutional Neural Network Algorithm. *J. Dent.* **2018**, *77*, 106–111. [CrossRef]
26. Ekert, T.; Krois, J.; Meinhold, L.; Elhennawy, K.; Emara, R.; Golla, T.; Schwendicke, F. Deep Learning for the Radiographic Detection of Apical Lesions. *J. Endod.* **2019**, *45*, 917–922.e5. [CrossRef]
27. Krois, J.; Ekert, T.; Meinhold, L.; Golla, T.; Kharbot, B.; Wittemeier, A.; Dörfer, C.; Schwendicke, F. Deep Learning for the Radiographic Detection of Periodontal Bone Loss. *Sci. Rep.* **2019**, *9*, 8495. [CrossRef]
28. Geetha, V.; Aprameya, K.S.; Hinduja, D.M. Dental Caries Diagnosis in Digital Radiographs Using Back-Propagation Neural Network. *Health Inf. Sci. Syst.* **2020**, *8*, 8. [CrossRef]
29. Endres, M.G.; Hillen, F.; Salloumis, M.; Sedaghat, A.R.; Niehues, S.M.; Quatela, O.; Hanken, H.; Smeets, R.; Beck-Broichsitter, B.; Rendenbach, C.; et al. Development of a Deep Learning Algorithm for Periapical Disease Detection in Dental Radiographs. *Diagnostics* **2020**, *10*, 430. [CrossRef]
30. Lee, J.H.; Kim, D.H.; Jeong, S.N. Diagnosis of Cystic Lesions Using Panoramic and Cone Beam Computed Tomographic Images Based on Deep Learning Neural Network. *Oral Dis.* **2020**, *26*, 152–158. [CrossRef]
31. Hashem, M.; Youssef, A.E. Teeth Infection and Fatigue Prediction Using Optimized Neural Networks and Big Data Analytic Tool. *Clust. Comput.* **2020**, *23*, 1669–1682. [CrossRef]
32. Zheng, Z.; Yan, H.; Setzer, F.C.; Shi, K.J.; Mupparapu, M.; Li, J. Anatomically Constrained Deep Learning for Automating Dental CBCT Segmentation and Lesion Detection. *IEEE Trans. Autom. Sci. Eng.* **2021**, *18*, 603–614. [CrossRef]
33. Lee, D.W.; Kim, S.Y.; Jeong, S.N.; Lee, J.H. Artificial Intelligence in Fractured Dental Implant Detection and Classification: Evaluation Using Dataset from Two Dental Hospitals. *Diagnostics* **2021**, *11*, 233. [CrossRef] [PubMed]
34. Bui, T.H.; Hamamoto, K.; Paing, M.P. Deep Fusion Feature Extraction for Caries Detection on Dental Panoramic Radiographs. *Appl. Sci.* **2021**, *11*, 2005. [CrossRef]
35. Cha, J.Y.; Yoon, H.I.; Yeo, I.S.; Huh, K.H.; Han, J.S. Peri-Implant Bone Loss Measurement Using a Region-Based Convolutional Neural Network on Dental Periapical Radiographs. *J. Clin. Med.* **2021**, *10*, 1009. [CrossRef]
36. Kearney, V.P.; Yansane, A.I.M.; Brandon, R.G.; Vaderhobli, R.; Lin, G.H.; Hekmatian, H.; Deng, W.; Joshi, N.; Bhandari, H.; Sadat, A.S.; et al. A Generative Adversarial Inpainting Network to Enhance Prediction of Periodontal Clinical Attachment Level. *J. Dent.* **2022**, *123*, 104211. [CrossRef] [PubMed]
37. Li, S.; Liu, J.; Zhou, Z.; Zhou, Z.; Wu, X.; Li, Y.; Wang, S.; Liao, W.; Ying, S.; Zhao, Z. Artificial Intelligence for Caries and Periapical Periodontitis Detection. *J. Dent.* **2022**, *122*, 104107. [CrossRef]
38. Liu, J.; Liu, Y.; Li, S.; Ying, S.; Zheng, L.; Zhao, Z. Artificial Intelligence-Aided Detection of Ectopic Eruption of Maxillary First Molars Based on Panoramic Radiographs. *J. Dent.* **2022**, *125*, 104239. [CrossRef]
39. Aljabri, M.; Aljameel, S.S.; Min-Allah, N.; Alhuthayfi, J.; Alghamdi, L.; Alduhailan, N.; Alfehaid, R.; Alqarawi, R.; Alhareky, M.; Shahin, S.Y.; et al. Canine Impaction Classification from Panoramic Dental Radiographic Images Using Deep Learning Models. *Inform. Med. Unlocked* **2022**, *30*, 100918. [CrossRef]
40. Ying, S.; Wang, B.; Zhu, H.; Liu, W.; Huang, F. Caries Segmentation on Tooth X-Ray Images with a Deep Network. *J. Dent.* **2022**, *119*, 104076. [CrossRef] [PubMed]
41. Imak, A.; Celebi, A.; Siddique, K.; Turkoglu, M.; Sengur, A.; Salam, I. Dental Caries Detection Using Score-Based Multi-Input Deep Convolutional Neural Network. *IEEE Access* **2022**, *10*, 18320–18329. [CrossRef]
42. Tajima, S.; Okamoto, Y.; Kobayashi, T.; Kiwaki, M.; Sonoda, C.; Tomie, K.; Saito, H.; Ishikawa, Y.; Takayoshi, S. Development of an Automatic Detection Model Using Artificial Intelligence for the Detection of Cyst-like Radiolucent Lesions of the Jaws on Panoramic Radiographs with Small Training Datasets. *J. Oral Maxillofac. Surg. Med. Pathol.* **2022**, *34*, 553–560. [CrossRef]
43. Feher, B.; Krois, J. Emulating Clinical Diagnostic Reasoning for Jaw Cysts with Machine Learning. *Diagnostics* **2022**, *12*, 1968. [CrossRef]
44. Tsoromokos, N.; Parinussa, S.; Claessen, F.; Moin, D.A.; Loos, B.G. Estimation of Alveolar Bone Loss in Periodontitis Using Machine Learning. *Int. Dent. J.* **2022**, *72*, 621–627. [CrossRef] [PubMed]
45. Zhu, Y.; Xu, T.; Peng, L.; Cao, Y.; Zhao, X.; Li, S.; Zhao, Y.; Meng, F.; Ding, J.; Liang, S. Faster-RCNN Based Intelligent Detection and Localization of Dental Caries. *Displays* **2022**, *74*, 102201. [CrossRef]

46. Muhammed Sunnetci, K.; Ulukaya, S.; Alkan, A. Periodontal Bone Loss Detection Based on Hybrid Deep Learning and Machine Learning Models with a User-Friendly Application. *Biomed. Signal Process. Control* **2022**, *77*, 103844. [CrossRef]
47. Chen, L.C.; Papandreou, G.; Kokkinos, I.; Murphy, K.; Yuille, A.L. DeepLab: Semantic Image Segmentation with Deep Convolutional Nets, Atrous Convolution, and Fully Connected CRFs. *IEEE Trans. Pattern Anal. Mach. Intell.* **2018**, *40*, 834–848. [CrossRef]
48. He, K.; Gkioxari, G.; Dollár, P.; Girshick, R. Mask R-CNN. *IEEE Trans. Pattern Anal. Mach. Intell.* **2020**, *42*, 386–397. [CrossRef]
49. Huang, G.; Liu, Z.; Van Der Maaten, L.; Weinberger, K.Q. Densely Connected Convolutional Networks. In Proceedings of the 2017 IEEE Conference on Computer Vision and Pattern Recognition (CVPR), Honolulu, HI, USA, 21–26 July 2017; pp. 2261–2269. [CrossRef]
50. Redmon, J.; Divvala, S.; Girshick, R.; Farhadi, A. You Only Look Once: Unified, Real-Time Object Detection. In Proceedings of the 2016 IEEE Conference on Computer Vision and Pattern Recognition (CVPR), Las Vegas, NV, USA, 27–30 June 2016; pp. 779–788. [CrossRef]
51. Sandler, M.; Howard, A.; Zhu, M.; Zhmoginov, A.; Chen, L.C. MobileNetV2: Inverted Residuals and Linear Bottlenecks. In Proceedings of the 2018 IEEE/CVF Conference on Computer Vision and Pattern Recognition, Salt Lake City, UT, USA, 18–23 June 2018; pp. 4510–4520. [CrossRef]
52. Chen, J.; Lu, Y.; Yu, Q.; Luo, X.; Adeli, E.; Wang, Y.; Lu, L.; Yuille, A.L.; Zhou, Y. TransUNet: Transformers Make Strong Encoders for Medical Image Segmentation. *arXiv* **2021**, arXiv:2102.04306.
53. Schwendicke, F.; Golla, T.; Dreher, M.; Krois, J. Convolutional Neural Networks for Dental Image Diagnostics: A Scoping Review. *J. Dent.* **2019**, *91*, 103226. [CrossRef]
54. Collins, G.S.; Reitsma, J.B.; Altman, D.G.; Moons, K.G.M. Transparent Reporting of a Multivariable Prediction Model for Individual Prognosis or Diagnosis (TRIPOD): The TRIPOD Statement. *BMC Med.* **2015**, *13*, 1. [CrossRef]
55. Wolff, R.F.; Moons, K.G.M.; Riley, R.D.; Whiting, P.F.; Westwood, M.; Collins, G.S.; Reitsma, J.B.; Kleijnen, J.; Mallett, S. PROBAST: A Tool to Assess the Risk of Bias and Applicability of Prediction Model Studies. *Ann. Intern. Med.* **2019**, *170*, 51–58. [CrossRef] [PubMed]
56. Collins, G.S.; Dhiman, P.; Navarro, C.L.A.; Ma, J.; Hooft, L.; Reitsma, J.B.; Logullo, P.; Beam, A.L.; Peng, L.; Van Calster, B.; et al. Protocol for Development of a Reporting Guideline (TRIPOD-AI) and Risk of Bias Tool (PROBAST-AI) for Diagnostic and Prognostic Prediction Model Studies Based on Artificial Intelligence. *BMJ Open* **2021**, *11*, e048008. [CrossRef] [PubMed]
57. Schwendicke, F.; Singh, T.; Lee, J.H.; Gaudin, R.; Chaurasia, A.; Wiegand, T.; Uribe, S.; Krois, J. Artificial Intelligence in Dental Research: Checklist for Authors, Reviewers, Readers. *J. Dent.* **2021**, *107*, 103610. [CrossRef] [PubMed]
58. Norgeot, B.; Quer, G.; Beaulieu-Jones, B.K.; Torkamani, A.; Dias, R.; Gianfrancesco, M.; Arnaout, R.; Kohane, I.S.; Saria, S.; Topol, E.; et al. Minimum Information about Clinical Artificial Intelligence Modeling: The MI-CLAIM Checklist. *Nat. Med.* **2020**, *26*, 1320–1324. [CrossRef]
59. Bonfanti, S.; Gargantini, A.; Mashkoor, A. A Systematic Literature Review of the Use of Formal Methods in Medical Software Systems. *J. Softw. Evol. Process* **2018**, *30*, e1943. [CrossRef]

Disclaimer/Publisher's Note: The statements, opinions and data contained in all publications are solely those of the individual author(s) and contributor(s) and not of MDPI and/or the editor(s). MDPI and/or the editor(s) disclaim responsibility for any injury to people or property resulting from any ideas, methods, instructions or products referred to in the content.

MDPI
St. Alban-Anlage 66
4052 Basel
Switzerland
www.mdpi.com

Computation Editorial Office
E-mail: computation@mdpi.com
www.mdpi.com/journal/computation

Disclaimer/Publisher's Note: The statements, opinions and data contained in all publications are solely those of the individual author(s) and contributor(s) and not of MDPI and/or the editor(s). MDPI and/or the editor(s) disclaim responsibility for any injury to people or property resulting from any ideas, methods, instructions or products referred to in the content.